AF588231

Birkhäuser

Compact Textbooks in Mathematics

This textbook series presents concise introductions to current topics in mathematics and mainly addresses advanced undergraduates and master students. The concept is to offer small books covering subject matter equivalent to 2- or 3-hour lectures or seminars which are also suitable for self-study. The books provide students and teachers with new perspectives and novel approaches. They may feature examples and exercises to illustrate key concepts and applications of the theoretical contents. The series also includes textbooks specifically speaking to the needs of students from other disciplines such as physics, computer science, engineering, life sciences, and finance.

- **Compact:** Small books presenting the relevant knowledge
- **Learning made easy:** Examples and exercises illustrate the application of the contents
- **Useful for lecturers:** Each title can serve as basis and guideline for a semester course/lecture/seminar of 2–3 hours per week.

Jürgen Herzog · Somayeh Moradi ·
Masoomeh Rahimbeigi

Numerical Semigroups

A Commutative Algebra Approach

Jürgen Herzog
Fakultät für Mathematik
Universität Duisburg-Essen
Essen, Germany

Somayeh Moradi
Department of Mathematics
Faculty of Science
Ilam University
Ilam, Iran

Masoomeh Rahimbeigi
Fakultät für Mathematik
Universität Duisburg-Essen
Essen, Germany

ISSN 2296-4568 ISSN 2296-455X (electronic)
Compact Textbooks in Mathematics
ISBN 978-3-032-05423-4 ISBN 978-3-032-05424-1 (eBook)
https://doi.org/10.1007/978-3-032-05424-1

This book is published under the imprint Birkhäuser, www.birkhauser-science.com by the registered company Springer Nature Switzerland AG
The registered company address is: Gewerbestrasse 11, 6330 Cham, Switzerland

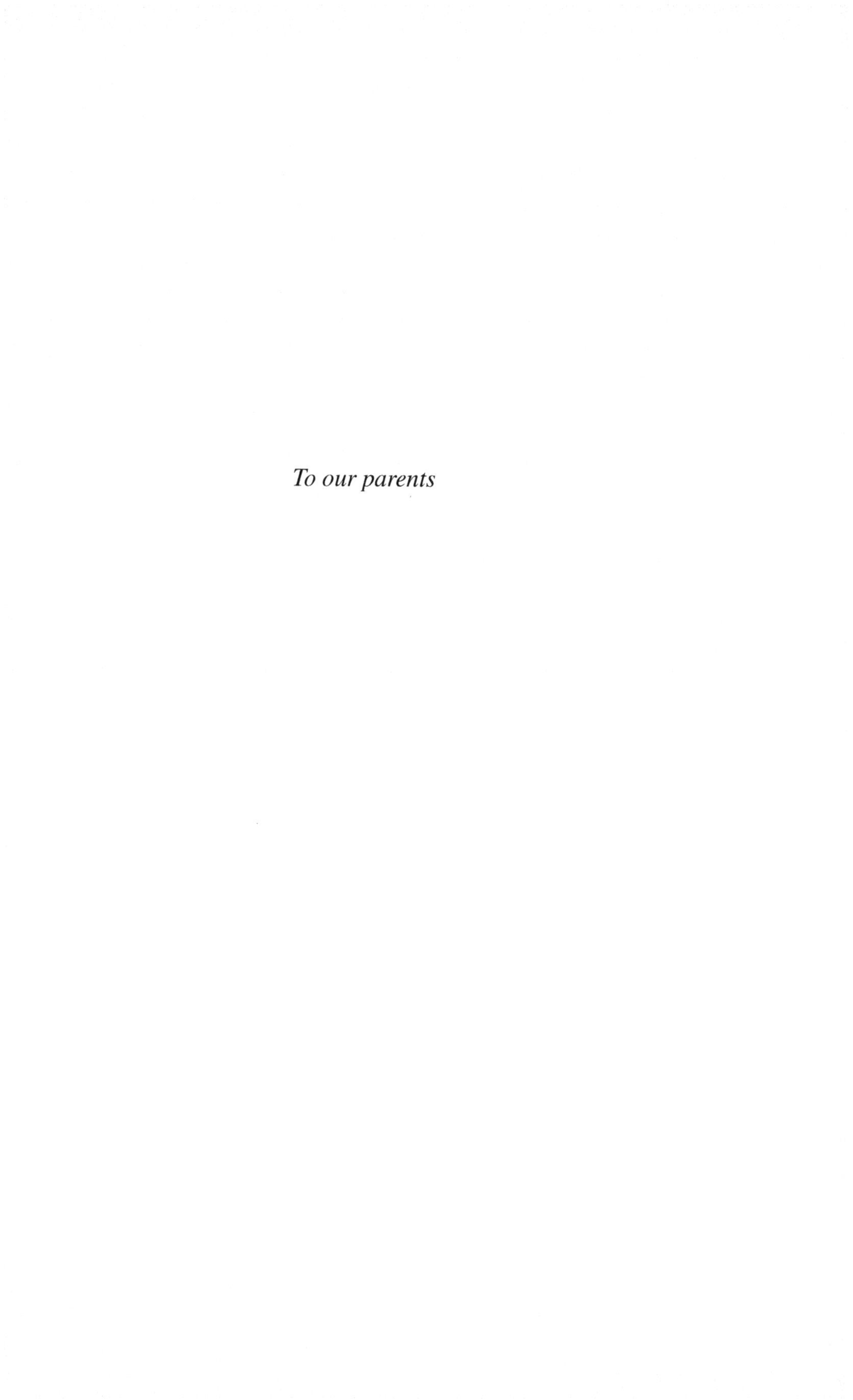

To our parents

Preface

The study of numerical semigroups has long held a central role at the intersection of algebra, number theory, and combinatorics. Despite their simple nature, numerical semigroups offer profound insights into many mathematical structures and have numerous applications in coding theory, algebraic geometry, number theory and even in the analysis of Diophantine equations. There are excellent resources on numerical semigroups and their applications in the above mentioned fields, see the books [2] and [65] and the references therein. However, few works bridge this area to commutative algebra in a systematic way. This book aims to fill that gap by providing a comprehensive introduction to the interplay between numerical semigroups and commutative algebra.

Given that the theory of numerical semigroup rings heavily relies on commutative algebra and homological algebra concepts, Chaps. 1–3 of the book provide a detailed treatment of the needed material from these fundamental areas. To this end, we focus primarily on graded modules over finitely generated graded K-algebras, where K is a field. These chapters together with the appendix serve as a self-contained introduction to basic topics in commutative and homological algebra, making them useful even for readers not specifically interested in numerical semigroups. They are prepared to be accessible to readers who may not have a specialized background in these areas but wish to gain a solid foundation to tackle the more specific and advanced material later in the book. For readers who wish to gain a deeper understanding of commutative algebra, the books [3, 15, 26, 51, 71] are invaluable. For a comprehensive introduction to homological algebra, we recommend the books [17, 68, 79].

Chapters 4–6 explore the deep interplay between numerical semigroups and commutative algebra. The main objective is to see how the structure of a numerical semigroup and its fundamental invariants can be read from the algebraic invariants of its associated semigroup ring mainly obtained from its minimal graded free resolution and its canonical module. Chapter 4 introduces basic concepts and properties of a numerical semigroup, its semigroup ring and the defining ideal of the semigroup ring. Well-structured numerical semigroups like complete intersection numerical semigroups, symmetric numerical semigroups, 3-generated numerical semigroups and numerical semigroups of maximal embedding dimension are more thoroughly discussed. Chapter 5 focuses on the Frobenius number, a central invariant in the study of numerical semigroups, and its determination via the minimal graded free resolution of the numerical semigroup ring. Other combinatorial invariants of numerical semigroups like pseudo-Frobenius numbers and the type, which can be deduced from the structure of the minimal graded free resolution are considered, as well. In Chap. 6, we turn to the study of the canonical ideal of a numerical semigroup along with the canonical trace, which measures how far a numerical semigroup is from being symmetric. These concepts lead to introduce interesting families of numerical semigroups, namely pseudo-symmetric, almost symmetric and nearly symmetric numerical semigroups.

The material in this book is intended for advanced undergraduate and graduate students, as well as researchers in algebra and related fields. Each chapter is accompanied by examples, detailed proofs, and exercises that encourage further exploration and understanding.

We gratefully acknowledge the University of Duisburg-Essen for hosting the second and third authors multiple times, enabling collaboration with the first author in the writing of this book. The second author's work was supported by a fellowship from the Alexander von Humboldt Foundation.

Jürgen Herzog unfortunately passed away before this book could be published, and his absence is deeply felt. It was a privilege to collaborate with him. He was not only a coauthor but also a mentor, a friend, and a source of constant inspiration. His sudden passing has left a deep void in the mathematical community and in our hearts. We remember him with gratitude for his generous spirit, intellectual brilliance, and passionate dedication to mathematics. This book is as much a testament to his ideas as it is a continuation of his lifelong dialogue with algebra.

Essen, Germany — Jürgen Herzog
Ilam, Iran — Somayeh Moradi
Essen, Germany — Masoomeh Rahimbeigi
March 2025

Competing Interests The authors have no competing interests to declare that are relevant to the content of this manuscript.

Contents

Basic Commutative Algebra

1

In this chapter, we present basic concepts of commutative algebra, which are needed in the study of numerical semigroup rings. Given that graded structures naturally emerge in semigroup rings, our emphasis will be on graded rings and modules. Noetherian rings and modules, associated prime ideals, dimension theory and Hilbert series for graded modules are presented. Throughout this book R is a commutative ring with identity and K is a field, unless specified otherwise.

1.1 Modules Over Finitely Generated Graded K-Algebras and Noetherian Rings

1.1.1 Graded Rings and Modules

We introduce positively graded K-algebras and discuss some fundamental properties of finitely generated graded modules over such algebras. Some of the statements and proofs are valid for more general commutative rings. In these cases we phrase and prove the results in this larger generality.

A ring R is called *graded*, if

(1) $R = \bigoplus_{i\in\mathbb{Z}} R_i$, as an abelian group;
(2) $R_i R_j \subseteq R_{i+j}$ for all i and j.

The elements $r \in R_i$ are called *homogeneous* of degree i, and we write $\deg r = i$, if $r \in R_i$. More generally, each element $r \in R$ can uniquely be written as a finite sum $r = \sum_i r_i$ with $r_i \in R_i$. The elements r_i in this sum are called the *homogeneous components* of r.

From (2) it can be seen that R_0 is a ring and each R_i is a R_0-module, because $R_0 R_i \subseteq R_i$ for all i. The R_0-module R_i is called the *ith homogeneous component* of R.

J. Herzog et al., *Numerical Semigroups*, Compact Textbooks in Mathematics,
https://doi.org/10.1007/978-3-032-05424-1_1

Note that the identity element of R coincides with the identity element of R_0. Indeed, we may write $1 = \sum_i r_i$, and hence for each j, we have $r_j = r_j 1 = \sum_i r_i r_j$. This implies that $r_j = r_0 r_j$ for all j. Therefore, $r_0 = r_0 1 = \sum_j r_0 r_j = \sum_j r_j = 1$. This in particular implies that R is an R_0-algebra. Since R has a graded structure whose 0th graded component is R_0, we call R a *graded R_0-algebra*. If $R_i = 0$ for $i < 0$, then R is called a *positively graded ring*.

The graded R_0-algebra R is called *standard graded*, if $R = R_0[R_1]$. The prototype of a standard graded R_0-algebra is the polynomial ring $S = K[x_1, \ldots, x_n]$ over a field K with $\deg x_i = 1$ for all i. The polynomial $f = x_1^3 - x_1x_2x_3$ is homogeneous of degree 3, while $g = x_3^2 - x_1^2x_2^4 + (3/4)x_1^3x_2^3 \in \mathbb{Q}[x_1, x_2, x_3]$ is not homogeneous. It has the homogeneous components x_3^2 and $-x_1^2x_2^4 + (3/4)x_1^3x_2^3$.

Note that the homogeneous component S_i of S is the K-vector space which has a K-basis consisting of all the monomials $x_1^{a_1}x_2^{a_2}\cdots x_n^{a_n}$ with $a_1 + a_2 + \cdots + a_n = i$. So, for example, $K[x_1, x_2, x_3]_2$ has the K-basis

$$x_1^2, x_2^2, x_3^2, x_1x_2, x_1x_3, x_2x_3.$$

In our applications $R_0 = K$ is a field and the graded K-algebra R is positively graded and *finitely generated*. In other words, there exist homogeneous elements $r_1, \ldots, r_m \in R$ such that $R = K[r_1, \ldots, r_m]$.

Lemma 1.1
Let R be a finitely generated graded K-algebra. Then each homogeneous component R_i of R is a finitely generated K-vector space.

Proof. Let $R = K[r_1, \ldots, r_m]$, where the r_i's are homogeneous. The component R_i is generated as a K-vector space by all products $r_{j_1}\cdots r_{j_k}$ with $1 \leq j_1 \leq j_2 \leq \cdots \leq j_k \leq m$ and $\sum_{l=1}^k \deg r_{j_l} = i$. Obviously, there exist only finitely many such products. □

Let $S = K[x_1, \ldots, x_n]$ be the polynomial ring in the variables $x_1, \ldots, x_n$. We may assign to each of the variables a degree, say $\deg x_i = a_i$, with $a_i \in \mathbb{Z}_{>0}$ for all i. With this assignment, the polynomial ring S becomes a finitely generated graded K-algebra. If $a_i = 1$ for $i = 1, \ldots, n$, then S is a standard graded K-algebra.

Exercise 1.2
(a) Let $S = K[x_1, x_2]$ be the graded polynomial ring in two variables over K with $\deg x_1 = 2$ and $\deg x_2 = 3$. Compute $\dim_K S_j$ for all j.

(b) Let $n \geq 2$, and let $S = K[x_1, \ldots, x_n]$ be the graded polynomial ring with $\deg x_i = a_i$ and $a_i \in \mathbb{Z}_{>0}$ for all i. Show that $\limsup_{i\to\infty} \dim_K S_i = \infty$.

Let R be a graded ring. An R-module M is called a *graded R-module*, if $M = \bigoplus_{j\in\mathbb{Z}} M_j$ as an abelian group, and $R_iM_j \subseteq M_{i+j}$ for all i and j. Note that each M_j is a R_0-module, since $R_0M_j = M_j$ for all j. Similarly as for R, we say that $x \in M$ is *homogeneous of degree* i, if $x \in M_i$, and we write $\deg x = i$.

Let M and N be graded R-modules. A map $\varphi\ M \to N$ is called a *graded R-module homomorphism*, if φ is an R-module homomorphism and $\varphi(M_i) \subseteq N_i$ for all i.

We denote by GMod_R the category of graded R-modules whose morphisms are the graded R-module homomorphisms.

Examples 1.1
Let R be a graded ring.

(a) R is a graded R-module.

(b) Let M be a graded R-module and $a \in \mathbb{Z}$. The graded R-module $M(a)$ is the R-module M equipped with the new grading $M(a)_j = M_{a+j}$ for all j. The module $M(a)$ is called the *module M shifted by a*.

Let R be a graded ring and let M be a graded R-module. A submodule $U \subseteq M$ is called a *graded submodule* of M, if U is a graded module, and $U_i \subseteq M_i$ for all i. The graded submodules of R are called the *graded ideals* of R. When R is a graded K-algebra, the ideal $\mathfrak{m} = \bigoplus_{i>0} R_i$ is the (unique) *graded maximal ideal* of R and $R/\mathfrak{m} \cong K$.

The following result tells us which submodules of M are graded submodules.

Proposition 1.3
Let M be a graded R-module and $U \subseteq M$ be a submodule of M. The following conditions are equivalent:

(i) U is a graded submodule of M.
(ii) $U = \sum_i U \cap M_i$.
(iii) Let $x \in M$. Then $x \in U$ if and only if all the graded components of x belong to U.

If the equivalent conditions hold, then $U = \bigoplus_{i\in\mathbb{Z}} U_i$, where $U_i = U \cap M_i$ for all i.

Proof. (i) $\Rightarrow$ (ii) Since U is a graded submodule of M, it follows that $U_i \subseteq M_i$ for all i. This implies that $U_i \subseteq U \cap M_i$. Since $U = \sum_i U_i$, we also have $U \subseteq \sum_i U \cap M_i$. The other inclusion is obvious.

(ii) $\Rightarrow$ (i) We set $U_i = U \cap M_i$. By assumption, $U = \sum_i U_i$. Since $U_i \subseteq M_i$ for all i and since $M = \bigoplus_i M_i$, it follows that $U = \bigoplus_i U_i$. Therefore, U is a graded submodule of M.

It is obvious that (ii) is equivalent to (iii). □

Proposition 1.4
Let $\varphi M \to N$ be a graded R-module homomorphism. Then $\operatorname{Ker}\varphi$ is a graded submodule of M and $\Im\varphi$ is a graded submodule of N.

Proof. $\operatorname{Ker}\varphi \subseteq M$ is a submodule of M. We show that $\operatorname{Ker}\varphi$ is graded. Let $m \in \operatorname{Ker}\varphi$, with $m = \sum_i m_i$ and $m_i \in M_i$. Then $0 = \sum_i \varphi(m_i)$ with $\varphi(m_i) \in N_i$. This implies that $\varphi(m_i) = 0$ for all i, and hence $m_i \in \operatorname{Ker}\varphi$ for all i. Thus the result follows from Proposition 1.3.

Now, set $U = \varphi(M)$, which is a submodule of N. Then for any $\varphi(m) \in U$ we have $\varphi(m) = \sum_i \varphi(m_i)$, where $m = \sum_i m_i$. Since $\varphi(m_i) \in N_i \cap U$, again by Proposition 1.3, it follows that $\Im\varphi$ is graded. □

Example 1.5
(a) Let $M \in \mathrm{GMod}_R$, and let U be a graded submodule of M. Then the inclusion map $U \to M$ is a morphism in the category GMod_R.
(b) Let M be a graded R-module and let $f \in R$ be homogeneous with $\deg(f) = a$. Then

$$M(-a) \to M, \ m \mapsto fm,$$

is a graded R-module homomorphism, that is, a morphism in the category GMod_R.

Exercise 1.6
Let M be a graded R-module, and let N be a graded submodule of M. Show that M/N is a graded R-module with the grading $(M/N)_i = (M_i + N)/N = M_i/N_i$.

Exercise 1.7
Let M be a graded R-module. Show that a submodule U of M is graded if and only if U is generated by homogeneous elements.

Many algebraic operations on graded modules and ideals provide modules and ideals which admit natural gradings. The problems in the next exercise give us some examples.

Exercise 1.8
Let R be a graded ring, and let M be a graded R-module.

(a) Let N and L be graded submodules of M. Show that $N : L = \{r \in R : rL \subseteq N\}$ is a graded ideal of R.

(b) Show that if $I \subseteq R$ is a graded ideal, then IM is a graded submodule of M.

(c) Let $\{U_\lambda\}_{\lambda\in\Lambda}$ be a family of graded submodules of M. Prove that the module $\bigcap_{\lambda\in\Lambda} U_\lambda$ is a graded submodule of M.

(d) Let $\{U_{\lambda\in\Lambda}\}$ be a family of graded R-modules. Show that $N = \bigoplus_{\lambda\in\Lambda} U_\lambda$ is a graded R-module with $N_i = \bigoplus_{\lambda\in\Lambda}(U_\lambda)_i$.

We call a graded R-module F *(finitely generated) graded free*, if F has a (finite) homogeneous basis. The definition of a basis is given in the beginning of Sect. 1.1.2. Exercise 1.8(d) implies that F is a graded free R-module if and only if it is of the form $F \cong \bigoplus_{\lambda \in \Lambda} Re_\lambda$ with each e_λ homogeneous of degree, say a_λ. Then for each $j \in \mathbb{Z}$, we have $F_j = \bigoplus_{\lambda \in \Lambda} R_{j-a_\lambda}$.

1.1.2 Noetherian (Graded) Rings and Modules

Let R be an arbitrary ring, and let M be an R-module. A subset $\mathcal{G} \subseteq M$ is called a *system of generators* of M, if each element of M can be written as a linear combination of elements of $\mathcal{G}$ with coefficients in R. The set $\mathcal{G}$ is called a *minimal system of generators* of M, if no proper subset of $\mathcal{G}$ generates M. We say that M is *finitely generated*, if there exists a finite set $\mathcal{G}$ of generators of M. A system of generators $\mathcal{B}$ of M is called a *basis* of M, if any element of M is a unique R-linear combination of elements in $\mathcal{B}$. If M has a basis, then M is called a *free* R-module.

The following result gives an important characterization of free R-modules.

Proposition 1.9
Let F be an R-module with $\mathcal{G}$ as a system of generators. Then the following conditions are equivalent:

(i) F is free with basis $\mathcal{G}$.
(ii) For any R-module M and for any sequence $\{m_g\}_{g\in\mathcal{G}}$ in M there exists a unique R-module homomorphism $\varphi : F \to M$ with $\varphi(g) = m_g$ for any $g \in \mathcal{G}$.

Proof. (i) $\Rightarrow$ (ii): We define the map $\varphi : F \to M$ with $\sum_g r_g g \mapsto \sum_g r_g m_g$, where almost all r_g are zero. Since any element in F is a unique R-linear combination of elements of $\mathcal{G}$, the map $\varphi : F \to M$ is well-defined. It can be easily seen φ is an R-module homomorphism.

(ii) $\Rightarrow$ (i): Let $M = \bigoplus_{g\in\mathcal{G}} Re_g$ be a free R-module with the basis $\{e_g : g \in \mathcal{G}\}$. Then by assumption there exists a unique R-module homomorphism $\varphi : F \to M$ with $\varphi(g) = e_g$ for any $g \in \mathcal{G}$. Suppose $\sum_g r_g g = 0$ for some $r_g \in R$. Then $\sum_g r_g e_g = \varphi(\sum_g r_g g) = 0$. Then $r_g = 0$ for all g. This implies that $\mathcal{G}$ is a basis of F. □

Exercise 1.10
Let M be a finitely generated free R-module. Show that

(a) F has a finite basis.
(b) Any two bases for F have the same cardinality.

A module M satisfying the condition that any submodule of M is finitely generated is called *Noetherian*. A ring R is called Noetherian if it is Noetherian as an R-module, which means that every ideal in R is finitely generated.

Noetherian modules can be characterized as follows:

Lemma 1.11
Let M be an R-module. The following conditions are equivalent:

(i) M is Noetherian.
(ii) Any ascending chain of submodules of M becomes stationary.

Proof. (i) $\Rightarrow$ (ii): Let $U_1 \subseteq U_2 \subseteq U_3 \subseteq \cdots$ be an ascending chain of submodules of M. Then $U = \bigcup_{i \geq 1} U_i$ is a submodule of M, which by our assumption is finitely generated. Let $u_1, \ldots, u_n$ be a set of generators of U. Then there exists an integer i such that $u_k \in U_i$ for all k. It follows that $U = U_i$. Therefore, $U_j = U_i$ for all $j \geq i$.

(ii) $\Rightarrow$ (i): Let $U \subseteq M$ be a submodule of M. Assume that U is not finitely generated. Then there exists an infinite sequence $u_1, u_2, \ldots$ in U such that $u_{i+1} \notin U_i$, where U_i is generated by $u_1, \ldots, u_i$. Indeed, suppose $u_1, \ldots, u_i$ is already constructed. Since we assume that U is not finitely generated, we have $U_i \neq U$. Hence we may choose $u_{i+1} \in U \setminus U_i$. This sequence of submodules U_i is strictly increasing and does not become stationary. This contradicts our assumption. □

We let $\mathcal{M}_R$ to be the full subcategory of GMod_R whose objects are the finitely generated graded R-modules.

Lemma 1.12
Let R be a graded ring, and let M be a graded R-module. Then the following statements hold:

(a) Suppose that $M \in \mathcal{M}_R$. Then M admits a finite system of homogeneous generators of M.
(b) Suppose R is a finitely generated K-algebra and M is finitely generated. Then $\dim_K M_i < \infty$ for all i.

Proof. (a) Let $\mathcal{G} = \{g_1, \ldots, g_m\}$ be a system of generators of M. Then $g_i = \sum_j g_{ij}$ with $g_{ij} \in M_j$, and each of these sums has finitely many summands. It is then clear that the finitely many homogeneous elements g_{ij} generate M, as well.

(b) According to the proof of (a) we may assume that $\mathcal{G} = \{g_1, \ldots, g_m\}$ is a homogeneous system of generators of M. Let $a_j = \deg g_j$ for $j = 1, \ldots, m$. Then for any integer i we have $M_i = \sum_{j=1}^m R_{i-a_j}g_j$. By Lemma 1.1, $\dim_K R_{i-a_j} < \infty$ for all j. This yields the desired conclusion. □

Exercise 1.13
Let M be a Noetherian R-module, and let N be a submodule of M. Show that for any system of generators $\mathcal{G}$ of N, there exists $g_1, \ldots, g_r \in \mathcal{G}$ such that N is generated by $g_1, \ldots, g_r$.

The following important result will imply that all finitely generated K-algebras are Noetherian.

Theorem 1.14
(Hilbert's basis theorem) Let R be a Noetherian ring. Then the polynomial ring $R[x]$ is also Noetherian.

Proof. Assume that $R[x]$ is not Noetherian. Then there exists an ideal $I \subsetneq R[x]$ which is not finitely generated. We claim that there exists a sequence of polynomials $f_1, f_2, \ldots$ in I with the following property: for all i, $f_{i+1} \notin (f_1, \ldots, f_i)$ and f_{i+1} is a polynomial of smallest degree with this property.

We prove this by induction on i. Let f_1 be a nonzero polynomial of smallest degree in I and suppose we constructed already $f_1, \ldots, f_i$ for some $i \geq 1$. Since I is not finitely generated, it follows that $(f_1, \ldots, f_i) \neq I$. Therefore, there exists a polynomial $f \in I \setminus (f_1, \ldots, f_i)$. We let f_{i+1} be such a polynomial with smallest degree.

For all i, let $d_i = \deg f_i$ and $a_i \in R$ be the leading coefficient of f_i. Then by the choice of the f_i we have $d_1 \leq d_2 \leq \cdots$. Indeed, if $d_{i+1} < d_i$ for some i, then one could replace f_i by f_{i+1} which contradicts to the choice of f_i. The ideal $J \subseteq R$, generated by the elements a_i, is finitely generated, because R is Noetherian. Thus there exists an integer n such that $J = (a_1, \ldots, a_n)$. In particular, $a_{n+1} = \sum_{j=1}^n b_j a_j$ with $b_j \in R$. Let $g = f_{n+1} - \sum_{j=1}^n b_j x^{d_{n+1}-d_j} f_j$. Then $\deg g < \deg f_{n+1}$ and $g \in I \setminus (f_1, \ldots, f_n)$, a contradiction. □

Let as before S be the polynomial ring over K in a finite number of variables. By using induction on the number of variables of S and by using the fact that $K[x_1, \ldots, x_{i-1}][x_i] = K[x_1, \ldots, x_i]$ for all i, it follows from Hilbert's basis theorem that all ideals in S are finitely generated. Hence S is Noetherian. This result has nothing to do with the grading of S which we may ignore in this context.

On the other hand, let $S = K[x_1, x_2, \ldots]$ be the polynomial ring in infinitely many variables. Then S is not Noetherian, because the ideal $I = (x_1, x_2, \ldots) \subsetneq S$ is not finitely generated.

Corollary 1.15
Let $S \subseteq R$ be a subring of R, and assume there exist elements $r_1, \dots, r_m \in R$ such that $R = S[r_1, \dots, r_m]$. If S is Noetherian ring, then R is a Noetherian ring. In particular, any finitely generated K-algebra R (not necessarily graded) is a Noetherian ring.

Proof. Let $T = S[x_1, \dots, x_m]$ be the polynomial ring over S. There exists a surjective S-algebra homomorphism $\varphi\ T \to R$ with $\varphi(x_i) = r_i$ for $i = 1, \dots, m$, see Exercise 1.16. Then $R \cong T/J$, where $J = \operatorname{Ker}\varphi$. Now let $I \subseteq R$ be an ideal, and let $L = \{f \in T\ \ \varphi(f) \in I\}$. Then $L \subseteq T$ is an ideal in T. By Theorem 1.14, T is Noetherian. Therefore, L is finitely generated. If $f_1, \dots, f_s$ is a set of generators of L, then $\varphi(f_1), \dots, \varphi(f_s)$ is a set of generators of I. □

Exercise 1.16
(a) Let $S \subset R$ be subring of R, and let $T = S[x_1, \dots, x_m]$ be the polynomial ring over the ring S. Futhermore, let $r_1, \dots, r_m \in R$. Then there exists a unique S-algebra homomorphism $\varphi\ T \to R$ with $\varphi(x_i) = r_i$ for $i = 1, \dots, m$. In particular, if $R = S[r_1, \dots, r_m]$, then $R \cong T/J$, where $J = \operatorname{Ker}\varphi$.
(b) Show that $K[x^2, xy, y^2] \cong K[x_1, x_2, x_3]/(x_2^2 - x_1x_3)$.

The following result is an analogous statement of Corollary 1.15 for the case of modules.

Theorem 1.17
Let R be a Noetherian ring, and let M be an R-module. Then M is finitely generated if and only if it is Noetherian.

Proof. We only need to show that M is Noetherian if M is finitely generated. Let $m_1, \dots, m_s$ be generators of M, and let F be the free R-module with basis $e_1, \dots, e_s$. We define the R-module homomorphism $\epsilon\ F \to M$ with $\epsilon(e_i) = m_i$ for $i = 1, \dots, s$, see Proposition 1.9. Let U be a submodule of M, and let $V = \{v \in F\ \ \epsilon(v) \in U\}$. Then $V \subseteq F$ is a submodule of F and $\epsilon(V) = U$. It is enough to show that V is finitely generated, because if $v_1, \dots, v_t$ is a system of generators of V, then $\epsilon(v_1), \dots, \epsilon(v_t)$ is a system of generators of U. By induction on s we show that V is finitely generated. If $s = 1$, then $V = Ie_1$, where I is an ideal in R. Since R is Noetherian, I and hence also V, is finitely generated.

Now, let $s > 1$. For any $v \in V$, we have $v = v(1)e_1 + \cdots + v(s)e_s$ with $v(i) \in R$. The set $I = \{v(1)\ \ v \in V\}$ is an ideal in R. Since R is Noetherian, I is finitely generated. Let $f_1, \dots, f_k$ be a system of generators of I, and let $w_1, \dots, w_k \in V$

with $w_i(1) = f_i$ for $i = 1, \ldots, k$. We let $W \subseteq V$ be the submodule of V generated by $w_1, \ldots, w_k$ and $V' = \{v \in V \ \ v(1) = 0\}$. By our induction hypothesis, V' is finitely generated, and moreover, $V = W + V'$. Indeed, for any $v \in V$, there exist $r_1, \ldots, r_k \in R$ such that $v(1) = \sum_{i=1}^{k} r_i f_i$. Then $v - \sum_{i=1}^{k} r_i w_i \in V'$, which implies that $v \in W + V'$. Since W is also finitely generated, it follows that V is finitely generated. □

Exercise 1.18
Let R be a graded K-algebra. Show that R is Noetherian if and only if R is a finitely generated K-algebra.

The following exercise provides another characterization of Noetherian modules.

Exercise 1.19
Let M be an R-module. Show that M is Noetherian if and only if any nonempty set of submodules of M has a maximal element (with respect to inclusion).

Let R be a finitely generated graded K-algebra. To show the Noetherian property for a graded R-module M with $\dim_K M_i < \infty$ for all i, it is enough to consider the graded submodules of M. This is shown in

Proposition 1.20
Let R be a finitely generated graded K-algebra, and let $M = \bigoplus_{i\in\mathbb{Z}} M_i$ be a graded R-module such that $\dim_K M_i < \infty$ for all $i \in \mathbb{Z}$. Then M is finitely generated if and only if any ascending chain of graded submodules of M becomes stationary.

Proof. Suppose that any ascending chain of graded submodules of M becomes stationary. First we show that $M = \bigoplus_{i=k}^{\infty} M_i$ for some k. Indeed, for any integer $j \geq 0$, set $N_j = \bigoplus_{i\geq -j} M_i$. Then the ascending chain $N_0 \subseteq N_1 \subseteq \cdots$ of graded submodules of M stops. Let $k \in \mathbb{Z}$ such that $N_k = N_\ell$ for all $\ell > k$. This implies that $M_i = 0$ for $i < k$. Now, for any $i \geq k$ let U_i be the graded submodule of M generated by $\bigcup_{j=k}^{i} M_j$. Then by our assumption, there exists an integer t such that $U_t = U_i$ for all $i > t$. This implies that $M_\ell \subseteq U_t$ for all $\ell \in \mathbb{Z}$. Hence $M = U_t$. Since each M_i has a finite K-basis B_i, we conclude that M is generated by $\bigcup_{j=k}^{t} B_i$. The converse statement is obvious. □

For an ideal I of a ring R, the *radical* of I is defined as

$$\sqrt{I} = \{r \in R : r^n \in I \text{ for some positive integer } n\}.$$

The radical of an ideal is again an ideal.

Exercise 1.21

Let R be a ring, and let I be an ideal of R. Show that

(a) If R is Noetherian, then $(\sqrt{I})^n \subseteq I$ for some positive integer n.

(b) If R is graded and I is a graded ideal of R, then $\sqrt{I}$ is a graded ideal, as well.

Krull's intersection theorem is one of the fundamental theorems in commutative algebra. Here, we present the graded version of this theorem for graded modules over graded K-algebras.

Observe that any graded K-algebra R has a unique graded maximal ideal, namely the ideal $\mathfrak{m} = \bigoplus_{i>0} R_i$.

Theorem 1.22

(Krull) Let R be a graded K-algebra with the graded maximal ideal $\mathfrak{m}$, and let M be a graded R-module with $M_i = 0$ for $i \ll 0$. Furthermore, let $U \subseteq M$ be a graded submodule of M, and let $I \subseteq \mathfrak{m}$ be a graded ideal. Then

$$\bigcap_{k\geq 0}(U + I^k M) = U.$$

In particular, one has $\bigcap_{k\geq 0} I^k M = 0$. The hypothesis on M is satisfied, if M is finitely generated.

Proof. Obviously, $U \subseteq \bigcap_{k\geq 0}(U + I^k M)$. Conversely, let $x \in \bigcap_{k\geq 0}(U + I^k M)$. By Exercise 1.8 this intersection is a graded R-module. Therefore, by Proposition 1.3 we may assume that x is homogeneous, say, $\deg x = a$. By assumption there exists an integer c such that $M_i = 0$ for $i < c$. Since $I \subseteq \mathfrak{m}$ and R is positively graded, it follows that there exists an integer $b > 0$ such that all elements of I have degree $\geq b$. Choose a number $\ell \geq 0$ big enough such that $\ell b + c > a$. Since $\deg x = a$, it follows that $x \in (U + I^\ell M)_a = U_a + (I^\ell M)_a = U_a \subseteq U$. □

One of the most important applications of Theorem 1.22 is the so-called Nakayama's lemma.

Theorem 1.23

(Nakayama) With the assumptions and notation as in Theorem 1.22, suppose that $U + IM = M$. Then $U = M$.

Proof. We first prove by induction on k that $U + I^k M = M$. For $k = 1$ this is our assumption. Now let $k \geq 1$. By induction hypothesis we may assume that $U + I^k M = M$. Then $M = U + I^k(U + IM) = U + I^k U + I^{k+1} M =$

$U + I^{k+1}M$. Now, we apply Krull's intersection theorem and obtain $M = \bigcap_{k\geq 0} (U + I^k M) = U$. □

Nakayama's lemma has the following remarkable consequences.

Corollary 1.24
Let R be a graded K-algebra with the graded maximal ideal $\mathfrak{m}$, let M be a graded R-module such that $M_i = 0$ for $i \ll 0$ (which is the case if M is finitely generated), and let $f_1, \ldots, f_r$ be homogeneous elements of M. Then $\{f_1, \ldots, f_r\}$ is a minimal set of homogeneous generators of M if and only if $f_1 + \mathfrak{m}M, \ldots, f_r + \mathfrak{m}M$ is a homogeneous K-basis of the K-vector space $M/\mathfrak{m}M$.

Proof. Set $v_i = f_i + \mathfrak{m}M$ for $i = 1, \ldots, r$. First suppose that $\{f_1, \ldots, f_r\}$ is a minimal set of homogeneous generators of M. Then $v_1, \ldots, v_r$ is a system of generators of the K-vector space $M/\mathfrak{m}M$. Now, let $\{v_{i_1}, \ldots, v_{i_s}\} \subseteq \{v_1, \ldots, v_r\}$ be a K-basis of $M/\mathfrak{m}M$. Then one can see that $M = \mathfrak{m}M + U$, where U is the submodule of M which is generated by $f_{i_1}, \ldots, f_{i_s}$. Theorem 1.23 implies that $U = M$. By our assumption on $f_1, \ldots, f_r$ we obtain $\{v_{i_1}, \ldots, v_{i_s}\} = \{v_1, \ldots, v_r\}$.

Conversely, suppose that $v_1, \ldots, v_r$ is a homogeneous K-basis of the K-vector space $M/\mathfrak{m}M$. Let U be the submodule of M which is generated by $f_1, \ldots, f_r$. Note that $U + \mathfrak{m}M = M$. Therefore, Theorem 1.23 implies that $U = M$. By the minimality of $v_1, \ldots, v_r$, the set $\{f_1, \ldots, f_r\}$ is indeed a minimal set of generators of M. □

Exercise 1.25
Let R be a graded K-algebra and M be a finitely generated graded R-module. Show that $M_i = 0$ for $i \ll 0$.

Suppose that M is a finitely generated graded R-module over the graded K-algebra R. Then by Corollary 1.24, $M/\mathfrak{m}M$ is a finite dimensional graded K-vector space and all the homogeneous minimal systems of generators of M have the same cardinality, namely, $\dim_K M/\mathfrak{m}M$. The minimal number of a homogeneous set of generators of M will be denoted by $\mu(M)$. In other words, $\mu(M) = \dim_K M/\mathfrak{m}M$.

Exercise 1.26
Let $S = K[x_1, \ldots, x_n]$ be the standard graded polynomial ring. Show that
(a) If $n = 1$ and $I \subseteq S$ is a nonzero graded ideal, then $\mu(I) = 1$.
(b) If $n > 1$, then for all $k \in \mathbb{Z}_{\geq 0}$ there exists a graded ideal I with $\mu(I) = k$.

Exercise 1.27
Let R be a graded K-algebra, let M be a finitely generated graded R-module and let $\mathscr{G}$ be a minimal homogeneous system of generators of M. We denote by β_{0i} the number of elements $f \in \mathscr{G}$ with $\deg f = i$.

Show that $\beta_{0i} = \dim_K(M/\mathfrak{m}M)_i$, and hence it does not depend on the particular chosen minimal set $\mathscr{G}$ of generators of M. Hint: use Nakayama's lemma.

We define the *embedding dimension* of a finitely generated graded K-algebra R with graded maximal ideal $\mathfrak{m}$ as the number

$$\operatorname{emb} R = \dim_K \mathfrak{m}/\mathfrak{m}^2.$$

By Corollary 1.24, $\operatorname{emb} R = \mu(\mathfrak{m})$.

The embedding dimension of a local ring $(R, \mathfrak{m})$ is defined similarly.

Note that if $R = K[f_1, \ldots, f_m]$, then $\mathfrak{m}$ is generated by $f_1, \ldots, f_m$. In particular, $m \geq \operatorname{emb} R$. If $m = \operatorname{emb} R$, then we say that R is minimally generated by $f_1, \ldots, f_m$ as a K-algebra.

Suppose R is generated by $f_1, \ldots, f_m$. Let $S = K[x_1, \ldots, x_m]$ be the polynomial ring over a field K with the grading $\deg(x_i) = \deg(f_i)$ for all i, and let $\varphi\ S \to R$ be the graded K-algebra homomorphism with $\varphi(x_i) = f_i$ for $i = 1, \ldots, m$. We call $S/\operatorname{Ker}\varphi$ a *minimal presentation* of R if $f_1, \ldots, f_m$ minimally generate R.

Exercise 1.28
Let $\varphi\ S \to R$ be as above. Show that $S/\operatorname{Ker}\varphi$ is a minimal presentation of R if and only if $\operatorname{Ker}\varphi \subseteq (x_1, \ldots, x_m)^2$.

Let M be a nonzero graded R-module. Then M is called a *simple* R-module if (0) and M are the only submodules of M.

Lemma 1.29
Let R be a graded K-algebra. A finitely generated graded R-module M is simple if and only if there exists an integer a such that $M \cong (R/\mathfrak{m})(a)$.

Proof. Suppose that M is simple. Then $\mu(M) = 1$. Indeed if $\mu(M) > 1$, and $f \in M$ be one of the minimal generators of M, then $(0) \neq Rf \subsetneq M$, which is not possible. Hence, $M = Rf$ for some homogeneous element $f \in M$ with $\deg f = -a$ for some $a \in \mathbb{Z}$. Therefore, $M \cong (R/I)(a)$, where $I \subseteq R$ is a graded ideal. If $I \neq \mathfrak{m}$, then R/I is not a field, and R/I contains a nonzero proper ideal, a contradiction.

Conversely, if $M \cong (R/\mathfrak{m})(a)$, then M is simple, because a field does not contain any proper submodules. □

Let M be an R-module. A *chain of submodules* of M of length m is a sequence of submodules

$$(0) = U_m \subsetneq \cdots \subsetneq U_1 \subsetneq U_0 = M.$$

The chain is called a *composition series* if U_{i-1}/U_i is a simple R-module for $1 \leq i \leq m$.

We say that M is of *finite length* if there exists an integer k such that the length of any chain of submodules of M is bounded by k.

Theorem 1.30

(Jordan-Hölder theorem) If a module has a composition series of length m, then the module is of finite length and all composition series of this module have length m.

Proof. We prove this theorem by induction on m. Let M be an R-module with composition series $(0) = U_m \subsetneq \cdots \subsetneq U_1 \subsetneq U_0 = M$. We want to show that the length of any chain of submodule of M is at most m. If $m = 1$, then M is simple and $(0) \subsetneq M$ is the only chain in M. So we may assume that $m > 1$. Let $(0) = V_s \subsetneq \cdots \subsetneq V_1 \subsetneq V_0 = M$ be a chain of submodules of M.

If $V_1 \subseteq U_1$, then $(0) = V_s \subsetneq \cdots \subsetneq V_2 \subsetneq U_1$ is a chain of submodules of U_1 of length $s - 1$. Since $(0) = U_m \subsetneq \cdots \subsetneq U_1$ is a composition series of U_1 of length $m - 1$, our induction hypothesis implies that $s - 1 \leq m - 1$, and we are done.

Now, assume that $V_1 \nsubseteq U_1$. Since M/U_1 is simple, we have $M = U_1 + V_1$. Therefore, $M/U_1 = (U_1 + V_1)/U_1 \cong V_1/(U_1 \cap V_1)$. This implies that $V_1/(U_1 \cap V_1)$ is simple. Note that $U_1 \cap V_1 \subsetneq U_1$. Indeed, if $U_1 \cap V_1 = U_1$, then $U_1 \subseteq V_1 \subseteq M$. Since M/U_1 is simple, it follows that $V_1 = U_1$ or $U_1 = M$, both is not possible. Therefore, any chain of submodule of $U_1 \cap V_1$ of length r can be extended to a chain of submodules of U_1 of length $r + 1$. Since U_1 has a composition series of length $m - 1$, the induction hypothesis implies that $r + 1 \leq m - 1$. Hence, $r \leq m - 2$. It follows that $U_1 \cap V_1$ has a composition series $\mathscr{C}$ of length $\leq m - 2$. The composition series $\mathscr{C}$ can be extended by $U_1 \cap V_1 \subsetneq V_1$ to obtain a composition series of V_1 of length $\leq m - 1$. Therefore, applying the induction hypothesis to V_1 we have $s - 1 \leq m - 1$, as desired. □

For a module M of finite length, the common length of composition series of M is called the *length* of M, denoted $\ell(M)$. If M is not of finite length, we set $\ell(M) = \infty$. A sequence

$$M_1 \xrightarrow{f_1} M_2 \xrightarrow{f_2} \cdots \xrightarrow{f_{n-1}} M_n \xrightarrow{f_n} M_{n+1}$$

of R-modules and R-module homomorphisms is called an *exact sequence* if $\Im f_i = \operatorname{Ker} f_{i+1}$ for any $1 \leq i \leq n-1$. Any exact sequence of the fxorm

$$0 \to M \to N \to P \to 0$$

is called a *short exact sequence*.

Exercise 1.31
Let $0 \to U \to M \to N \to 0$ be a short exact sequence of R-modules. Show that $\ell(M) = \ell(U) + \ell(N)$.

Proposition 1.32
Let R be a finitely generated graded K-algebra, and let M be a finitely generated graded R-module. Then

(a) M is of finite length if and only if there exists a positive integer k such that $\mathfrak{m}^k M = 0$.

(b) $\ell(M) = \dim_K M$. Moreover, $\ell(M) < \infty$ if and only if $M_i = 0$ for all but finitely many i.

Proof. (a) By an iterated use of Exercise 1.31,

$$\ell(M/\mathfrak{m}^k M) = \sum_{i=0}^{k-1} \ell(\mathfrak{m}^i M/\mathfrak{m}^{i+1} M)$$

for any positive integer k. Now, assume that $\ell(M) < \infty$. If $\mathfrak{m}^i M \neq 0$ for all i, then by Nakayama's lemma $\mathfrak{m}^i M/\mathfrak{m}^{i+1} M \neq 0$ for all i, which implies that $\ell(\mathfrak{m}^i M/\mathfrak{m}^{i+1} M) > 0$. Hence, $\ell(M/\mathfrak{m}^k M) \geq k$ and then $\ell(M) > \ell(M/\mathfrak{m}^k M) \geq k$ for all k, a contradiction.

Conversely, assume that $\mathfrak{m}^k M = 0$ for some k. Then $\ell(M) = \ell(M/\mathfrak{m}^k M)$. Note that each $\mathfrak{m}^i M/\mathfrak{m}^{i+1} M$ is a finite dimensional K-vector space, because by Nakayama's lemma $\dim_K(\mathfrak{m}^i M/\mathfrak{m}^{i+1} M) = \mu(\mathfrak{m}^i M)$. On the other hand, since R is Noetherian and M is finitely generated, we have $\mu(\mathfrak{m}^i M) < \infty$. So $\mathfrak{m}^i M/\mathfrak{m}^{i+1} M$ is of finite length. Thus, M is of finite length, as well.

(b) Suppose that $\ell(M) < \infty$. Then (a) implies that there exists a positive integer k with $\mathfrak{m}^k M = 0$. Then $\ell(M) = \sum_{i=0}^{k-1} \ell(\mathfrak{m}^i M/\mathfrak{m}^{i+1} M) = \sum_{i=0}^{k-1} \dim_K(\mathfrak{m}^i M/\mathfrak{m}^{i+1} M) = \dim_K M = \sum_i \dim_K M_i$. This implies that $\dim_K M_i = 0$, except finitely many i. Finally, assume that $\ell(M) = \infty$. Then by (a), $\mathfrak{m}^i M \neq 0$ for all i and therefore, $\dim_K(\mathfrak{m}^i M/\mathfrak{m}^{i+1} M) > 0$. Hence, $\dim_K M \geq \dim_K M/\mathfrak{m}^k M \geq k$ for all k. This shows that $\dim_K M = \infty$. □

Exercise 1.33

Let $S = K[x_1, \ldots, x_n]$ be the polynomial ring with the standard grading, and let $I \subseteq S$ be a graded ideal.

(a) Show that $\ell(S/I) < \infty$ if and only if for all i there exists a positive integer a_i such that $x_i^{a_i} \in I$.

(b) Compute $\ell(S/(x_1^{a_1}, \ldots, x_n^{a_n}))$.

For later applications we also need the local version of Nakayama's lemma.

Theorem 1.34

Let $(R, \mathfrak{m})$ be a local ring, $I \subseteq \mathfrak{m}$ be an ideal, M be an R-module and let N be a submodule of M such that M/N is finitely generated. Suppose that $M = N + IM$. Then $M = N$.

Proof. Suppose $M \neq N$, and set $W = M/N$. Then $W \neq 0$. Let $w_1, \ldots, w_m$ be a minimal system of generators of W. Since $W = IW$, it follows that $w_m = \sum_{i=1}^{m} a_i w_i$ with $a_i \in I$. Therefore, $(1 - a_m)w_m = \sum_{i=1}^{m-1} a_i w_i$. The element $1 - a_m$ is a unit in R, since R is local and $a_m \in \mathfrak{m}$. This implies that $w_m \in \sum_{i=1}^{m-1} Rw_i$, contradicting the assumption that $w_1, \ldots, w_m$ is a minimal system of generators. $\square$

We conclude this section by considering an important invariant of modules. Let R be any domain, and let M be a finitely generated R-module. Then $S^{-1}M$ is a finitely generated $Q(R)$-vector space, where $S = R \setminus \{0\}$ and $Q(R)$ is the quotient field of R. The $Q(R)$-vector space dimension of $S^{-1}M$ is called the *rank* of the module M, denoted rank M. The module M is called a *torsion module*, if rank $M = 0$. This is the case if and only if $S^{-1}M = 0$.

Exercise 1.35

Let F be a free R-module with basis $e_1, \ldots, e_r$. Then we have rank $F = r$.

The following lemma shows that the rank function is additive on short exact sequences.

Lemma 1.36

Let R be a ring, and let $0 \to U \to M \to N \to 0$ be a short exact sequence of finitely generated R-modules. Then rank $M =$ rank $U +$ rank N.

Proof. We notice that localization is an exact functor. Thus localization of the given exact sequence yields the exact sequence $0 \to S^{-1}U \to S^{-1}M \to S^{-1}N \to 0$ of finite dimensional $Q(R)$-vector spaces. The desired conclusion follows from linear algebra, since a basis of $S^{-1}U$ together with the preimage of a basis of $S^{-1}N$ establishes a basis of $S^{-1}M$. □

The set $\operatorname{Ann}(M) = \{r \in R : rM = 0\}$ is called the set of *annihilators* of M. It is easily seen that $\operatorname{Ann}(M)$ is an ideal of R. This ideal is called the *annihilator* of M.

The following lemma states some other useful properties of the rank invariant.

Proposition 1.37
Let R be a domain, and let $M \neq 0$ be a finitely generated R-module. Then the following statements hold:

(a) M is a torsion module if and only if $\operatorname{Ann}(M) \neq 0$.
(b) There exists a free submodule F of M with rank F = rank M such that M/F is a torsion module.
(c) There exists a free R-module F with rank F = rank M and a homomorphism $M \to F$, whose cokernel is a torsion module.

In the graded case, the free module F in (b) and (c) can be chosen to be graded and the maps between F and M to be graded homomorphisms.

Proof. (a) Let $\{m_1, \ldots, m_s\}$ be a set of generators of M, and suppose that M is a torsion module. Then $m_i/1 = 0$ for $i = 1, \ldots, s$. Thus there exist $r_i \in S = R \setminus \{0\}$ such that $r_i m_i = 0$ for $i = 1, \ldots, s$. Let $r = r_1 \cdots r_s$. Since R is a domain, it follows that $r \neq 0$. Moreover, $rm_i = 0$ for all i. Hence, $rM = 0$. This shows that $\operatorname{Ann}(M) \neq 0$.

Conversely, suppose that $\operatorname{Ann}(M) \neq 0$. Then there exists $r \in R$, $r \neq 0$, such that $rM = 0$. This implies that $(r/1)S^{-1}M = 0$. Since $r/1$ is a unit in $Q(R)$, we conclude that $S^{-1}M = 0$. In other words, M is a torsion module.

(b) Let t be the rank of M, and let $\{m_1, \ldots, m_s\}$ be system of generators of M. The module $S^{-1}M$ is a $Q(R)$-vector space of dimension t which is generated by the elements $m_1/1, \ldots, m_s/1$. From linear algebra we know that a basis of $S^{-1}M$ can be chosen as a subset of $\{m_1/1, \ldots, m_s/1\}$. We may assume that $m_1/1, \ldots, m_t/1$ with $t \leq s$ is a basis of $S^{-1}M$. Let $F \subseteq M$ be the R submodule of M generated by $m_1, \ldots, m_t$. Then F is free with basis $m_1, \ldots, m_t$. Indeed, suppose that $a_1m_1 + \cdots + a_rm_t = 0$. Then $(a_1/1)(m_1/1) + \cdots + (a_r/1)(m_t/1) = 0$. Since $m_1/1, \ldots, m_t/1$ is a basis of $S^{-1}M$, it follows that $a_i = 0$ for $i = 1, \ldots, t$.

The exact sequence $0 \to F \to M \to M/F \to 0$ yields the exact sequence $0 \to S^{-1}F \to S^{-1}M \to S^{-1}(M/F) \to 0$. By the construction of F, the inclusion map $F \to M$ becomes equality after localization. This implies that $S^{-1}(M/F) = 0$, and shows that M/F is a torsion module.

(c) Let F be the free module as it is defined in the proof of (b). Since M/F is a torsion module, there exists $r \in R$ such that $rM \subseteq F$. We define the R-module homomorphism $M \to F$ by setting $m \mapsto rm$ for all $m \in M$. Then the induced homomorphism $S^{-1}M \to S^{-1}F$ becomes an isomorphism, since r is a unit after localization. This shows that the cokernel of $M \to F$ is a torsion module, as desired.

In the graded case M is generated by homogeneous elements, and the element r in the proof of (c) can also be chosen to be homogeneous. It follows that the homomorphism $F \to M$ constructed in the proof of (b) is graded, and that the homomorphism $M \to F$ constructed in the proof of (c) is graded, as well. □

1.1.3 Basic Properties of Integral Extensions

Let $S \subseteq R$ be a ring extension. An element $x \in R$ is called *integral* over S, if there exists an equation

$$x^n + a_1x^{n-1} + \cdots + a_{n-1}x + a_n = 0$$

for some positive integer n and $a_i \in S$. Such an equation is called an *integral equation* for x over S.

The ring extension $S \subseteq R$ is called an *integral extension* when any element of R is integral over S.

Lemma 1.38
Let $S \subseteq R$ be a ring extension, where S is Noetherian, and R is a finitely generated S-module. Then $S \subseteq R$ is an integral extension.

Proof Let $x \in R$ and consider the S-submodules $U_i = \sum_{k=0}^{i} Sx^k$ of R. Then $U_0 \subseteq U_1 \subseteq U_2 \subseteq \cdots$. Since S is Noetherian and R is a finitely generated R-module, by Theorem 1.17 there exists an integer n such that $U_{n-1} = U_n$. This implies that $x^n \in U_{n-1}$, from which one obtains immediately the desired conclusion. □

Corollary 1.39
Let S be a Noetherian ring, and let $R = S[f_1, \ldots, f_m]$. Then R is an integral extension of S if and only if each f_i is integral over S.

Proof. Suppose that $f_1, \ldots, f_m$ are integral over S. Let $g_i(x) = 0$ be an integral equation of f_i for all $1 \leq i \leq m$, and let $n_i = \deg(g_i)$. We claim that the set $\mathcal{G}$ which consists of elements of the form $f_1^{i_1} \cdots f_m^{i_m}$ with $0 \leq i_j < n_j$ for $j = 1, \ldots, m$ is a set of generators for R as an S-module. Indeed, let f be an arbitrary element in R. Then $f = \sum_{i_1,\ldots,i_m \in \mathbb{Z}_{\geq 0}} s_{i_1,\ldots,i_m} f_1^{i_1} \cdots f_m^{i_m}$, where all $s_{i_1,\ldots,i_m} \in S$. If $i_j \geq n_j$ for some j, then using the equation $g_j(x) = 0$, the element $f_j^{i_j}$ may be written as S-linear combination of $f_j, f_j^2, \ldots, f_j^{n_j-1}$. This can be done for each factor $f_\ell^{i_\ell}$. Then each $f_1^{i_1} \cdots f_m^{i_m}$ appearing in the expression of f may be rewritten as S-linear combination of elements of $\mathcal{G}$. The desired conclusion follows from Lemma 1.38. On the other hand if R is integral over S, then by definition each f_i is integral over S. □

Proposition 1.40
Let $S \subseteq R$ be a ring extension, where S is a Noetherian ring, and let $T \subseteq R$ be the set of elements in R which are integral over S. Then T is a ring extension of S.

Proof. Let $f, g \in T$. By Corollary 1.39, the ring $S[f, g]$ is an integral extension of S. This implies that $S[f, g] \subseteq T$. In particular, $f + g, fg \in T$. This shows that T is a ring containing S as a subring. □

The ring T in Proposition 1.40 is called the *integral closure* of S in R.

Lemma 1.41
Let $S \subseteq R$ be an integral extension, and let T be a multiplicatively closed subset of S. Then the followings hold:

(a) For any ideal $J \subsetneq R$ and $I = J \cap S$, the ring extension $S/I \subseteq R/J$ is integral.
(b) $T^{-1}S \subseteq T^{-1}R$ is an integral extension.
(c) Suppose in addition that S and R are integral domains and let $T = S \setminus \{0\}$. Then $T^{-1}R = Q(R)$ and the field extension $Q(S) \subseteq Q(R)$ is algebraic.
(d) If R is a field, then S is a field.

Proof. (a) Let $\overline{x} = x + J \in R/J$, and let $x^n + a_1x^{n-1} + \cdots + a_{n-1}x + a_n = 0$ be an integral equation for x over S. Then $\overline{x}^n + \overline{a}_1\overline{x}^{n-1} + \cdots + \overline{a}_{n-1}\overline{x} + \overline{a}_n = 0$ is an integral equation for $\overline{x}$ over S/I. Here, $\overline{a_i} = a_i + I$ for all i.

(b) Let $x/s \in T^{-1}R$, and let $x^n + a_1x^{n-1} + \cdots + a_{n-1}x + a_n = 0$ be an integral equation for x over S. Then $(x/s)^n + (a_1/s)(x/s)^{n-1} + \cdots + (a_n/s^n) = 0$ is an integral equation of x/s over $T^{-1}S$.

(c) It follows from (b) that all elements of $T^{-1}R$ satisfy an algebraic equation over the field $Q(S)$. This shows that $T^{-1}R$ is an algebraic field extension of $Q(S)$. In particular, we must have that $T^{-1}R = Q(R)$.

(d) Consider a nonzero element x in S. Then $x^{-1} \in R$, and hence there exists an equation $x^{-n} + s_1x^{-n+1} + \cdots + s_{n-1}x^{-1} + s_n = 0$ with $s_i \in S$. Multiplying this equation with x^{n-1} we see that $x^{-1} \in S$. □

Lemma 1.42

Let $S \subseteq R$ be a ring extension of integral domains, and suppose that $Q(S) \subseteq Q(R)$ is an algebraic field extension. Then for any nonzero ideal $I \subseteq R$ we have $I \cap S \neq 0$.

Proof. It is enough to show that if $x \in R$ is a nonzero element, then $(x) \cap S \neq 0$. Since $Q(S) \subseteq Q(R)$ is an algebraic field extension, there exists an equation $(x/1)^n + (a_1/b_1)(x/1)^{n-1} + \cdots + (a_{n-i}/b_{n-i})(x/1)^i = 0$ with $a_j, b_j \in S$ for all j and $a_{n-i}/b_{n-i} \neq 0$. By clearing denominators we obtain an equation $c_0x^n + c_1x^{n-1} + \cdots + c_{n-i}x^i = 0$ with $c_j \in S$ for all j and $c_{n-i} \neq 0$. Then $x^i(c_0x^{n-i} + c_1x^{n-i-1} + \cdots + c_{n-i}) = 0$. Since $x \neq 0$ and since S is an integral domain, it follows that $c_0x^{n-i} + c_1x^{n-i-1} + \cdots + c_{n-i} = 0$. Hence $c_{n-i} \in (x) \cap S$. □

The following exercise shows that the assumption $Q(S) \subseteq Q(R)$ is an algebraic field extension can not be omitted in Lemma 1.42.

Exercise 1.43

Let K be a field, and let $S = K[x_1]$ and $R = K[x_1, x_2]$ be polynomial rings.

(a) Show that $Q(S) \subseteq Q(R)$ is not an algebraic field extension.

(b) Consider the ideal $I = (x_2) \subsetneq R$. Show that $I \cap S = 0$.

Lemma 1.44

Let $S \subseteq R$ be an integral extension, and let P be a prime ideal in S. Then there exists a prime ideal Q in R such that $Q \cap S = P$.

Proof. Let $T = S \setminus P$. Then by Lemma 1.41(b), $S_P \subseteq T^{-1}R$ is again an integral extension. Let Q' be a maximal ideal of $T^{-1}R$. Lemma 1.41(a) implies that $S_P/(Q' \cap S_P) \subseteq (T^{-1}R)/Q'$ is an integral extension. We also have that $(T^{-1}R)/Q'$ is field. Hence Lemma 1.41(d) implies that $S_P/(Q' \cap S_P)$ is a field. This is only possible if $PS_P = Q' \cap S_P$. Thus, if $Q \subsetneq R$ is the prime ideal with $QT^{-1}R = Q'$, then $Q \cap S = P$. □

1.1.4 Associated Prime Ideals

Associated prime ideals determine the zerodivisors of a module. They also show up in irredundant primary decompositions. We mainly present the theory for general Noetherian rings, but also pay attention to graded rings.

Let R be a ring, and let M be an R-module. An element $r \in R$ is called a *zerodivisor* of M, when there exists a nonzero element $x \in M$ such that $rx = 0$. The set of all zerodivisors of M is denoted by $Z(M)$.

Let R be a Noetherian ring and M be an R-module. A prime ideal P of R is called an *associated prime ideal* of M if $P = \text{Ann}(x)$ for some $x \in M$. The set of all associated prime ideals of M is denoted by $\text{Ass}_R(M)$ or simply $\text{Ass}(M)$. For an ideal $I \subsetneq R$, by abuse of notation we denote $\text{Ass}(R/I)$ simply by $\text{Ass}(I)$, and we call it the *set of associated prime ideals* of I. So any associated prime ideal of I is of the form $P = \text{Ann}(x + I) = I : (x)$ for some $x \in R$.

Exercise 1.45
The prime ideal P is an associated prime ideal of M if and only if there exists a submodule of M which is isomorphic to R/P. In particular, if $P \in \text{Ass}(M)$, then R/P may be considered as a submodule of M.

The following lemma shows the existence of associated prime ideals in Noetherian rings.

Lemma 1.46
Let M be a nonzero R-module. Then any maximal element (with respect to inclusion) of the set $\mathscr{A} = \{\text{Ann}(x) : x \in M, x \neq 0\}$ is a prime ideal. Moreover, if R is a Noetherian ring, then $\text{Ass}(M) \neq \emptyset$.

Proof. Let $P = \text{Ann}(x)$ be a maximal element of $\mathscr{A}$. Let $r, s \in R$ with $rs \in P$ and $s \notin P$. Then $sx \neq 0$ and $P \subseteq \text{Ann}(sx)$. Since $\text{Ann}(sx) \in \mathscr{A}$, we should have $P = \text{Ann}(sx)$. Note that $r \in \text{Ann}(sx)$ which then means $r \in P$, as desired. Now, if R is a Noetherian ring, then by Exercise 1.19, the set $\mathscr{A}$ has a maximal element which by the first statement belongs to $\text{Ass}(M)$. □

Lemma 1.47
Let R be a Noetherian ring and M be an R-module. Then

$$Z(M) = \bigcup_{P \in \mathrm{Ass}(M)} P.$$

Proof. Clearly $P \subseteq Z(M)$ for any $P \in \mathrm{Ass}(M)$. Let $r \in Z(M)$. Then $r \in \mathrm{Ann}(x)$ for some nonzero element x in M. Since R is Noetherian, $\mathrm{Ann}(x) \subseteq \mathrm{Ann}(y_0)$, where $\mathrm{Ann}(y_0)$ is a maximal element of $\{\mathrm{Ann}(y) : y \in M, y \neq 0\}$. By Lemma 1.46, $\mathrm{Ann}(y_0)$ is a prime ideal. So it belongs to $\mathrm{Ass}(M)$ and then $r \in \bigcup_{P \in \mathrm{Ass}(M)} P$. □

Lemma 1.48
Let N be a submodule of M. Then $\mathrm{Ass}(N) \subseteq \mathrm{Ass}(M) \subseteq \mathrm{Ass}(N) \cup \mathrm{Ass}(M/N)$.

Proof. The first inclusion is clear from the definition. Now, let $P \in \mathrm{Ass}(M)$. If $P \in \mathrm{Ass}(N)$, we are done. So we may assume that $P \notin \mathrm{Ass}(N)$. Then $P = \mathrm{Ann}(x)$ for some $x \in M \setminus N$. We show that $P = \mathrm{Ann}(\overline{x})$, where $\overline{x}$ denotes the image of x in M/N. Obviously, $P \subseteq \mathrm{Ann}(\overline{x})$. Now, let $r \in \mathrm{Ann}(\overline{x})$, and by contradiction assume that $r \notin P$. Then $rx \in N$ and $rx \neq 0$. For any $s \in \mathrm{Ann}(rx)$, we have $rs \in P$. Since $r \notin P$, we have $s \in P$. This shows that $\mathrm{Ann}(rx) = P$. Hence, $P \in \mathrm{Ass}(N)$, a contradiction. Therefore, $P = \mathrm{Ann}(\overline{x})$ which implies that $P \in \mathrm{Ass}(M/N)$. □

Exercise 1.49
Let $M_1, \ldots, M_n$ be R-modules, and let $M = \bigoplus_{i=1}^n M_i$. Show that $\mathrm{Ass}(M) = \bigcup_{i=1}^n \mathrm{Ass}(M_i)$.

A proper submodule N of M is called *primary*, if the following condition is satisfied: if $rx \in N$ for some $r \in R$ and some $x \in M \setminus N$, then there exists a positive integer k such that $r^k M \subseteq N$. In other words, N is a primary submodule of M, whenever $Z(M/N) \subseteq \sqrt{\mathrm{Ann}(M/N)}$.

Exercise 1.50
Let N be a primary submodule of M. Show that $\sqrt{\mathrm{Ann}(M/N)}$ is a prime ideal.

A primary submodule N of M with $P = \sqrt{\mathrm{Ann}(M/N)}$ is called a *P-primary submodule* of M.

Exercise 1.51
Let R be a Noetherian ring, and let M be a finitely generated R-module. Show that for a proper submodule N of M the following conditions are equivalent:

(i) N is P-primary.
(ii) $Z(M/N) = \sqrt{\mathrm{Ann}(M/N)} = P$.
(iii) $\mathrm{Ass}(M/N) = \{P\}$.

In particular, an ideal $I \subsetneq R$ is primary if and only if $|\,\mathrm{Ass}(I)| = 1$.

Exercise 1.52
Let $N_1, \ldots, N_k$ be P-primary submodules of M. Show that $\bigcap_{i=1}^k N_i$ is P-primary.

A *primary decomposition* for a submodule N of M is an expression $N = \bigcap_{i=1}^k N_i$, where the N_i's are P_i-primary submodules of M. Moreover, if $P_1, \ldots, P_k$ are distinct and $\bigcap_{j \neq i} N_j \nsubseteq N_i$ for $1 \leq i \leq k$, we call $\bigcap_{i=1}^k N_i$, an *irredundant* primary decomposition of N.

Irreducible submodules naturally appear in the theory of primary decompositions. A proper submodule N of M is called *irreducible* if N can not be expressed as $N_1 \cap N_2$, where $N_1, N_2 \neq N$ are submodules of M. Otherwise, it is called *reducible*.

Lemma 1.53
Let M be a Noetherian R-module. Then any irreducible submodule of M is primary.

Proof. For a submodule N of M and $r \in R$, we set $N :_M (r) = \{x \in M : rx \in N\}$. Now, let N be an irreducible submodule of M, and let $rx \in N$ for some $x \in M \setminus N$ and $r \in R$. Since M is Noetherian, the ascending chain

$$N :_M (r) \subseteq N :_M (r^2) \subseteq \cdots \subseteq N :_M (r^k) \subseteq \cdots$$

of submodules of M becomes stationary. Hence there exists an integer k such that $N :_M (r^k) = N :_M (r^i)$ for any $i > k$. We show that $N = (N + r^k M) \cap (N + (x))$. It is obvious that $N \subseteq (N + r^k M) \cap (N + (x))$. Conversely, let $y = z_1 + r^k m = z_2 + sx$ be an element in the intersection, where $z_1, z_2 \in N$, $m \in M$ and $s \in R$. Then $ry = rz_1 + r^{k+1} m = rz_2 + rsx$. Since $rz_1, rz_2, rsx \in N$, we have $m \in N :_M (r^{k+1}) = N :_M (r^k)$. Hence $r^k m \in N$ and then $y \in N$, as desired. Since we assume that N is irreducible and $x \notin N$, we obtain $N = N + r^k M$, from which we get $r^k M \subseteq N$. $\square$

Lemma 1.54
Let M be a Noetherian R-module. Then any proper submodule of M is a finite intersection of irreducible submodules of M.

Proof. Let $\mathscr{S}$ be the set of proper submodules of M which can not be written as a finite intersection of irreducible submodules of M. Note that each module belonging to $\mathscr{S}$ is reducible.

By contradiction assume that $\mathscr{S} \neq \emptyset$. Since M is Noetherian, $\mathscr{S}$ has a maximal element, say N (see Exercise 1.19). Since N is reducible, we have $N = N_1 \cap N_2$, where N_1 and N_2 are submodules of M which properly contain N. The maximality of N implies that $N_1, N_2 \notin \S$, and hence both can be written as finite intersections of irreducible submodules of M. Then so does N, which is a contradiction. □

By using the above lemmata we are prepared to prove

Theorem 1.55
Any proper submodule N of a Noetherian R-module M has an irredundant primary decomposition.

Proof. By Lemmas 1.53 and 1.54, N has a primary decomposition, say $N = \bigcap_{i=1}^{k} N_i$. By the following two operations we can get an irredundant primary decomposition of N. First we discard those N_i's which contain $\bigcap_{j \neq i} N_j$ and next we intersect the N_i's which are P-primary for the same P, see Lemma 1.52. □

Theorem 1.56
Let R be a Noetherian ring, let M be a finitely generated R-module, and let N be a submodule of M with the irredundant primary decomposition $N = \bigcap_{i=1}^{k} N_i$, where N_i is P_i-primary for $1 \leq i \leq k$. Then $\operatorname{Ass}(M/N) = \{P_1, \ldots, P_k\}$. In particular, the prime ideals $P_1, \ldots, P_k$ are uniquely determined. In other words, they do not depend on the choice of irredundant primary decomposition of N.

Proof. Let $\varphi : M/\bigcap_{i=1}^{k} N_i \to \bigoplus_{i=1}^{k} M/N_i$ be the map $\varphi(x + \bigcap_{i=1}^{k} N_i) = (x + N_1, \ldots, x + N_k)$ for any $x \in M$. Then φ is a monomorphism of R-modules. This together with Lemma 1.48 and Exercise 1.49 implies that $\operatorname{Ass}(M/N) \subseteq$

$\bigcup_{i=1}^{k} \mathrm{Ass}(M/N_i)$. By Exercise 1.51, $\mathrm{Ass}(M/N_i) = \{P_i\}$. Hence $\mathrm{Ass}(M/N) \subseteq \{P_1, \ldots, P_k\}$.

For each i, we set $L_i = \bigcap_{j \neq i} N_j$. Then $N \subsetneq L_i$, and by the Second Isomorphism Theorem, $L_i/N \cong (L_i + N_i)/N_i \subseteq M/N_i$. So by Lemma 1.48, $\mathrm{Ass}(L_i/N) \subseteq \mathrm{Ass}(M/N_i) = \{P_i\}$. Since $L_i/N \neq 0$, by Lemmas 1.46 and 1.48 we have $\{P_i\} = \mathrm{Ass}(L_i/N) \subseteq \mathrm{Ass}(M/N)$. This completes the proof. □

If in above theorem we choose $N = 0$, then we obtain

Corollary 1.57
Let R be a Noetherian ring, and let M be a finitely generated R-module. Then $\mathrm{Ass}(M)$ is a finite set.

The set $\mathrm{Spec}(R)$ denotes the set of prime ideals of R. It is called the *spectrum* of R. The *support* of an R-module M is defined as

$$\mathrm{Supp}_R(M) = \{P \in \mathrm{Spec}(R) : \ M_P \neq 0\}.$$

Here M_P denotes the module of fractions of M with respect to the multiplicatively closed set $S = R \setminus P$. If there is no ambiguity about the ring R we drop the index R and simply write $\mathrm{Supp}(M)$.

For an ideal I of R, we set $V(I) = \{P \in \mathrm{Spec}(R) : \ I \subseteq P\}$. Observe that $\mathrm{Supp}_R(R/I) = V(I)$.

Exercise 1.58
Let $0 \to U \to M \to N \to 0$ be a short exact sequence of R-module. Then $\mathrm{Supp}(M) = \mathrm{Supp}(U) \cup \mathrm{Supp}(N)$. Hint: use the fact that localization is an exact functor.

Lemma 1.59
Let M be a finitely generated R-module. Then $\mathrm{Supp}(M) = V(\mathrm{Ann}(M))$.

Proof. Let M be generated by $x_1, \ldots, x_n$, and let P be a prime ideal with $\mathrm{Ann}(M) \subseteq P$. By contradiction assume that $M_P = 0$. Then there exists $s \in R \setminus P$ such that $sM = 0$, which means that $s \in \mathrm{Ann}(M) \setminus P$. This is a contradiction. Hence, $P \in \mathrm{Supp}(M)$. Conversely, assume that P is a prime ideal of R which does not contain $\mathrm{Ann}(M)$. Let $s \in \mathrm{Ann}(M) \setminus P$. Then, $sM = 0$ which implies that $M_P = 0$. This shows that $\mathrm{Supp}(M) \subseteq V(\mathrm{Ann}(M))$. □

Exercise 1.60
Let M be an R-module. Prove that
(a) $\mathrm{Ass}(M) \subseteq \mathrm{Supp}(M)$.
(b) $\mathrm{Supp}(M/IM) = \mathrm{Supp}(M) \cap V(I)$.

Minimal elements (with respect to inclusion) of $\mathrm{Supp}(M)$ are called *minimal prime ideals* of M. The set of minimal prime ideals of M is denoted by $\mathrm{Min}(M)$. For an ideal $I \subsetneq R$, by abusing the notation, we denote $\mathrm{Min}(R/I)$ simply by $\mathrm{Min}(I)$ and call it the set of minimal prime ideals of I.

Proposition 1.61
Let R be a Noetherian ring, and let M be a finitely generated R-module. Then

(a) $\mathrm{Min}(M) \subseteq \mathrm{Ass}(M)$. In particular, the set of minimal elements of $\mathrm{Ass}(M)$ (with respect to inclusion) is precisely the set $\mathrm{Min}(M)$.
(b) $\mathrm{Min}(M)$ is finite.
(c) $\sqrt{\mathrm{Ann}(M)} = \bigcap_{P \in \mathrm{Ass}(M)} P = \bigcap_{P \in \mathrm{Min}(M)} P$.

Proof. (a) Let $P \in \mathrm{Min}(M)$. Then $M_P \neq 0$ and in view of Exercise 1.60, $\mathrm{Ass}_{R_P}(M_P) \subseteq \mathrm{Supp}_{R_P}(M_P) = \{PR_P\}$. By Lemma 1.46, $\mathrm{Ass}_{R_P}(M_P)$ is not empty. Thus $\mathrm{Ass}_{R_P}(M_P) = \{PR_P\}$. Hence $P \in \mathrm{Ass}(M)$.

(b) follows from part (a) and Corollary 1.57.

(c) By Lemma 1.59, Proposition 1.62 and part (a), the result holds. □

Proposition 1.62
Let R be a ring, and let I be an ideal of R. Then
(a) $\sqrt{I} = \bigcap_{P \in V(I)} P$.
(b) If R is Noetherian and I is a radical ideal, then $\mathrm{Min}(I) = \mathrm{Ass}(I)$.

Proof. (a) Let $P \in V(I)$. Then clearly $\sqrt{I} \subseteq \sqrt{P} = P$ and hence $\sqrt{I} \subseteq \bigcap_{P \in V(I)} P$. Now, let $x \in \bigcap_{P \in V(I)} P$. By contradiction assume that $x \notin \sqrt{I}$. Then $S = \{x^n : n \geq 0\}$ is a multiplicatively closed subset of R with $I \cap S = \emptyset$. Hence $S^{-1}I$ is a proper ideal of $S^{-1}R$. Thus it is contained in a maximal ideal of $S^{-1}R$, say $S^{-1}Q$, where Q is a prime ideal of R with $Q \cap S = \emptyset$ and $I \subseteq Q$. Hence, $Q \in V(I)$. But by our assumption $x \in Q \cap S$, a contradiction.

(b) Since R is Noetherian, I has a minimal primary decomposition and $\operatorname{Min}(I) \subseteq \operatorname{Ass}(I)$. On the other hand by (a) we have $I = \sqrt{I} = \bigcap_{P \in \operatorname{Min}(I)} P$, which is indeed a minimal primary decomposition of I, see Exercise 1.63. Hence, $\operatorname{Min}(I) = \operatorname{Ass}(I)$. □

Exercise 1.63
Let $I_1, \ldots, I_n$ and P be ideals of a ring R, where P is a prime ideal and $\bigcap_{i=1}^{n} I_i \subseteq P$. Show that $I_i \subseteq P$ for some i.

Example 1.64
Let $I = (xy, xz) \subsetneq K[x, y, z]$. Then $I = (x) \cap (y, z)$. Indeed, it is clear that $I = (xy, xz) \subseteq (x) \cap (y, z)$. Conversely, let $f \in (x) \cap (y, z)$. Then $f = xg$ with $g \in K[x, y, z]$ and $xg \in (y, z)$. Consider the map $\varphi\, K[x, y, z] \to K[x]$ with $x \mapsto x$ and $y, z \mapsto 0$. Then $\operatorname{Ker} \varphi = (y, z)$, and we have $0 = \varphi(xg) = \varphi(x)\varphi(g) = x\varphi(g)$. This implies that $\varphi(g) = 0$. Therefore, $g \in (y, z)$ and $f \in x(y, z) = (xy, xz)$.

It follows from Proposition 1.62 that $\operatorname{Min}(I) = \operatorname{Ass}(I) = \{(x), (y, z)\}$.

Next we study how the graded structure of a module affects its associated prime ideals.

Proposition 1.65
Let M be a graded R-module, and let $P \in \operatorname{Ass}(M)$. Then $P = \operatorname{Ann}(y)$ for some homogeneous element $y \in M$ and P is a graded prime ideal. In particular, any minimal prime ideal of M is graded.

Proof. Let $P = \operatorname{Ann}(x)$, where $x = x_i + x_{i+1} + \cdots + x_d$ with $x_\ell \in M_\ell$ for all ℓ and $x_i \neq 0$. Moreover, let $r = r_j + r_{j+1} + \cdots + r_k \in P$, with $r_\ell \in R_\ell$ for all ℓ. Since $r_j x_i$ is the $(i + j)$th graded component of rx, we have $r_j x_i = 0$. Similarly, $r_{j+1} x_i + r_j x_{i+1} = 0$, which implies that $r_j^2 x_{i+1} = 0$. Proceeding the same way we have $r_j x_i = r_j^2 x_{i+1} = \cdots = r_j^{d-i+1} x_d = 0$. Hence, $r_j^{d-i+1} x = 0$, which means that $r_j \in \sqrt{\operatorname{Ann}(x)} = \sqrt{P} = P$. Thus $r - r_j = r_{j+1} + \cdots + r_k \in P$. Using a similar argument as above we get $r_{j+1} \in P$, and step by step we get $r_\ell \in P$ for all ℓ. This shows that P is a graded ideal.

Next we find a homogeneous element $y \in M$ with $P = \operatorname{Ann}(y)$. For any homogeneous element $r \in P$ we have $rx_i = rx_{i+1} = \cdots = rx_d = 0$, and then $r \in \bigcap_{\ell=i}^{d} \operatorname{Ann}(x_\ell)$. Since P is graded, this means that $P \subseteq \bigcap_{\ell=i}^{d} \operatorname{Ann}(x_\ell)$. On the other hand, it is clear that $\bigcap_{\ell=i}^{d} \operatorname{Ann}(x_\ell) \subseteq P$. So by Exercise 1.63 we have $\operatorname{Ann}(x_s) \subseteq P$ for some $i \leq s \leq d$. So $P = \operatorname{Ann}(x_s)$. □

1.2 Dimension Theory

In this section we present dimension theory for general Noetherian rings as much as it is required for later applications. In particular, the Krull dimension for graded rings and modules is considered.

1.2.1 Dimension of Finitely Generated Graded K-Algebras

We first consider an arbitrary Noetherian ring R. The *Krull dimension* of R, denoted by $\dim R$, is the supremum of the numbers n for which there exists a chain

$$P_0 \subsetneq P_1 \subsetneq \cdots \subsetneq P_n \tag{1.1}$$

of prime ideals in R. The length of the chain (1.1) is defined to be n. Let $P \in \operatorname{Spec}(R)$. Then the *height* of P is the supremum of lengths of chains as in (1.1) with $P_n = P$. More generally, if $I \subsetneq R$ is an ideal, one defines the height of I as

$$\text{height } I = \inf\{\text{height } P : P \in V(I)\}.$$

By Proposition 1.61, $\operatorname{Min}(I)$ is finite. Thus, if $\operatorname{Min}(I) = \{P_1, \ldots, P_m\}$, then

$$\text{height } I = \min\{\text{height } P_1, \ldots, \text{height } P_m\}.$$

You may ask whether height I is a finite number. We will see below that this is indeed the case, if R is Noetherian, as we assume here. The proof of this fact follows from one of the most important theorems in dimension theory, which is known as Krull's generalized principal ideal theorem. We first begin with Krull's principal ideal theorem. For its proof we have to consider symbolic powers.

Let $P \in \operatorname{Spec}(R)$, and let $k \geq 1$ be an integer. Then the preimage of $P^k R_P$ in R with respect to the canonical homomorphism $R \to R_P$ with $r \mapsto r/1$, is denoted by $P^{(k)}$ and is called the k-th *symbolic power* of P. Note that if $r \in P^{(k)}$, then $r/1 = s/1$ for some $s \in P^k$. Hence there exists $t \in R \setminus P$ such that $t(r - s) = 0$. This implies that $tr \in P^k$. Conversely, if $tr \in P^k$ for some $t \in R \setminus P$, then $(t/1)(r/1) \in P^k R_P$. This implies that $r/1 \in P^k R_P$, since $t/1$ is a unit in R_P. It follows that $r \in P^{(k)}$. Thus we have shown that

$$P^{(k)} = \{r \in R \;\; tr \in P^k \text{ for some } t \in R \setminus P\}.$$

Obviously, one has $P^k \subseteq P^{(k)}$ for all k.

Theorem 1.66
(Krull's principal ideal theorem) Let R be a Noetherian ring, let $x \in R$ be a nonunit, and let P be a minimal prime ideal of (x). Then height $P \leq 1$.

Proof. Since P is a minimal prime ideal of (x), it follows that PR_P is a minimal prime ideal of xR_P. Replacing R by R_P, we may therefore assume that R is local with maximal ideal $\mathfrak{m}$ and that $\mathfrak{m}$ is a minimal prime ideal of (x). It is enough to show that for any prime ideal $Q \subsetneq \mathfrak{m}$, height $Q = 0$.

We consider the preimage of $Q^i R_Q$ under the canonical map $R \to R_Q$, and denote it by $Q^{(i)}$. Then $Q^{(i)} \subseteq R$ and we obtain the chain of ideals $(x) + Q^{(1)} \supseteq (x) + Q^{(2)} \supseteq \cdots$. By Proposition 1.62 and Exercise 1.21, we have $\mathfrak{m}^k \subseteq (x)$ for some k. Therefore, $\ell(R/(x)) < \infty$. Hence, there exists an integer k such that $(x) + Q^{(k)} = (x) + Q^{(k+1)}$.

Let $y \in Q^{(k)}$, Then $y = rx + z$ with $z \in Q^{(k+1)}$. Hence $rx \in Q^{(k)}$. Note that $Q^{(k)}$ is Q-primary. Since $x \notin Q$, it follows that $r \in Q^{(k)}$. Thus, $Q^{(k)} = xQ^{(k)} + Q^{(k+1)}$. Now we use Theorem 1.34, the local version of Nakayama's lemma, and obtain that $Q^{(k)} = Q^{(k+1)}$. This implies that $Q^k R_Q = Q^{k+1} R_Q$. Applying again Theorem 1.34 we conclude that $Q^k R_Q = (0)$. This implies that $\operatorname{Spec}(R_Q) = \{QR_Q\}$, and hence height $Q = 0$, as desired. □

Our next goal is to prove Krull's generalized principal ideal theorem. For its proof we need the so-called prime avoidance lemma.

Lemma 1.67

Let R be an arbitrary commutative ring, let $P_1, \ldots, P_s \in \operatorname{Spec}(R)$, and let $I \subsetneq R$ be an ideal. If $I \subseteq \bigcup_{i=1}^s P_i$, then $I \subseteq P_i$ for some i. Moreover, if R is a graded K-algebra, I is a graded ideal and all P_i are graded prime ideals, then r can be chosen to be homogeneous.

Proof. We consider only the graded case. The nongraded case is similar and even easier to prove. We proceed by induction on s. Suppose $s = 1$. Since $I \not\subseteq P_1$, there exists $r \in I \setminus P_1$. Since $r \notin P_1$, at least one homogeneous component r' of r does not belong to P_1. On the other hand, since I is a graded ideal, all homogeneous components of r belong to I, see Proposition 1.3. Thus, $r' \in I \setminus P_1$.

Now let $s > 1$. We may assume that there is no containment among the P_i's. Then $I \cap P_j \not\subseteq P_i$ for $j \neq i$, because P_i is a prime ideal, see Exercise 1.63. Since $I \cap P_j$ is a graded ideal, by induction hypothesis, for each j there exists a homogeneous element $r_j \in (I \cap P_j) \setminus \bigcup_{i, i \neq j} P_i$. We may assume that all r_j have the same degree. Otherwise we replace each of them by a suitable power. Then we set $r = \sum_{i=1}^s \prod_{j, j \neq i} r_j$. Obviously, $r \in I$ and $r \notin P_i$ for $i = 1, \ldots, s$. □

Krull's principal ideal theorem together with the prime avoidance lemma gives us

Lemma 1.68

Let R be a Noetherian ring, and let $P' \subsetneq Q' \subsetneq P$ be a chain of prime ideals in R. Furthermore, let $P_1, \ldots, P_s \in \operatorname{Spec}(R)$ with $P \not\subseteq P_i$ for all i. Then we can find $Q \in \operatorname{Spec}(R)$ with $P' \subsetneq Q \subsetneq P$ and $Q \not\subseteq P_i$ for all i.

Proof. By Lemma 1.67 there exists $x \in P$ with $x \notin P_i$ for all i and $x \notin P'$. A minimal prime ideal of $xR_P + P'R_P$ is of the form QR_P for some $Q \in \operatorname{Spec}(R)$ with $P' \subseteq Q \subseteq P$. By Theorem 1.66, $\operatorname{height}(Q/P') = 1$. Hence, since by assumption $\dim R_P/P'R_P \geq 2$, it follows that $Q \subsetneq P$. Furthermore, since $x \in Q$, we have $P' \subsetneq Q$ and $Q \not\subseteq P_i$ for all i. □

Now we can prove

Theorem 1.69

(Krull's generalized principal ideal theorem) Let R be a Noetherian ring, and let $I \subsetneq R$ be an ideal generated by m elements. Then $\operatorname{height} P \leq m$ for any minimal prime ideal P of I. In particular, $\operatorname{height} I \leq m$.

Proof. We proceed by induction on m. The case $m = 1$ is covered by Theorem 1.66. Now let $m > 1$ and $I = (r_1, \ldots, r_m)$. Let $Q_1, \ldots, Q_s$ be the minimal prime ideals of $J = (r_1, \ldots, r_{m-1})$. By induction hypothesis, $\operatorname{height} Q_i \leq m - 1$ for all i.

Let P be a minimal prime ideal of I, and let $P_0 \subsetneq P_1 \subsetneq \cdots \subsetneq P_h = P$ be a chain of prime ideals. If $h \leq 1$ for all such chains, we are done. Hence we may assume that $h \geq 2$. Furthermore, we may assume that $P \not\subseteq \bigcup_{i=1}^{s} Q_i$, because otherwise $P \subseteq Q_i$ for some i which implies that $\operatorname{height} P \leq m - 1$, and we are done.

Applying Lemma 1.68 repeatedly, we find a chain as above with $P_1 \not\subseteq \bigcup_{i=1}^{s} Q_i$. We let $\overline{R} = R/J$ and mark the images of the elements and ideals of R in $\overline{R}$ by an overline. Note that $\overline{P}$ is a minimal prime ideal of $(\overline{r}_m)$. Thus, $\operatorname{height} \overline{P} \leq 1$ by Theorem 1.66, and we also have $J \subseteq P_1 + J \subseteq P$. We claim that P is a minimal prime ideal of $P_1 + J$. Suppose that is not the case. Then there exists $P' \in \operatorname{Spec}(R)$ with $P_1 + J \subseteq P' \subsetneq P$. Since $\operatorname{height} \overline{P} \leq 1$, it follows that P' is a minimal prime ideal of J, and hence $P' = Q_i$ for some i, which implies that $P_1 \subseteq Q_i$, a contradiction.

The claim implies that P/P_1 is a minimal prime ideal of the ideal $(P_1 + J)/P_1$ in R/P_1. Since $(P_1 + J)/P_1$ is generated by $m - 1$ elements, our induction hypothesis implies that $\operatorname{height}(P/P_1) \leq m - 1$. Since $h - 1 \leq \operatorname{height}(P/P_1)$, we conclude that $h \leq m$. □

Example 1.70
Let $P = (x_{i_1}, x_{i_2}, \ldots, x_{i_k}) \subsetneq S = K[x_1, x_2, \ldots, x_n]$. Then P is a prime ideal, because S/P is a polynomial ring over K in the variables x_j which do not belong to P. We claim that height $P = k$. Indeed, by Theorem 1.69 we have height $P \leq k$. On the other hand, we have the following chain of prime ideal

$$(0) \subsetneq (x_{i_1}) \subsetneq (x_{i_1}, x_{i_2}) \subsetneq \cdots \subsetneq (x_{i_1}, x_{i_2}, \ldots, x_{i_k}) = P,$$

which shows that height $P \geq k$.

Corollary 1.71
Let R be a Noetherian ring and let $I \subsetneq R$ be an ideal. Then height $I < \infty$.

Proof. Since R is Noetherian, I is finitely generated. Therefore, the desired result follows from Theorem 1.69. □

Since we are mainly interested in graded K-algebras, the following result is very important for us.

Theorem 1.72
Let K be a field, and let R be a finitely generated graded K-algebra with graded maximal ideal $\mathfrak{m}$. Then

$$\dim R = \text{height}\,\mathfrak{m}.$$

In particular, $\dim R < \infty$.

We postpone the proof of this theorem for the moment, and first consider some consequences.

Corollary 1.73
Let $S = K[x_1, \ldots, x_n]$ be the polynomial ring. Then $\dim S = n$.

Proof. Let $\mathfrak{m} = (x_1, \ldots, x_n)$. Then $\mathfrak{m}$ is the graded maximal ideal of S, and by Example 1.70 we have height $\mathfrak{m} = n$. Hence the desired result follows from Theorem 1.72. □

When R is a Noetherian local ring, it follows from Theorem 1.69 that $\dim R \leq \operatorname{emb} R$. A similar inequality holds for graded K-algebras.

Corollary 1.74
Let R be a finitely generated graded K-algebra. Then the following hold:

(a) $\dim R \leq \operatorname{emb} R$.
(b) $\dim R = \operatorname{emb} R$ if and only if R is isomorphic to the polynomial ring.

Proof. (a) follows from Theorems 1.69 and 1.72.

(b) If R is a polynomial ring, then Corollary 1.73 implies that $\dim R = \operatorname{emb} R$. Conversely, let $n = \operatorname{emb} R$. Then $R \cong S/I$, where $S = K[x_1, \ldots, x_n]$ is the polynomial ring and $I \subseteq (x_1, \ldots, x_n)^2$ is a graded ideal, see Exercise 1.28. Since we assume that $\dim R = n$, there exists a chain of prime ideals $P_0 \subsetneq P_1 \subsetneq \cdots \subsetneq P_n = (x_1, \ldots, x_n)$ in S such that $I \subseteq P_0$. Suppose that R is not a polynomial ring. Then $I \neq (0)$ and $(0) \subsetneq P_0 \subsetneq P_1 \subsetneq \cdots \subsetneq P_n = (x_1, \ldots, x_n)$ is a chain of prime ideals in S of length $n + 1$. This contradicts Theorem 1.72. □

Exercise 1.75
Let $S = K[x, y]$ be the polynomial ring, and let $I = (x^2, xy) \subsetneq S$. Show that $\operatorname{height} I = 1$, and that I cannot be generated by 1 element. Which are the associated prime ideals of I?

For a submodule N of a graded R-module M, we denote by N^* the submodule of M generated by all homogeneous elements of N. Indeed, N^* is the largest graded submodule of M contained in N.

Example 1.76
Let $S = K[x, y]$ be a polynomial ring with the standard grading, and let $I = (x^2, x + y^2)$ be an ideal. Then, $x + y^2 \notin I^*$, but $x^2, xy^2, y^4 \in I^*$.Therefore, $J \subseteq I^*$ where $J = (x^2, xy^2, y^4)$. We claim that $J = I^*$. Indeed, let $f \in I$ be homogeneous of degree d. We want to show that $f \in J$. We write $f = x^2 g + h$ where $g \in K[x, y]$ and $h = axy^{d-1} + by^d \in I$ with $a, b \in K$. We may assume that $h \neq 0$. Now we consider the K-algebra homomorphism φ $K[x, y] \to K[y]$ with $\varphi(x) = -y^2$ and $\varphi(y) = y$. Then, $\varphi(I) = (y^4)$ and $-ay^{d+1} + by^d = \varphi(h) \in \varphi(I) = (y^4)$. If $b \neq 0$, then $d \geq 4$, which implies that $h \in J$, and if $b = 0$, then $ay^{d+1} \in (y^4)$. This implies that $d \geq 3$. It follows again that $h \in J$.

Lemma 1.77
Let R be a graded ring. If P is a prime ideal of R, then so is P^*.

Proof. Let x and y be two elements of R such that $xy \in P^*$ and $x \notin P^*$. We may write $x = \sum_i x_i$ and $y = \sum_i y_i$, where $x_i, y_i \in R_i$ for all i. Suppose $x_i \in P$ for all i. Then $x_i \in P^*$ for all i, and hence $x \in P^*$. This is a contradiction. Therefore, there exists i such that $x_i \notin P$. Let j be the smallest integer such that $x_j \notin P$. Since $\sum_{i<j} x_i \in P^*$, we may replace x by $\sum_{i \geq j} x_i$, and hence may assume that $x = x_j + x_{j+1} + \cdots + x_d$ such that $x_j \notin P$. We may assume that $y \neq 0$. Let k be the smallest integer such that $y_k \neq 0$. Since P^* is a graded ideal, $x_j y_k \in P^*$, because $(xy)_{j+k} = x_j y_k$. Thus, $x_j y_k \in P$ and then $y_k \in P$. Now, set $y' = y - y_k$. Then $xy' = xy - xy_k \in P^*$. By the same argument we get $y_{k+1} \in P$ and step by step we obtain $y_i \in P$ for all i. Hence, $y_i \in P^*$ for all i, and then $y \in P^*$, as desired. □

Exercise 1.78
Let M be a graded R-module. Show that if P be a prime ideal in $\mathrm{Supp}(M)$, then $P^* \in \mathrm{Supp}(M)$.

Theorem 1.72 is a simple consequence of

Theorem 1.79
Let R be a finitely generated graded K-algebra, and let $P \in \mathrm{Spec}(R)$. Then the following hold:

(a) height $P/P^* = 1$, if P is not a graded ideal.
(b) If P is a graded ideal of height h, then there exists a chain

$$P_0 \subsetneq P_1 \subsetneq \cdots \subsetneq P_h = P$$

of graded prime ideals.
(c) If P is not a graded ideal, then height $P =$ height $P^* + 1$.

Proof. (a) If we replace R by R/P^* and P by P/P^*, we may assume that R is a graded domain and P is a nonzero prime ideal in R which contains no nonzero homogeneous element of R, and we have to show that height $P = 1$. Clearly height $P \geq 1$. Thus it remains to be shown that height $P \leq 1$. To this end, we consider the set S of nonzero homogeneous elements of R. The set S is multiplicatively closed and $P \cap S = \emptyset$.

Therefore, height P = height $S^{-1}P$. We claim that $\dim S^{-1}R \leq 1$, and this will then imply that height $P \leq 1$.

We first note that $S^{-1}R$ is a graded ring (not positively graded). The grading is given by $(S^{-1}R)_i = \{r/s \ \ r, s \text{ homogeneous}, s \neq 0, \ \deg r - \deg s = i\}$. The 0th graded component $(S^{-1}R)_0$ is a field which we denote by L. If $S^{-1}R = L$, then $\dim S^{-1}R = 0$ and hence height $P = 0$, which is a contradiction. Therefore, $S^{-1}R \neq L$. Since any nonzero homogeneous element of $S^{-1}R$ is invertible, by Exercise 1.80, $S^{-1}R = L[t, t^{-1}]$ for some homogeneous element t of positive degree in R which is transcendental over L. Hence, $\dim S^{-1}R = \dim L[t, t^{-1}] = 1$.

(b) Let $P_0 \subsetneq P_1 \subsetneq \cdots \subsetneq P_h = P$ be a chain of prime ideals. Since P_0 is a minimal prime ideal of R, it follows from Lemma 1.77 that $P_0^* = P_0$. In other words, P_0 is a graded prime ideal. Thus, if $h = 1$, then the statement holds. If $h > 1$, we may assume by induction that $P_0, \ldots, P_{i-1}$ are graded prime ideals for some $i > 0$. If $i = h$, we are done. Thus, we may assume that $i < h$.

If P_{i+1} is not graded we may replace P_i by P_{i+1}^*. Indeed, if $P_{i-1} = P_{i+1}^*$, then (a) implies that height $P_{i+1}/P_{i-1} = 1$, contradicting the fact $P_{i-1} \subsetneq P_i \subsetneq P_{i+1}$. On the other hand, suppose there is a prime ideal Q with $P_{i-1} \subsetneq Q \subsetneq P_{i+1}^*$. Then we obtain the chain of prime ideals

$$P_0 \subsetneq \cdots \subsetneq P_{i-1} \subsetneq Q \subsetneq P_{i+1}^* \subsetneq P_{i+1} \subsetneq \cdots \subsetneq P_h = P$$

of length $h + 1$, which is again a contradiction.

If P_{i+1} is graded, then there exists a homogeneous element $r \in P_{i+1} \setminus P_{i-1}$. Let Q be a minimal prime ideal of $(r) + P_{i-1}$ contained in P_{i+1}. By Theorem 1.66 we have height $Q/P_{i-1} \leq 1$ and this implies that $Q \neq P_{i+1}$. Furthermore, Q is graded, since Q is a minimal prime ideal of the graded ideal $(r) + P_{i-1}$, see Proposition 1.65. Thus, we may replace P_i by Q, and in the new chain the prime ideals $P_0, \ldots, P_i$ are all graded.

(c) Let h = height P. It is clear that $h \geq$ height $P^* + 1$. In order to prove the opposite inequality, we observe that by the proof of (b) there exists a chain of prime ideals $P_0 \subsetneq P_1 \subsetneq \cdots \subsetneq P_h = P$ for which $P_1, \ldots, P_{h-2}$ are graded and $P_{h-2} \neq P^*$. This shows that height $P^* \geq h - 1$ = height $P - 1$. □

Exercise 1.80

Let R be a graded K-algebra. The following conditions are equivalent:

(i) Every nonzero homogeneous element in R is invertible.
(ii) $R = K$ or $R = K[t, t^{-1}]$ for some homogeneous element t of positive degree in R which is transcendental over K.

▶ **Remark 1.81** If I is not a prime ideal, then one may have height I = height I^*, see Example 1.76.

Proof (of Theorem 1.72). It is clear that height $\mathfrak{m} \leq \dim R$. In order to prove the opposite inequality, let $d = \dim R$, and let $P \in \operatorname{Spec}(R)$ be a prime ideal of height d. If P is the graded maximal ideal $\mathfrak{m}$ of R, we are done. Otherwise, P is not graded, and by Lemma 1.77 and Theorem 1.79(c) P^* is a graded prime ideal with height $P^* = d - 1$. Since P^* is graded, it follows that $P^* \subseteq \mathfrak{m}$, and since P^* is not maximal we see that $P^* \subsetneq \mathfrak{m}$. By Theorem 1.79(b) there is a chain of graded prime ideals $P_0 \subsetneq P_1 \subsetneq \cdots \subsetneq P_{d-1} = P^*$. Then $P_0 \subsetneq P_1 \subsetneq \cdots \subsetneq P_{d-1} \subsetneq \mathfrak{m}$ is a chain of graded prime ideals. Hence, height $\mathfrak{m} \geq d = \dim R$. □

1.2.2 Krull Dimension of Modules and Systems of Parameters

Let R be a Noetherian ring, and let M be a finitely generated R-module. In Lemma 1.59 we have seen that $\operatorname{Supp}(M)$ coincides with the set of prime ideals containing the annihilator of M. Thus, it is natural to set

$$\dim M = \dim(R/\operatorname{Ann}(M)).$$

Let R be a finitely generated graded K-algebra. If $M \in \mathcal{M}_R$, then $\operatorname{Ann}(M)$ is a graded ideal and hence $R/\operatorname{Ann}(M)$ is again a finitely generated graded K-algebra. By Theorem 1.72, $\dim M < \infty$, and $\dim M$ is equal to the longest chain of graded prime ideals

$$P_0 \subsetneq P_1 \subsetneq \cdots \subsetneq P_n = \mathfrak{m}$$

in the support of M. Of course, if $P_0 \in \operatorname{Supp}(M)$, then all the other P_i also belong to the support of M.

Exercise 1.82
Let $0 \to U \to M \to N \to 0$ be an exact sequence in the category $\mathcal{M}_R$. Show that $\dim M = \max\{\dim U, \dim N\}$.

In Proposition 1.32 we have seen equivalent conditions for a module $M \in \mathcal{M}_R$ to be of finite length. In the following theorem we show that modules of finite length in $\mathcal{M}_R$ are precisely those which are zero dimensional. We will also see that these conditions are satisfied if and only if M is Artinian. An R-module M is called *Artinian*, if any descending chain of submodules of M becomes stationary.

Theorem 1.83
Let R be a finitely generated graded K-algebra with graded maximal ideal $\mathfrak{m}$ and $M \in \mathcal{M}_R$. Show that the following conditions are equivalent:

(i) $\dim M = 0$.

(ii) There exists a positive integer k such that $\mathfrak{m}^k M = 0$.
(iii) $M_i \neq 0$ for only finitely many i.
(iv) M has finite length with $\ell_R(M) = \dim_K M$.
(v) M is Artinian.

Proof. (i) $\iff$ (ii): Note that $\dim M = 0$ if and only if $V(\mathrm{Ann}(M)) = \{\mathfrak{m}\}$. By Proposition 1.62 the latter equality is equivalent to $\sqrt{\mathrm{Ann}(M)} = \mathfrak{m}$. Using this and Exercise 1.21, we conclude that $\dim M = 0$ if and only if $\mathfrak{m}^k \subseteq \mathrm{Ann}(M)$ for some positive integer k.

(ii) $\iff$ (iii) $\iff$ (iv) follow from Proposition 1.32.

(iv) $\Rightarrow$ (v) follows from the fact that any chain of submodules of M has finite length.

(v) $\Rightarrow$ (iii): Let M be an Artinian R-module. For any integer $i \geq 0$, set $N_i = \bigoplus_{j=i}^{\infty} M_j$. Then $N_0 \supseteq N_1 \supseteq \cdots$ is a descending chain of submodules of M. Hence there exists an integer k such that $N_i = N_k$ for all $i \geq k$. Then $M_i \subseteq N_i = N_{i+1}$ for any $i \geq k$. This implies that $M_i = 0$ for $i \geq k$. This together with Exercise 1.25 gives us the desired result.

□

Now we present another important characterization of the Krull dimension of a module, based on the following

Theorem 1.84

Let R be a local ring (or a graded K-algebra) with the (graded) maximal ideal $\mathfrak{m}$, and let M be a finitely generated (graded) R-module with $\dim M = d$. Then the following hold:

(a) Let $x_1, \ldots, x_m \in R$ be (homogeneous) elements such that

$$\dim M/(x_1, \ldots, x_m)M = 0.$$

Then $m \geq d$.

(b) There exist (homogeneous) elements $x_1, \ldots, x_d \in R$ with

$$\dim M/(x_1, \ldots, x_d)M = 0.$$

Moreover, if K is infinite and R is standard graded, then these elements can be chosen such that $\deg x_i = 1$ for all i.

With the above assumptions a sequence $x_1, \dots, x_d \in R$ of (homogeneous) elements is called a *system of parameters* of M, if $\dim M/(x_1, \dots, x_d)M = 0$, where $d = \dim M$.

Proof (of Theorem 1.84). In this proof we consider only the graded case. The local case is shown in a similar way.

(a) Replacing R by $R/\operatorname{Ann}(M)$ we may assume that $\operatorname{Supp}(M) = \operatorname{Spec}(R)$. We claim that $\dim R/(x_1, \dots, x_m) = 0$. Suppose that this is not the case. Then there exists a graded prime ideal $P \neq \mathfrak{m}$ with $(x_1, \dots, x_m) \subseteq P$. Then $\dim M/PM = 0$, and hence $M_P/PM_P = (M/PM)_P = 0$, since $P \neq \mathfrak{m}$. Therefore, $M_P = PM_P$. Thus, Theorem 1.34 implies that $M_P = 0$, so that $P \notin \operatorname{Supp}(M)$, a contradiction.

Now, Theorem 1.69 shows that $m \geq \operatorname{height} \mathfrak{m}$. By Theorem 1.72, $\operatorname{height} \mathfrak{m} = \dim R$. Therefore, $m \geq \dim R$. Since $\dim R = \dim M$, we are done.

(b) As in part (a) we may assume that $\operatorname{Supp}(M) = \operatorname{Spec}(R)$. We proceed by induction on d. If $d = 0$, there is nothing to prove. Assume now that $d > 0$, and let $P_1, \dots, P_r$ be the minimal prime ideals of R. Suppose that $\mathfrak{m} \subseteq \bigcup_{i=1}^r P_i$. Then $\mathfrak{m} \subseteq P_i$ for some i, see Lemma 1.67. Therefore, $\mathfrak{m} = P_i$ and $\dim R = 0$. Since $\dim R = \dim M$, this is a contradiction. Now, as we know that $\mathfrak{m} \not\subseteq P_i$ for all i, Lemma 1.67 implies that there exists a homogeneous element $x_1 \in \mathfrak{m}$ which is not contained in any minimal prime ideal of R. This implies that $\dim R/x_1R < d$. Since M/x_1M is an R/x_1R-module, it follows that $\dim M/x_1M < d$. Say, $\dim M/x_1M = c$. By our induction hypothesis there exist homogeneous elements $x_2, \dots, x_{c+1} \in R$ such that $\dim(M/x_1M)/(x_2, \dots, x_{c+1})(M/x_1M) = 0$. Since $(M/x_1M)/(x_2, \dots, x_{c+1})(M/x_1M) \cong M/(x_1, \dots, x_{c+1})M$, part (a) implies that $c \geq d - 1$. Since $c < d$ we get $c = d - 1$. This yields $\dim M/(x_1, \dots, x_d)M = 0$, as desired.

Suppose in addition, that K is infinite and R is standard graded. We may assume that $\dim M = \dim R > 0$, and we want to find x_1 as before but with $\deg x_1 = 1$. Since $\dim R > 0$, no P_i contains $\mathfrak{m}$. Therefore, each of the K-vector spaces $R_1 \cap P_i$ is a proper K-subspace of R_1. Since K is infinite, R_1 cannot be covered by finitely many proper K-subspaces, see Lemma 1.85. Thus, we may choose $x_1 \in R_1 \setminus \bigcup_{i=1}^r P_i$, which is an element of degree 1. This argument can be used for the choice of the other x_i's, as well. □

Lemma 1.85
Let K be an infinite field, let V be a finitely generated K-vector space, and let $V_1, \dots, V_m$ be proper K-subspaces of V. Then $\bigcup_{i=1}^m V_i \neq V$.

Proof. By induction on m we prove the assertion. If $m = 1$, there is nothing to prove. Let $m > 1$ and by contradiction assume that $V = \bigcup_{i=1}^m V_i$. By our induction hypothesis we may assume that $V_1 \not\subseteq \bigcup_{i=2}^m V_i$. Indeed, if $V_1 \subseteq \bigcup_{i=2}^m V_i$, then $V = \bigcup_{i=2}^m V_i$, a contradiction.

Now, let $v \in V_1$ such that $v \neq 0$, and $w \in V \setminus V_1$. Then $v + \lambda w \notin V_1$ for all $\lambda \in K \setminus \{0\}$. Since $|K| = \infty$ and $V = \bigcup_{i=1}^{m} V_i$, there exists j such that $v + \lambda w \in V_j$ for infinitely many λ. Choose λ_1 and λ_2 in $K \setminus \{0\}$ with $\lambda_1 \neq \lambda_2$ and $v + \lambda_i w \in V_j$ for $i = 1, 2$. Then $(\lambda_1 - \lambda_2)w = (v + \lambda_1 w) - (v + \lambda_2 w) \in V_j$. This implies that $w \in V_j$, and hence also $v \in V_j$. Since $v \in V_1$ was chosen arbitrarily, it follows that $V_1 \subseteq \bigcup_{i=2}^{m} V_i$, a contradiction. □

As a complement to Theorem 1.84 we have

Corollary 1.86
Let R be a local ring or a graded K-algebra with the (graded) maximal ideal $\mathfrak{m}$. Let $M \in \mathscr{M}_R$, and let $x_1, \ldots, x_i \in \mathfrak{m}$ be a sequence of (homogeneous) elements. Then

$$\dim M/(x_1, \ldots, x_i)M \geq \dim M - i.$$

Equality holds if and only if $x_1, \ldots, x_i$ is part of a system of parameters of M.

Proof. Set $N = M/(x_1, \ldots, x_i)M$, and suppose that $c = \dim N$. By Theorem 1.84(b), there exist (homogeneous) elements $x_{i+1}, \ldots, x_{i+c} \in \mathfrak{m}$ such that $\dim N/(x_{i+1}, \ldots, x_{i+c})N = 0$.

Since $N/(x_{i+1}, \ldots, x_{i+c})N \cong M/(x_1, \ldots, x_{i+c})M$, Theorem 1.84(a) implies that $i + c \geq \dim M$, as desired.

If $c = \dim M - i$, then $x_1, \ldots, x_{i+c}$ is a system of parameters of M, and hence $x_1, \ldots, x_i$ is part of a system of parameters of M. Conversely, assume that $x_1, \ldots, x_i$ is part of the system of parameters $x_1, \ldots, x_d$ of M. Then $\dim N/(x_{i+1}, \ldots, x_d)N = 0$. Thus, Theorem 1.84(a) implies that $\dim M - i = d - i \geq \dim N$. The other inequality is always valid, as we have seen before. Thus $\dim N = \dim M - i$. □

In the next proposition we consider a case where equality holds in Corollary 1.86.

Let R be an arbitrary ring and let M be an R-module. The sequence $\mathbf{f} = f_1, \ldots, f_m$ in R is called *regular* on M, or simply an *M-sequence*, if $M/(\mathbf{f})M \neq 0$ and f_i is *regular* on $M/(f_1, \ldots, f_{i-1})M$ for all $i = 1, \ldots, m$, which means that the multiplication map

$$M/(f_1, \ldots, f_{i-1})M \xrightarrow{f_i} M/(f_1, \ldots, f_{i-1})M$$

is injective. (Here $f_0 = 0$.) In other words, f_i is a non-zerodivisor on $M/(f_1, \ldots, f_{i-1})M$.

Example 1.87
Let K be a field, and let $S = K[x_1, \ldots, x_n]$ be the polynomial ring in the variables $x_1, \ldots, x_n$. Then $\mathbf{x} = x_1, \ldots, x_n$ is an S-sequence. This follows from the fact that $S/(x_1, \ldots, x_i) \cong K[x_{i+1}, \ldots x_n]$ for all i.

Exercise 1.88
Let $f_1, \ldots, f_m$ be an M-sequence. Show that for any positive integers $a_1, \ldots, a_m$, the sequence $f_1^{a_1}, \ldots, f_m^{a_m}$ is also an M-sequence.

Proposition 1.89
Let R be a Noetherian local ring or a graded K-algebra, let M be a finitely generated (graded) R-module, and let $\mathbf{f} = f_1, \ldots, f_r$ be a (homogeneous) M-sequence. Then $\dim M/(\mathbf{f})M = \dim M - r$. In particular, any (homogeneous) M-sequence is part of a system of parameters of M.

Proof. We prove the statement by induction on r. Let $r = 1$. By Corollary 1.86, $\dim M/f_1M \geq \dim M - 1$. Assume that $\dim M/f_1M = d$ and let $P_0 \subsetneq P_1 \subsetneq \cdots \subsetneq P_d$ be a chain of prime ideals in $\mathrm{Supp}(M/f_1M)$. Since P_0 contains f_1 and f_1 is a non-zerodivisor of M, P_0 is not a minimal prime ideal of M. This implies that $\dim M \geq d + 1$. Hence $\dim M/f_1M = \dim M - 1$. Now, let $r > 1$ and set $M' = M/(f_1, \ldots, f_{r-1})M$. By using the induction hypothesis, $\dim M/(\mathbf{f})M = \dim M'/f_rM' = \dim M' - 1 = \dim M - (r-1) - 1 = \dim M - r$. □

We conclude this section with the graded version of Noether's normalization theorem.

Theorem 1.90
(Noether normalization theorem) Let R be a graded K-algebra of dimension d with graded maximal ideal $\mathfrak{m}$, and let $x_1, \ldots, x_d \in \mathfrak{m}$ be a homogeneous system of parameters of R. Then the graded K-subalgebra $S = K[x_1, \ldots, x_d]$ of R is a d-dimensional polynomial ring and R is a finitely generated S-module. Moreover, if K is infinite and R is standard graded, then the x_i can be chosen such that $\deg x_i = 1$ for all i.

Proof. We denote the graded maximal ideal of S by $\mathfrak{n}$, and notice that R is a graded S-module. Since $x_1, \ldots, x_d$ is a system of parameters of R, it follows that $\dim R/\mathfrak{n}R = 0$. Hence by Theorem 1.83, $\dim_K R/\mathfrak{n}R < \infty$. Since $\dim_K R_i = 0$ for $i < 0$, we may apply Corollary 1.24 and deduce that R is a finitely generated S-module. The additional statement about the x_i, when K is infinite, follows from Theorem 1.84.

It remains to show that $S = K[x_1, \ldots, x_d]$ is a polynomial ring over K. By Theorem 1.72, $\dim S \leq d = \dim R$. By Lemma 1.38, the ring extension $S \subseteq R$ is integral since R is a finite S-module. Thus, Theorem 1.91 implies that $\dim S \geq d$. Hence, we have $\dim S = d$. Now the desired conclusion follows from Corollary 1.74(b). □

Theorem 1.91
Let $S \subseteq R$ be an integral extension, and let $P_0 \subsetneq P_1 \subsetneq \cdots \subsetneq P_n$ be a chain of prime ideals in R. Then $P_0 \cap S \subsetneq P_1 \cap S \subsetneq \cdots \subsetneq P_n \cap S$ is a chain of prime ideals in S. In particular, $\dim S \geq \dim R$.

Proof. It is enough to show that if $P_1 \subsetneq P_2$ are prime ideals in R, then $P_1 \cap S \neq P_2 \cap S$. Assume that this is not the case. Then $P_1 \cap S = P_2 \cap S$. We denote this prime ideal by L. Let $P = P_2/P_1$. Then P is a nonzero prime ideal in R/P_1 with $P \cap (S/L) = (0)$. By Lemma 1.41(a), $S/L \subseteq R/P_1$ is an integral extension. Therefore, Lemma 1.41(c) implies that $Q(S/L) \subseteq Q(R/P_1)$ is an algebraic field extension. Hence we may apply Lemma 1.42, and deduce that $P \cap (S/L) \neq (0)$, a contradiction. □

Let L/K be a field extension. We remind you that the *transcendence degree* of L/K, denoted $\operatorname{trdeg}(L/K)$, is defined as the largest cardinality of an algebraically independent subset of L over K.

Corollary 1.92
Let R be a graded K-algebra which is an integral domain. Then $\operatorname{trdeg}(Q(R)/K) = \dim R$.

Proof. Let $S = K[x_1, \ldots, x_d] \subseteq R$ be a normalization of R. In other words, S is a polynomial ring, $d = \dim R$ and R is a finitely generated S-module. Since S is a polynomial ring, we have $\operatorname{trdeg}(Q(S)/K) = d$. Furthermore, Lemma 1.41(c) implies that the field extension $Q(R)/Q(S)$ is algebraic. It follows that $\operatorname{trdeg}(Q(R)/K) = \operatorname{trdeg}(Q(S)/K) = d$, as desired. □

Example 1.93
Let $R = K[x^4, x^3y, xy^3, y^4]$. The quotient field $Q(R)$ of R contains the element $x/y = x^4/x^3y$. Since $x^3y = (y/x)(x^4)$, $xy^3 = (y/x)^3(x^4)$ and $y^4 = (y/x)^4(x^4)$, it follows that $Q(R) = K(x^4, x/y)$. Since $x^4, x/y \in K(x, y)$ are algebraically independent over K, we see that $\operatorname{trdeg}(Q(R)/K) = 2$. Thus Corollary 1.92 implies that $\dim R = 2$.

The next result, which is known as the "Going-up Theorem", complements Theorem 1.91.

Theorem 1.94
Let $S \subseteq R$ be an integral extension, let $P_0 \subsetneq P_1 \subsetneq \cdots \subsetneq P_n$ be a chain of prime ideals in S. Then there exists a chain of prime ideals $Q_0 \subsetneq Q_1 \subsetneq \cdots \subsetneq Q_n$ in R with $Q_i \cap S = P_i$ for all i.

Proof. We proof the theorem by induction on n. If $n = 0$, then by Lemma 1.44 there exists a prime ideal Q_0 in R with $Q_0 \cap S = P_0$. Suppose now that we found already a chain of prime ideals $Q_0 \subsetneq Q_1 \subsetneq \cdots \subsetneq Q_{n-1}$ in R with $Q_i \cap S = P_i$ for $i = 0, \ldots, n-1$. By Lemma 1.41(a), $S/P_{n-1} \subseteq R/Q_{n-1}$ is an integral extension. Hence by Lemma 1.44 there exists a prime ideal Q in R/Q_{n-1} such that $Q \cap (S/P_{n-1}) = P_n/P_{n-1}$. Let Q_n be the preimage of Q in R. Then $Q_{n-1} \subsetneq Q_n$ and $Q_n \cap S = P_n$. □

Corollary 1.95
Let $S \subseteq R$ be an integral extension. Then $\dim R = \dim S$.

Proof. By Theorem 1.91 we know already that $\dim S \geq \dim R$. The inequality $\dim R \geq \dim S$ follows from Theorem 1.94. □

1.3 Hilbert Series and Hilbert Functions

In this section we prove basic properties of Hilbert series of graded modules and present Hilbert's famous theorem which says that for any $M \in \mathscr{M}_R$, the dimension of the K-vector space M_i, as a function of i, is a polynomial function for $i \gg 0$. We also consider Hilbert-Samuel functions defined over local rings.

1.3.1 Hilbert Series of Graded Modules

Let R be a finitely generated graded K-algebra, and let $M \in \mathscr{M}_R$. By Lemma 1.12(b) we have $\dim_K M_i < \infty$ for all i. The numerical function

$$H(M, -)\ \mathbb{Z} \to \mathbb{Z}_{\geq 0} \quad \text{with} \quad H(M, i) = \dim_K M_i$$

is called the *Hilbert function* of M.

For example consider the polynomial ring $S = K[x, y]$ in two variables with the standard grading, and let $M = S(5)$. Then $H(M, i) = \dim_K S(5)_i = \dim_K S_{5+i} = i + 6$, because $\dim_K S_j = j + 1$.

A series $\sum_{i\in\mathbb{Z}} a_i t^i$ with indeterminate t and coefficients a_i in some field is called a *formal Laurent series* if $a_i = 0$ for $i \ll 0$, and it is called a *Laurent polynomial* if $a_i = 0$ for $i \ll 0$ and $i \gg 0$.

The formal Laurent series

$$\mathrm{Hilb}_M(t) = \sum_i H(M, i)t^i$$

is called the *Hilbert series* of M. In our example, $\mathrm{Hilb}_M(t) = \sum_{i\geq -5}(i + 6)t^i$.

The following easy observations are very useful.

Lemma 1.96

(a) Let $M \in \mathcal{M}_R$, and let j be an integer. Then $\mathrm{Hilb}_{M(-j)}(t) = t^j \,\mathrm{Hilb}_M(t)$.

(b) Let $0 \to M_1 \to M_2 \to M_3 \to 0$ be a short exact sequence in $\mathcal{M}_R$. Then

$$\mathrm{Hilb}_{M_2}(t) = \mathrm{Hilb}_{M_1}(t) + \mathrm{Hilb}_{M_3}(t).$$

Proof. (a) We have

$$\begin{aligned}\mathrm{Hilb}_{M(-j)}(t) &= \sum_i H(M(-j), i)t^i = \sum_i \dim_K M(-j)_i t^i = \sum_i \dim_K M_{i-j} t^i\\ &= \sum_i \dim_K M_i t^{i+j} = t^j \sum_i \dim_K M_i t^i = t^j \,\mathrm{Hilb}_M(t).\end{aligned}$$

(b) We use the fact that if $0 \to U \to V \to W \to 0$ is a short exact sequence of K-vector spaces, then $\dim_K V = \dim_K U + \dim_K W$.

Since $0 \to (M_1)_i \to (M_2)_i \to (M_3)_i \to 0$ is an exact sequence of K-vector spaces, we get

$$\begin{aligned}\mathrm{Hilb}_{M_2}(t) &= \sum_i \dim_K (M_2)_i t^i = \sum_i (\dim_K (M_1)_i + \dim_K (M_3)_i)t^i\\ &= \sum_i \dim_K (M_1)_i t^i + \sum_i \dim_K (M_3)_i t^i = \mathrm{Hilb}_{M_1}(t) + \mathrm{Hilb}_{M_3}(t).\end{aligned}$$ □

Exercise 1.97

Let $M \in \mathcal{M}_R$, let $f_1, \ldots, f_m$ be a homogeneous M-sequence with $\deg f_i = a_i$ for all i, and let $\overline{M} = M/(f_1, \ldots, f_m)M$. Show that

$$\mathrm{Hilb}_{\overline{M}}(t) = \mathrm{Hilb}_M(t)(1 - t^{a_1}) \cdots (1 - t^{a_m}).$$

Example 1.98
Let $S = K[x_1, \ldots, x_n]$ be the polynomial ring with the standard grading. Then $H(S, i) = \dim_K S_i$ is equal to the number of monomials $x_1^{b_1} \cdots x_n^{b_n}$ of degree i in S. A simple inductive argument shows that

$$H(S, i) = \binom{n+i-1}{i}.$$

It follows that

$$\mathrm{Hilb}_S(t) = \sum_{i \geq 0} \binom{n+i-1}{i} t^i = \frac{1}{(1-t)^n}.$$

More generally, we may consider the polynomial ring $S = K[x_1, \ldots, x_n]$ with grading given by $\deg x_i = a_i$ with $a_i \in \mathbb{Z}_{>0}$. Then $H(S, i)$ is the number of monomials $x_1^{b_1} \cdots x_n^{b_n}$ of degree $\sum_{j=1}^n b_j a_j = i$. By applying induction on n one can give a recursive formula for $H(S, i)$ which is not as explicit as we have it in the standard graded case. But we have

Lemma 1.99
Let $S = K[x_1, \ldots, x_n]$ be the graded polynomial ring with $\deg x_i = a_i$ for $i = 1, \ldots, n$. Then

$$\mathrm{Hilb}_S(t) = \frac{1}{(1-t^{a_1}) \cdots (1-t^{a_n})}.$$

Proof. We note that $1/(1-t^a) = \sum_{j \geq 0} t^{ja}$ for any integer $a > 0$. Hence

$$\frac{1}{(1-t^{a_1}) \cdots (1-t^{a_n})} = \prod_{k=1}^{n} (\sum_{j \geq 0} t^{ja_k}) = \sum_{i \geq 0} c_i t^i,$$

where $c_i = |\{(b_1, \ldots, b_n) \ b_j \in \mathbb{Z}_{\geq 0}, \ \sum_{j=1}^n b_j a_j = i\}|$. This shows that $c_i = H(S, i)$ for all i. □

Exercise 1.100
Let $S = K[x_1, \ldots, x_n]$ be the graded polynomial ring with $\deg x_i = a_i \in \mathbb{Z}_{>0}$ for $i = 1, \ldots, n$. Show that $S_j \neq 0$ for all $j \gg 0$ if and only if $\gcd(a_1, \ldots, a_n) = 1$. Hint: use the argument given in the proof of Theorem 4.1.

Now we turn to the most general case.

Theorem 1.101
Let R be a finitely generated graded K-algebra, generated by the homogeneous generators $r_1, \ldots, r_n$ with $\deg r_i = a_i > 0$ for $i = 1, \ldots, n$, and let $M \in \mathcal{M}_R$. Then there exists a Laurent polynomial $Q(t)$ with integer coefficients such that

$$\mathrm{Hilb}_M(t) = \frac{Q(t)}{(1-t^{a_1})\cdots(1-t^{a_n})}.$$

Proof. Let $S = K[x_1, \ldots, x_n]$ be the graded polynomial ring with $\deg x_i = a_i$ for $i = 1, \ldots, n$, and let $\epsilon\, S \to R$ be the K-algebra homomorphism with $\epsilon(x_i) = r_i$ for all i. Via the map ϵ we may view M as a graded S-module which is again finitely generated. For the further discussions we need to consider graded free resolutions. If you want to recall this concept, we advice you to go to Sect. A in Chap. 2. There you find that the S-module M has a graded free resolution $F_{\bullet} : 0 \to F_n \to \cdots \to F_0 \to 0$.

For $i = 0, \ldots, n-1$ we may split this resolution into short exact sequences $0 \to Z_{i+1} \to F_i \to Z_i \to 0$, where $Z_0 = M$, $Z_n = F_n$ and $Z_{n+1} = 0$. It follows from Lemma 1.96(b) that $\mathrm{Hilb}_{F_i}(t) = \mathrm{Hilb}_{Z_{i+1}}(t) + \mathrm{Hilb}_{Z_i}(t)$ for all i. This gives us

$$\begin{aligned}\sum_{i=0}^{n}(-1)^i \,\mathrm{Hilb}_{F_i}(t) &= \sum_{i=0}^{n}(-1)^i (\mathrm{Hilb}_{Z_{i+1}}(t) + \mathrm{Hilb}_{Z_i}(t)) \\ &= \mathrm{Hilb}_{Z_0}(t) = \mathrm{Hilb}_M(t).\end{aligned}$$

Next we observe that $F_i = \bigoplus_j S(-j)^{b_{ij}}$ with $b_{ij} \in \mathbb{Z}_{\geq 0}$ and $b_{ij} = 0$ for all but finitely many j. Since $\mathrm{Hilb}_{S(-j)}(t) = t^j \,\mathrm{Hilb}_S(t)$, see Lemma 1.96(a), it follows together with Lemma 1.99 that

$$\mathrm{Hilb}_{F_i}(t) = \sum_j b_{ij} t^j \,\mathrm{Hilb}_S(t) = \frac{\sum_j b_{ij} t^j}{(1-t^{a_1})\cdots(1-t^{a_n})}.$$

Thus, we obtain the desired formula for $\mathrm{Hilb}_M(t)$ with

$$Q(t) = \sum_{i=0}^{n}(-1)^i \sum_j b_{ij} t^j. \tag{1.2}$$

□

Example 1.102
Consider the K-subalgebra $R = K[t^2, t^3]$ of the polynomial ring $K[t]$. Then $R = S/(f)$ where $S = K[x_1, x_2]$ is the polynomial ring and $f = x_1^3 - x_2^2$. The polynomial f is homogeneous of degree 6 if we set $\deg x_1 = 2$ and $\deg x_2 = 3$. Then $0 \to S(-6) \to S \to R \to 0$ is the free resolution of R as an S-module, where the map

$S(-6) \to S$ is multiplication by f. It follows from (1.2) that $Q(t) = 1 - t^6$ and hence $\mathrm{Hilb}_R(t) = (1-t^6)/(1-t^2)(1-t^3) = (1+t^3)/(1-t^2) = 1 + \sum_{i \geq 2} t^i$.

Suppose now that R is standard graded, the embedding dimension of R is n and $M \in \mathscr{M}_R$. Then Theorem 1.101 implies that $\mathrm{Hilb}_M(t) = Q(t)/(1-t)^n$, where $Q(t)$ is a Laurent polynomial with integer coefficients. If it happens that $Q(1) = 0$, then $Q(t) = (1-t)\widetilde{Q}(t)$, where $\widetilde{Q}(t)$ is again a Laurent polynomial, and it follows that $\mathrm{Hilb}_M(t) = \widetilde{Q}(t)/(1-t)^{n-1}$. By repeating this argument we arrive at a unique presentation of $\mathrm{Hilb}_M(t)$ as a fraction $Q_M(t)/(1-t)^m$ with $Q_M(1) \neq 0$ and $m \leq n$. Since this presentation is unique, the exponent m and $Q_M(1)$ are invariants of M. What is the meaning of the exponent m? Theorem 1.104 gives us the answer. First we need to prove

Lemma 1.103
Let R be a finitely generated graded K-algebra with the graded maximal ideal $\mathfrak{m}$, let M be a finitely generated R-module, and let $N = \bigcup_{k>0}(0 :_M \mathfrak{m}^k)$. Then

(a) The R-module N has finite length.
(b) $\mathfrak{m} \notin \mathrm{Ass}(M/N)$.

Proof. (a) By Theorem 1.17, M is Noetherian. Therefore there exists an integer $j > 0$ such that $N = 0 :_M \mathfrak{m}^j$. Then $\mathfrak{m}^j N = 0$ and hence by Proposition 1.32, N has finite length.

(b) Assume that $\mathfrak{m} \in \mathrm{Ass}(M/N)$ and $\mathfrak{m} = \mathrm{Ann}(\overline{y})$ for some $\overline{y} \in M/N$. This means that $y \notin N$ and $\mathfrak{m}y \subseteq N$. Then

$$y \in (0 :_M \mathfrak{m}^j) :_M \mathfrak{m} = (0 :_M \mathfrak{m}^{j+1}) = N,$$

which is a contradiction. □

Theorem 1.104
Let R be a standard graded K-algebra, and let $M \in \mathscr{M}_R$ with $M \neq 0$ and $\dim M = d$. Then the following holds:

(a) $\mathrm{Hilb}_M(t) = Q_M(t)/(1-t)^d$.
(b) $Q_M(1) > 0$.

Proof. We proceed by induction on d. Suppose $d = 0$. Then Theorem 1.83 tells us that $M_i = 0$ for all but finitely many i. Let $\mathrm{Hilb}_M(t) = Q_M(t)/(1-t)^m$. We want to show that $m = 0$. Suppose this is not the case. Then $\mathrm{Hilb}_M(t) = \sum_i H(M,i)t^i$ has infinitely many nonzero terms, contradicting the fact that $M_i = 0$ for almost all i. Moreover, $Q_M(1) = \sum_i H(M,i) > 0$. This proves (a) and (b) when $d = 0$.

Now let $d > 0$. Without restriction we may assume that K is infinite. Indeed, if L be a field extension of K, then one can see that $\mathrm{Hilb}_M(t) = \mathrm{Hilb}_{L\otimes_K M}(t)$. Suppose first that $\mathfrak{m} \notin \mathrm{Ass}(M)$, and let $\mathrm{Ass}(M) = \{P_1, \ldots, P_m\}$. Because R_1 generates $\mathfrak{m}$, the K-vector spaces $R_1 \cap P_i$ are proper subspaces of R_1. By Exercise 1.85, there exists an element $x \in R_1$ which does not belong to any P_i. Thus, Lemma 1.47 implies that x is a non-zerodivisor. Therefore, $(1-t)\,\mathrm{Hilb}_M(t) = \mathrm{Hilb}_{M/xM}(t)$, see Exercise 1.97. By our induction hypothesis we have $\mathrm{Hilb}_{M/xM}(t) = Q_{M/xM}(t)/(1-t^{d-1})$ with $Q_{M/xM}(1) > 0$. This implies that $\mathrm{Hilb}_M(t) = Q_M(t)/(1-t)^d$ with $Q_M(t) = Q_{M/xM}(t)$. This proves (a) and (b) in this case. Here we used the fact that $\dim M/xM = d-1$. Indeed, by Theorem 1.86, $\dim M/xM \geq d-1$. On the other hand, since x belongs to no minimal prime ideal of M, it follows that $\dim M/xM < d$, and hence $\dim M/xM = d-1$.

Now we deal with the case that $\mathfrak{m} \in \mathrm{Ass}(M)$. Then $U = \bigcup_{k>0}(0 :_M \mathfrak{m}^k)$ is a nonzero submodule of M of finite length, see Lemma 1.103(a). Thus, Exercise 1.83 implies that $\dim U = 0$. Hence, $\mathrm{Hilb}_U(t) = Q_U(t)$. From the exact sequence $0 \to U \to M \to M/U \to 0$ we deduce that $\mathrm{Hilb}_M(t) = \mathrm{Hilb}_{M/U}(t) + Q_U(t)$. By Lemma 1.103(b), $\mathfrak{m} \notin \mathrm{Ass}(M/U)$. Thus, by the first part of the proof, $\mathrm{Hilb}_{M/U}(t) = Q_{M/U}(t)/(1-t)^m$ where $m = \dim M/U$. Since $\dim U = 0$, Exercise 1.82 implies that $\dim M = \dim M/U$. Hence, $\mathrm{Hilb}_{M/U}(t) = Q_{M/U}(t)/(1-t)^d$. Now it follows that $\mathrm{Hilb}_M(t) = Q_M(t)/(1-t)^d$ with $Q_M(t) = Q_{M/U}(t) + (1-t)^d Q_U(t)$. Indeed, $Q_M(1) = Q_{M/U}(1) > 0$. □

1.3.2 Hilbert Function and Hilbert-Samuel Function

We maintain the assumptions and notation of Theorem 1.104. The positive number $Q_M(1)$ is called the *multiplicity* of M and is denoted by $e(M)$. This is a very important invariant of M, and in geometric and combinatorial contexts it has a concrete interpretation. It follows from Theorem 1.104 that $e(M)$ is a positive integer.

For the K-algebra R in Example 1.102 we have $\mathrm{Hilb}_R(t) = Q_R(t)/(1-t)$, where $Q_R(t) = 1 + t + \cdots + t^{d-1}$. Therefore, $\dim R = 1$ and $e(R) = d$.

The multiplicity of M appears also in the following result.

Theorem 1.105

Let R be a standard graded K-algebra, and let $M \in \mathscr{M}_R$ with $\dim M = d > 0$. Then the Hilbert function $H(M,i)$ is a polynomial function on i for $i \gg 0$.

More precisely, for $i \gg 0$ we have

$$H(M,i) = \frac{e(M)}{(d-1)!} i^{d-1} + \text{lower terms}.$$

Proof. We know from Example 1.98 that

$$\frac{1}{(1-t)^d} = \sum_{k\geq 0} \binom{d+k-1}{k} t^k = \sum_{k\geq 0} \binom{d+k-1}{d-1} t^k.$$

Now let $Q_M(t) = \sum_{j=a}^{b} h_j t^j$. Then Theorem 1.104 implies that

$$\mathrm{Hilb}_M(t) = (\sum_{j=a}^{b} h_j t^j)(\sum_{k\geq 0} \binom{d+k-1}{d-1} t^k) = \sum_{i\geq a} (\sum_{\substack{j+k=i \\ k\geq 0, a\leq j\leq b}} h_j \binom{d+k-1}{d-1}) t^i.$$

This shows that

$$H(M,i) = \sum_{j=a}^{b} h_j \binom{d+i-j-1}{d-1} = \frac{e(M)}{(d-1)!} i^{d-1} + \text{lower terms}$$

for all $i \geq b$. □

The polynomial in Theorem 1.105 related to M is called the *Hilbert polynomial* of M.

The theorem tells us that if $\dim M > 0$, then $\dim_K M_i$ is constant for $i \gg 0$ if and only if $\dim M = 1$, and this constant number is the multiplicity of M.

Exercise 1.106
Let $S = K[x_1, \ldots, x_n]$ be the standard graded polynomial ring, and let $f_1, \ldots, f_m \in S$ be a regular sequence of homogeneous polynomials with $\deg f_i = a_i$. Show that $e(S/(f_1, \ldots, f_m)) = \prod_{i=1}^{m} a_i$.

For the results in Theorems 1.104 and 1.105 there exist also local versions: let $(R, \mathfrak{m})$ be a Noetherian local ring with residue field $K = R/\mathfrak{m}$, and let M be a finitely generated R-module. We define the standard graded K-algebra

$$\mathrm{gr}_{\mathfrak{m}}(R) = \bigoplus_{i\geq 0} \mathfrak{m}^i/\mathfrak{m}^{i+1},$$

and the finitely generated graded $\mathrm{gr}_{\mathfrak{m}}(R)$-module

$$\mathrm{gr}_{\mathfrak{m}}(M) = \bigoplus_{i \geq 0} \mathfrak{m}^i M/\mathfrak{m}^{i+1} M.$$

$\mathrm{gr}_{\mathfrak{m}}(R)$ is called the *associated graded ring* of R, and $\mathrm{gr}_{\mathfrak{m}}(M)$ is called the *associated graded module* of M.

The Hilbert function of the graded module $\mathrm{gr}_{\mathfrak{m}}(M)$ over the standard graded K-algebra $\mathrm{gr}_{\mathfrak{m}}(R)$ is called the *Hilbert-Samuel function* of M, and is denoted $HS(M, -)$. Thus $HS(M, i) = \dim_K(\mathfrak{m}^i M/\mathfrak{m}^{i+1} M)$. It follows from Theorem 1.34 that $HS(M, i) = \mu(\mathfrak{m}^i M)$. In particular, $HS(R, i) = \mu(\mathfrak{m}^i)$. It is also common to consider the series $\sum_{i \geq 0} HS(M, i)t^i$. This series is called the *Hilbert-Samuel series* of M. We denote this series by $\mathrm{HilbS}_M(t)$.

The proofs of the next two theorems follow immediately from Theorems 1.104 and 1.105, provided we know that $\dim M = \dim \mathrm{gr}_{\mathfrak{m}}(M)$. We give a reference to this important and non-trivial fact in the notes to this section. In particular, the local ring R and the associated graded ring $\mathrm{gr}_{\mathfrak{m}}(R)$ have the same Krull dimension.

Theorem 1.107
Let $(R, \mathfrak{m})$ be a Noetherian local ring, and let M be a finitely generated R-module of dimension d. Then

$$\mathrm{HilbS}_M(t) = Q_M(t)/(1-t)^d \quad \text{with} \quad Q_M(1) > 0.$$

We call $Q_M(1)$ the multiplicity of M and denote it by $e(M)$. Thus $e(M) = e(\mathrm{gr}_{\mathfrak{m}}(M))$. We also have

Theorem 1.108
Let $(R, \mathfrak{m})$ be a Noetherian local ring, and let M be a finitely generated R-module of dimension d. Then for $i \gg 0$, $HS(M, i)$ is a polynomial function on i of the form $\frac{e(M)}{(d-1)!} i^{d-1}$ + lower terms.

When R is a finitely generated graded K-algebra with graded maximal ideal $\mathfrak{m}$, we can define the Hilbert-Samuel function $HS(M, -)$ for $M \in \mathscr{M}_R$ similarly as in the local case. We set $HS(M, i) = \dim_K \mathfrak{m}^i M/\mathfrak{m}^{i+1} M$ for all $i \geq 0$, and let $\mathrm{HilbS}_M(t) = \sum_{i \geq 0} HS(M, i)t^i$ be the Hilbert-Samuel series of M. Then for $M \in \mathscr{M}_R$ we have the analogue statements as in the local case in Theorems 1.107 and 1.108.

Note that if R is standard graded, then $R \cong \mathrm{gr}_{\mathfrak{m}}(R)$, and hence $\mathrm{Hilb}_R(t) = \mathrm{HilbS}_R(t)$. But this equality does not hold, if R is not standard graded.

Consider for example the graded K-algebra $K[t^2, t^3] \subsetneq K[t]$ from Example 1.102. There we observed that $\mathrm{Hilb}_R(t) = (1 + t^3)/(1 - t^2)$. On the other hand, the graded maximal ideal of R is $\mathfrak{m} = (t^2, t^3)$, and it is easily seen that $\mathfrak{m}^i = (t^{2i}, t^{2i+1})$ and then $\mu(\mathfrak{m}^i) = 2$ for all $i \geq 1$. This implies that $\mathrm{HilbS}_R(t) = 1 + 2\sum_{i\geq 1} t^i = (1 + t)/(1 - t)$.

Homological Methods (on Graded Structures) 2

Important tools in the study of graded modules over graded K-algebras are considered including the functors Hom and tensor in the category of graded R-modules and functors derived from them, Koszul complexes, minimal graded free resolutions and graded injective resolutions.

2.1 Hom and Tensor for Graded Modules

The concepts of Hom and tensor functors in the category of all R-modules are recalled in Sect. A.3. In this section we study the restriction of these functors to the category of graded modules over finitely generated graded K-algebras.

2.1.1 Hom for Graded Modules

In this section, unless otherwise stated, we will always assume that R is a finitely generated graded K-algebra. Let as before, GMod_R be the category whose objects are graded R-modules and whose morphisms are the graded R-module homomorphisms. For $M, N \in \mathrm{GMod}_R$, we denote by $\mathrm{Hom}_R(M, N)_0$, the set of morphisms from M to N. The set $\mathrm{Hom}_R(M, N)_0$ is actually a K-vector space. Indeed, if $\varphi, \psi \in \mathrm{Hom}_R(M, N)_0$ and $a, b \in K$, then $\gamma = a\varphi + b\psi$ is the R-module homomorphism with $\gamma(m) = a\varphi(m) + b\psi(m)$ for all $m \in M$. It is a graded R-module homomorphism, and hence a morphism.

Let $M, N \in \mathrm{GMod}_R$. Then the set $\mathrm{Hom}_R(M, N)$ of R-module homomorphisms from M to N is an R-module which contains $\mathrm{Hom}_R(M, N)_0$ as a K-vector space.

J. Herzog et al., *Numerical Semigroups*, Compact Textbooks in Mathematics,
https://doi.org/10.1007/978-3-032-05424-1_2

More generally, for all integers j we define the K-vector spaces

$$\mathrm{Hom}_R(M, N)_j = \{\varphi \in \mathrm{Hom}_R(M, N) \quad \varphi(M_i) \subseteq N_{i+j} \text{ for all } i\}.$$

The elements of $\mathrm{Hom}_R(M, N)_j$ are called graded R-module homomorphisms of degree j. We set $\deg \varphi = j$ if $\varphi \in \mathrm{Hom}_R(M, N)_j$. Note that

$$\mathrm{Hom}_R(M, N)_j = \mathrm{Hom}_R(M(-j), N)_0 = \mathrm{Hom}_R(M, N(j))_0.$$

Exercise 2.1
Let $M, N \in \mathrm{GMod}_R$. Show that

$$R_i \,\mathrm{Hom}_R(M, N)_j \subseteq \mathrm{Hom}_R(M, N)_{i+j}$$

for all i and j.

It follows from Exercise 2.1 that the direct sum of the graded R-module homomorphisms is a graded R-module, whose graded components are the K-vector spaces $\mathrm{Hom}_R(M, N)_j$. To give this module a name, we set

$$\underline{\mathrm{Hom}}_R(M, N) = \bigoplus_j \mathrm{Hom}_R(M, N)_j,$$

as an interior direct sum of K-vector spaces. Let as before, $\mathscr{M}_R$ be the category whose objects are finitely generated graded R-modules and whose morphisms are the graded R-module homomorphisms, and let $M, N \in \mathscr{M}_R$. Since $\mathrm{Hom}_R(M, N)$ is a finitely generated R-module and since $\underline{\mathrm{Hom}}_R(M, N)$ is an R-submodule of $\mathrm{Hom}_R(M, N)$, Theorem 1.17 implies that $\underline{\mathrm{Hom}}_R(M, N)$ is also finitely generated. It follows that $\underline{\mathrm{Hom}}_R(M, N) \in \mathscr{M}_R$.

Let $\varphi \; M \to M'$ be a morphism in GMod_R. Furthermore, let $\alpha \; M' \to N$ be a graded R-module homomorphism of degree j. Then the map

$$\underline{\mathrm{Hom}}_R(\varphi, N) \; \underline{\mathrm{Hom}}_R(M', N) \to \underline{\mathrm{Hom}}_R(M, N), \quad \alpha \mapsto \alpha \circ \varphi$$

is a morphism in the category GMod_R, and it follows that $\underline{\mathrm{Hom}}_R(-, N) \; \mathrm{GMod}_R \to \mathrm{GMod}_R$ is a contravariant functor. Similarly,

$$\underline{\mathrm{Hom}}_R(M, -) \; \mathrm{GMod}_R \to \mathrm{GMod}_R$$

is a covariant functor. Both of these functors induce by restrictions to $\mathscr{M}_R$ functors from $\mathscr{M}_R$ to $\mathscr{M}_R$.

Theorem 2.2
Let $M, N \in \mathscr{M}_R$. Then we have $\underline{\mathrm{Hom}}_R(M, N) = \mathrm{Hom}_R(M, N)$.

Proof. Note that $\underline{\mathrm{Hom}}_R(M, N) \subseteq \mathrm{Hom}_R(M, N)$. Therefore, it is enough to show that any $\varphi \in \mathrm{Hom}_R(M, N)$ is a finite sum of homogeneous R-module homomorphisms.

We first consider the case that $M = F$ is free. Let $F = \bigoplus_{i=1}^r Re_i$ with e_i homogeneous of degree a_i for $i = 1, \ldots, r$, and let $\varphi \in \mathrm{Hom}_R(F, N)$. Since F is free, φ is uniquely determined by the elements $\varphi(e_1), \ldots, \varphi(e_r) \in N$. We write $\varphi(e_i) = \sum_j n_{ij}$ with $n_{ij} \in N_j$ for all i and j, and consider the R-module homomorphisms $\varphi_{ij}\ F \to N$ defined by

$$\varphi_{ij}(e_k) = \begin{cases} n_{ij}, & \text{if } k = i, \\ 0, & \text{if } k \neq i. \end{cases}$$

Note that φ_{ij} is a graded R-module homomorphism of degree $j - a_i$. Moreover, only finitely many φ_{ij} are different from 0, since only finitely many n_{ij} are different from 0. Thus, φ can be written as the finite sum $\varphi = \sum_{i,j} \varphi_{ij}$, and this shows that $\underline{\mathrm{Hom}}_R(F, N) = \mathrm{Hom}_R(F, N)$.

Now let $M \in \mathcal{M}_R$ be arbitrary and let $m_1, \ldots, m_r$ be a homogeneous system of generators of M. We let $F = \bigoplus_{i=1}^r Re_i$ be the free R module with $\deg(e_i) = \deg(m_i)$ for $i = 1, \ldots, r$, and let $\alpha : F \to M$ be the morphism with $\alpha(e_i) = m_i$ for $i = 1, \ldots, r$. Let $C \subseteq F$ be the kernel of α. By Proposition 1.4, $C \in \mathcal{M}_R$. Similarly as for M, there exists a free module $G \in \mathcal{M}_R$ and a morphism $G \to C$. Composing this morphism with the inclusion map $C \to F$, we obtain a morphism $\beta\ G \to F$. By our construction,

$$G \xrightarrow{\beta} F \xrightarrow{\alpha} M \longrightarrow 0 \tag{2.1}$$

is an exact sequence in the category $\mathcal{M}_R$ which by the left exactness of $\mathrm{Hom}_R(-, N)$ induces the exact sequence

$$0 \longrightarrow \mathrm{Hom}_R(M, N) \xrightarrow{\mathrm{Hom}_R(\alpha, N)} \mathrm{Hom}_R(F, N) \xrightarrow{\mathrm{Hom}_R(\beta, N)} \mathrm{Hom}_R(G, N).$$

Similarly, we obtain a sequence

$$0 \longrightarrow \underline{\mathrm{Hom}}_R(M, N) \xrightarrow{\underline{\mathrm{Hom}}_R(\alpha, N)} \underline{\mathrm{Hom}}_R(F, N) \xrightarrow{\underline{\mathrm{Hom}}_R(\beta, N)} \underline{\mathrm{Hom}}_R(G, N).$$

Indeed, in exactly the same way as for $\mathrm{Hom}_R(-, N)$, left exactness of $\underline{\mathrm{Hom}}_R(-, N)$ is shown. Hence this sequence is also exact.

By the first part of our proof we have $\mathrm{Hom}_R(F, N) = \underline{\mathrm{Hom}}_R(F, N)$ and $\mathrm{Hom}_R(G, N) = \underline{\mathrm{Hom}}_R(G, N)$. Then it follows that $\mathrm{Hom}_R(\beta, N) = \underline{\mathrm{Hom}}_R(\beta, N)$, and hence

$$\mathrm{Hom}_R(M, N) = \mathrm{Ker}(\mathrm{Hom}_R(\beta, N)) = \mathrm{Ker}(\underline{\mathrm{Hom}}_R(\beta, N)) = \underline{\mathrm{Hom}}_R(M, N). \quad \square$$

By Theorem 2.2, if $M, N \in \mathcal{M}_R$, then $\mathrm{Hom}_R(M, N) \in \mathcal{M}_R$, and for each $j \in \mathbb{Z}$, the jth graded component $\mathrm{Hom}_R(M, N)_j$ of $\mathrm{Hom}_R(M, N)$ consists of all graded R-module homomorphisms of degree j. Moreover, the functor $\mathrm{Hom}_R(-, N)$ in the category of finitely generated R-modules can be restricted to the category $\mathcal{M}_R$, and its restriction coincides with $\underline{\mathrm{Hom}}_R(-, N)$. A similar statement holds for the functor $\mathrm{Hom}_R(M, -)$.

Exercise 2.3
Let $M, N \in \mathcal{M}_R$ and $a \in \mathbb{Z}$. Show that

$$\underline{\mathrm{Hom}}_R(M(-a), N) = \underline{\mathrm{Hom}}_R(M, N(a)) = \underline{\mathrm{Hom}}_R(M, N)(a).$$

Exercise 2.4
Let H be a free R-module with basis $e_1, \dots, e_n$, and $H^* = \mathrm{Hom}_R(H, R)$. The module H^* is called the R-*dual* of H. We let $e_i^* \in H^*$ be the element which is uniquely determined by the condition that

$$e_i^*(e_j) = \begin{cases} 1, & \text{if } i = j, \\ 0, & \text{if } i \neq j. \end{cases}$$

(a) Show that H^* is a free R-module and that $e_1^*, \dots, e_n^*$ is a basis of H^*. This basis is called the *dual basis* of the basis $e_1, \dots, e_n$.

(b) Assume in addition that $H \in \mathcal{M}_R$ with $\deg(e_i) = a_i$ for $i = 1, \dots, n$. Show that $H^* \in \mathcal{M}_R$ and $\deg(e_i^*) = -a_i$ for $i = 1, \dots, n$.

Let F be a free R-module with basis $\mathcal{F} = f_1, \dots, f_n$ and G be a free R-module with basis $\mathcal{G} = g_1, \dots, g_m$, and let $\varphi : F \to G$ be an R-module homomorphism with $\varphi(f_j) = \sum_{i=1}^m a_{ij} g_i$. Then $A = (a_{ij})$ is called the matrix representing φ with respect to $\mathcal{F}$ and $\mathcal{G}$. Set $\varphi^* = \mathrm{Hom}_R(\varphi, R) : G^* \to F^*$.

Exercise 2.5
The matrix representing φ^* with respect to $\mathcal{G}^*$ and $\mathcal{F}^*$ is the transpose of A, where $\mathcal{G}^*$ and $\mathcal{F}^*$ are the dual bases of $\mathcal{G}$ and $\mathcal{F}$, respectively.

2.1.2 Tensor Products for Graded Modules

This subsection is devoted to introduce and study the tensor product for graded modules. Let $M, N, W \in \mathrm{GMod}_R$. A map $\alpha\ M \times N \to W$ is called R-*bilinear* if for all $r \in R$, $m, m' \in M$ and $n, n' \in N$, we have

(1) $\alpha(rm, n) = \alpha(m, rn) = r\alpha(m, n)$
(2) $\alpha(m + m', n) = \alpha(m, n) + \alpha(m', n)$
(3) $\alpha(m, n + n') = \alpha(m, n) + \alpha(m, n')$.

We denote by $\underline{\mathrm{Bihom}}_R(M \times N, W)$ the set of R-bilinear maps $\varphi\ M \times N \to W$ with the property that $\varphi(m, n) \in W_{i+j}$ for all $m \in M_i$ and $n \in N_j$. The elements of $\underline{\mathrm{Bihom}}_R(M \times N, W)$ are called graded R-bilinear maps.

Let $M, N \in \mathrm{GMod}_R$. An object T in GMod_R is called the tensor product of M and N in the category GMod_R, if T satisfies the following universal property:

There exists a graded R-bilinear map $\alpha_{M,N} : M \times N \to T$ such that for any $W \in \mathrm{GMod}_R$ and all graded R-bilinear maps $M \times N \to W$ there exists a unique morphism $\varphi'\ T \to W$ such that $\varphi = \varphi' \circ \alpha_{M,N}$.

If the object T exists, by the universal property it is uniquely determined up to isomorphism in GMod_R. We show that this object exists and we denote it by $M\underline{\otimes}N$. To prove the existence consider the tensor product $M \otimes_R N$ in Mod_R (see Sect. A.3). We show that there exists a grading on $M \otimes_R N$ such that for any homogeneous elements $m \in M$ and $n \in N$,

$$\deg(m \otimes n) = \deg(m) + \deg(n).$$

Let us first assume that M is a graded free R-module of rank 1 with homogeneous basis element e. Then $M = Re$. It follows that the map

$$N \longrightarrow Re \otimes_R N, \quad n \mapsto e \otimes n$$

is an isomorphism of R-modules. In particular, each element in $Re \otimes N$ can be uniquely written as $e \otimes n$ for some n. Thus, if n is homogeneous we may define $\deg(e \otimes n) = \deg(e) + \deg(n)$. This makes $Re \otimes N$ a graded R-module, and $N(-\deg(e)) \to Re \otimes N$, $n \mapsto e \otimes n$, is an isomorphism in the category GMod_R. Now, let M be a graded free R-module in GMod_R, say $M = \bigoplus_{\lambda \in \Lambda} Re_\lambda$ with $\deg(e_\lambda) = a_\lambda$. Then $M \otimes N \cong \bigoplus_{\lambda \in \Lambda} N(-a_\lambda)$ is graded with $\deg(m \otimes n) = \deg(m) + \deg(n)$ for any homogeneous elements $m \in M$ and $n \in N$.

Next, let $M \in \mathrm{GMod}_R$ be arbitrary and choose a graded free presentation of R-modules

$$G \longrightarrow F \longrightarrow M \longrightarrow 0$$

as in (2.1). Tensorizing with N we obtain the exact sequence

$$G \otimes_R N \xrightarrow{\alpha} F \otimes_R N \longrightarrow M \otimes_R N \longrightarrow 0.$$

Here, α is a morphism in GMod_R, and therefore, $M \otimes_R N \in \mathrm{GMod}_R$. The grading defined in $M \otimes_R N$ via this exact sequence is such that $\deg(m \otimes n) = \deg(m) + \deg(n)$ for all homogeneous $m \in M$ and $n \in N$. This shows that the grading of $M \otimes_R N$ induced by this exact sequence, does not depend on the particular chosen graded free presentation of M. One can see that $M \otimes_R N$ together with this grading satisfies the universal property of the tensor product of M and N in GMod_R with respect to the graded R-bilinear map $\alpha_{M,N} : M \times N \to M \otimes N$, $(m, n) \mapsto m \otimes n$.

This universal property has the consequence that the morphisms $\varphi\ M \to M'$ and $\psi : N \to N'$ in GMod_R induce morphisms $\varphi \otimes N\ \ M \otimes_R N \to M' \otimes_R N$ with $(\varphi \otimes N)(m \otimes n) = \varphi(m) \otimes n$, and $M \otimes \psi\ \ M \otimes_R N \to M \otimes_R N'$ with $(M \otimes \psi)(m \otimes n) = m \otimes \psi(n)$. This makes $\otimes\ \mathrm{GMod}_R \times \mathrm{GMod}_R \to \mathrm{GMod}_R$ a functor in two variables which is covariant in both variables.

Exercise 2.6
Let R be a graded K-algebra, $M \in \mathrm{GMod}_R$ and $a, b \in \mathbb{Z}$. Show that $R(a)\underline{\otimes}M \cong M(a)$ and $R(a)\underline{\otimes}R(b) \cong R(a+b)$.

In Sect. A.4, Tor- and Ext-functors are defined in the category of all R-modules. These functors can be defined in the category GMod_R exactly as in the category of all R-modules, replacing $\otimes_R$ by $\underline{\otimes}_R$ and Hom_R by $\underline{\mathrm{Hom}}_R$. These functors will be denoted by $\underline{\mathrm{Tor}}$ and $\underline{\mathrm{Ext}}$, respectively. All the properties that are proved for Tor and Ext in Sect. A.4 are valid correspondingly for $\underline{\mathrm{Tor}}$ and $\underline{\mathrm{Ext}}$ with similar arguments. We denote the restriction of these two functors to the category $\mathscr{M}_R$ again by $\underline{\mathrm{Tor}}$ and $\underline{\mathrm{Ext}}$.

2.2 The Koszul Complex

2.2.1 Definition and Properties of Koszul Complexes

Let R be a ring. Associated with a sequence $\mathbf{f} = f_1, \ldots, f_m$ of elements in R one defines the *Koszul complex* $K_\bullet(\mathbf{f}; R)$, which is defined as follows: let F be the free R-module with basis $e_1, \ldots, e_m$. We set $K_i(\mathbf{f}; R) = \bigwedge^i F$, which by definition is the *ith exterior power* of F. Then $K_i(\mathbf{f}; R)$ is a free R-module with basis consisting of the wedge products $e_{j_1} \wedge e_{j_2} \wedge \cdots \wedge e_{j_i}$ with $1 \leq j_1 < j_2 < \cdots < j_i \leq m$. In particular, $K_0(\mathbf{f}; R) = R$, $K_1(\mathbf{f}; R) = F$ and $\operatorname{rank} K_i(\mathbf{f}; R) = \binom{m}{i}$. For the moment you may view these wedge products simply as formal expressions. Later we introduce the exterior algebra structure on $\bigwedge^\bullet F = \bigoplus_{i=0}^m \bigwedge^i F$. In this algebra the wedge products are indeed products.

For $i = 1, \ldots, m$, the R-module homomorphism $\partial_i\ K_i(\mathbf{f}; R) \longrightarrow K_{i-1}(\mathbf{f}; R)$ is defined by

$$\partial_i(e_{j_1} \wedge e_{j_2} \wedge \cdots \wedge e_{j_i}) = \sum_{k=1}^{i} (-1)^{k+1} f_{j_k} e_{j_1} \wedge \cdots \wedge \widehat{e}_{j_k} \wedge \cdots \wedge e_{j_i}. \quad (2.2)$$

Here $\widehat{e}_{j_k}$ indicates that the factor e_{j_k} is omitted from the wedge product.

Exercise 2.7
Show that $K_\bullet(\mathbf{f}; R)$ together with the differential given in (2.2) is indeed a complex.

The complex $K_\bullet(\mathbf{f}; M) = K_\bullet(\mathbf{f}; R) \otimes_R M$ is called the *Koszul complex of M with respect to* $\mathbf{f}$.

In the case that R is a graded K-algebra and the elements f_i in the sequence are homogeneous, say $\deg f_i = a_i$, then we set

$$\deg(e_{j_1} \wedge e_{j_2} \wedge \cdots \wedge e_{j_i}) = a_{j_1} + a_{j_2} + \cdots + a_{j_i}.$$

With this definition, the differentials ∂_i are homogeneous of degree 0, and if M is graded, then $K_\bullet(\mathbf{f}; M)$ is a complex of graded R-modules.

Let $J \subseteq [m] = \{1, \ldots, m\}$ with $J = \{j_1 < j_2 < \cdots < j_i\}$. We set

$$\mathbf{e}_J = e_{j_1} \wedge e_{j_2} \wedge \cdots \wedge e_{j_i}.$$

Then the elements of $K_i(\mathbf{f}; M)$ can be identified with linear combinations of the form

$$\sum_{J \subseteq [m], |J|=i} w_J \mathbf{e}_J, \tag{2.3}$$

where $w_J \in M$ for all J.

In the case that R is a graded K-algebra, M is a graded R-module and $\mathbf{f}$ is a homogeneous sequence as above, the kth graded component of $K_i(\mathbf{f}; M)$ consists of linear combinations as in (2.3) with w_J homogeneous and $\deg w_J + \deg \mathbf{e}_J = k$ for all J, where for $J = \{j_1 < j_2 < \cdots < j_i\}$, $\deg \mathbf{e}_J = \sum_{r=1}^{i} a_{j_r}$.

Before we proceed, we use the fact that $(\bigwedge^{\bullet} F, \wedge)$ is a graded, alternating, associative and distributive R-algebra with graded components $\bigwedge^i F$ for $i = 0, \ldots, m$. The elements of degree 1 generate the algebra, and its relations are given by

$$u \wedge u = 0 \quad \text{for all} \quad u \in \bigwedge^1 F = F.$$

In particular, for all i and j we have

$$\begin{aligned} 0 &= (e_i + e_j) \wedge (e_i + e_j) = e_i \wedge e_i + e_i \wedge e_j + e_j \wedge e_i + e_j \wedge e_j \\ &= e_i \wedge e_j + e_j \wedge e_i, \end{aligned}$$

so that $e_i \wedge e_j = -e_j \wedge e_i$.

By using induction on the number of factors and by using the associativity of the wedge product, you can easily check the following rule. Let $1 \leq j_1 < j_2 < \cdots < j_i \leq m$, and let π be a permutation of the set $\{1, \ldots, i\}$. Then

$$e_{j_{\pi(1)}} \wedge e_{j_{\pi(2)}} \wedge \cdots \wedge e_{j_{\pi(i)}} = \operatorname{sgn}(\pi) e_{j_1} \wedge e_{j_2} \wedge \cdots \wedge e_{j_i}, \tag{2.4}$$

where $\operatorname{sgn}(\pi)$ denotes the *signature* of the permutation π.

Let M be an R-module. We denote by $Z_i(\mathbf{f}; M) \subseteq K_i(\mathbf{f}; M)$ the kernel of $\partial_i \otimes M$ and by $B_i(\mathbf{f}; M) \subseteq Z_i(\mathbf{f}; M)$ the image of $\partial_{i+1} \otimes M$. The R-module $Z_i(\mathbf{f}; M)$ is called the module of ith *Koszul cycles*, the R-module $B_i(\mathbf{f}; M)$ is called the module of ith *Koszul boundaries*, and $H_i(\mathbf{f}; M) = Z_i(\mathbf{f}; M)/B_i(\mathbf{f}; M)$ is called the ith Koszul homology of M with respect to $\mathbf{f}$. Note that if M is finitely generated or graded, then $H_i(\mathbf{f}; M)$ has the same property for all i.

As was seen before, $K_{\bullet}(\mathbf{f}; R)$ is a graded algebra. We give $K_{\bullet}(\mathbf{f}; M) = K_{\bullet}(\mathbf{f}; R) \otimes_R M$ a graded $K_{\bullet}(\mathbf{f}; R)$-module structure by defining the scalar multiplication which is uniquely defined by the requirement that for each $e_I \in K_i(\mathbf{f}; R)$ and $me_J \in K_j(\mathbf{f}; M)$, $e_I \wedge me_J = m(e_I \wedge e_J)$.

Exercise 2.8

Let R be a commutative ring, let $\mathbf{f} = f_1, \ldots, f_m$ be a sequence of elements of R and let M be an R-module.

(a) Show that $u \wedge v = (-1)^{ij} v \wedge u$ for $u \in K_i(\mathbf{f}; R)$ and $v \in K_j(\mathbf{f}; R)$, and that $u \wedge u = 0$ if i is odd.

(b) Let M be an R-module, and let $u \in K_i(\mathbf{f}; R)$ and $v \in K_j(\mathbf{f}; M)$. Show that

$$\partial_{i+j}(u \wedge v) = \partial_i(u) \wedge v + (-1)^i u \wedge \partial_j(v). \tag{2.5}$$

(c) Use (b) to show that if $u \in Z_i(\mathbf{f}; R)$ and $v \in Z_j(\mathbf{f}; R)$, then $u \wedge v \in Z_{i+j}(\mathbf{f}; R)$.

(d) Show that $Z_\bullet(\mathbf{f}; R)$ is a graded subalgebra of $K_\bullet(\mathbf{f}; R)$, and that $Z_\bullet(\mathbf{f}; M)$ is a graded $Z_\bullet(\mathbf{f}; R)$-module. Also show that $B_\bullet(\mathbf{f}; R) \subseteq Z_\bullet(\mathbf{f}; R)$ is a graded ideal in $Z_\bullet(\mathbf{f}; R)$.

(e) Deduce from (d) that $H_\bullet(\mathbf{f}; R)$ has a natural structure of a graded, alternating associative algebra and that $H_\bullet(\mathbf{f}; M)$ is a graded $H_\bullet(\mathbf{f}; R)$-algebra.

In general it is not easy to compute the Koszul homology. But, as we shall see, $H_0(\mathbf{f}; M)$ and $H_m(\mathbf{f}; M)$ can easily be interpreted. Here m is the length of the sequence $\mathbf{f}$.

Let $(R, \mathfrak{m})$ be a Noetherian local ring or a graded K-algebra with (graded) maximal ideal $\mathfrak{m}$, and let M be a (graded) R-module. The *socle* of M is defined to be

$$\operatorname{Soc}(M) = 0 :_M \mathfrak{m}. \tag{2.6}$$

In other words, $\operatorname{Soc}(M)$ is the submodule of M consisting of all elements of M, which are annihilated by $\mathfrak{m}$. If R is a graded K-algebra and $M \in \mathscr{M}_R$, then $\operatorname{Soc}(M)$ is a finitely generated graded K-vector space.

Exercise 2.9

Let $S = K[x_1, \ldots, x_n]$ be the polynomial ring over the field K, and let I be the ideal in S generated by $x_1^{a_1}, \ldots, x_n^{a_n}$ with $a_i > 0$ for all i. Furthermore, let $R = S/I$. Show that $\operatorname{Soc}(R) = x_1^{a_1-1} \cdots x_n^{a_n-1} R$.

Proposition 2.10

Let R be a Noetherian ring, let M be an R-module, and let $I \subseteq R$ be an ideal generated by $\mathbf{f} = f_1, \ldots, f_m$. Then $H_0(\mathbf{f}; M) = M/IM$ and $H_m(\mathbf{f}; M) \cong 0 :_M I$.

In particular, if R is a graded K-algebra and $I = \mathfrak{m}$ is the graded maximal ideal of R, then

$$H_m(\mathbf{f}; M) \cong \operatorname{Soc}(M)(-a_1 - a_2 - \cdots - a_m),$$

where $\deg f_i = a_i$ for $i = 1, \ldots, m$. Similar statements hold when $(R, \mathfrak{m})$ is local.

Proof. We observe that $Z_m(\mathbf{f}; M) = H_m(\mathbf{f}; M)$. So let $z \in Z_m(\mathbf{f}; M)$. Then $z = w\mathbf{e}_{[m]}$ with $w \in M$ and $0 = \partial_m(z) = \sum_{j=1}^{m} w(-1)^{j+1} f_j \mathbf{e}_{[m]\setminus\{j\}}$. This implies $w \in 0 :_M I$. Hence $0 :_M I \to H_m(\mathbf{f}; M)$, $w \mapsto w\mathbf{e}_{[m]}$, yields the desired isomorphism.

In the graded case, $Z_m(\mathbf{f}; M)$ is graded, and $w\mathbf{e}_{[m]}$ is homogeneous of degree k if w is homogeneous and $\deg w + \deg \mathbf{e}_{[m]} = k$, where $\deg \mathbf{e}_{[m]} = a_1 + a_2 + \cdots + a_m$. This yields the desired conclusion in the graded case. □

For the computation of Koszul homology two fundamental exact sequences are important. The first of them is pretty standard: let

$$0 \to U \longrightarrow M \longrightarrow N \longrightarrow 0$$

be an exact sequence of R-modules, and let $\mathbf{f} = f_1, \ldots, f_m$ be a sequence of elements of R. Then one obtains the short exact sequence of complexes

$$0 \to K_\bullet(\mathbf{f}; U) \longrightarrow K_\bullet(\mathbf{f}; M) \longrightarrow K_\bullet(\mathbf{f}; N) \to 0.$$

This sequence is indeed exact, since $\operatorname{Tor}_1^R(K_i(\mathbf{f}; R), N) = 0$ for all i, because $K_i(\mathbf{f}; R)$ is a free R-module, see Theorem A.41 and Exercise A.34. By Theorem A.4, the short exact sequence of Koszul complexes gives rise to the long exact sequence

$$\begin{aligned} 0 \to H_m(\mathbf{f}; U) \to H_m(\mathbf{f}; M) \to H_m(\mathbf{f}; N) \to H_{m-1}(\mathbf{f}; U) \to \cdots \\ \cdots \to H_{i+1}(\mathbf{f}; N) \to H_i(\mathbf{f}; U) \to H_i(\mathbf{f}; M) \to H_i(\mathbf{f}; N) \to \cdots \\ \cdots \to H_1(\mathbf{f}; N) \to H_0(\mathbf{f}; U) \to H_0(\mathbf{f}; M) \to H_0(\mathbf{f}; N) \to 0. \end{aligned} \tag{2.7}$$

This long exact sequence is called the *first fundamental long exact sequence* of Koszul homology. More interesting is the *second fundamental long exact sequence* of Koszul homology, which relates the Koszul homology of M with respect to $\mathbf{f}' = f_1, \ldots, f_{m-1}$ to the Koszul homology of M with respect to $\mathbf{f}$.

Theorem 2.11

With the notation and hypotheses introduced, one has the following long exact sequence

$$\begin{aligned} 0 \to H_m(\mathbf{f}; M) \to H_{m-1}(\mathbf{f}'; M) \to H_{m-1}(\mathbf{f}'; M) \to H_{m-1}(\mathbf{f}; M) \to \cdots \\ \cdots \to H_{i+1}(\mathbf{f}; M) \to H_i(\mathbf{f}'; M) \to H_i(\mathbf{f}'; M) \to H_i(\mathbf{f}; M) \to \cdots \\ \cdots \to H_1(\mathbf{f}; M) \to H_0(\mathbf{f}'; M) \to H_0(\mathbf{f}'; M) \to H_0(\mathbf{f}; M) \to 0, \end{aligned}$$

where for each i, the map $H_i(\mathbf{f}'; M) \to H_i(\mathbf{f}'; M)$ is multiplication by $(-1)^{i-1} f_m$. Suppose in addition that R is graded, M is a graded R-module, $\mathbf{f}$ is a sequence of homogeneous elements and $\deg f_m = a$. Then the long exact sequence in an exact sequence of graded R-modules, where $H_i(\mathbf{f}'; M) \to H_i(\mathbf{f}'; M)$ has to be replaced by $H_i(\mathbf{f}'; M)(-a) \to H_i(\mathbf{f}'; M)$.

Proof. Let $u \in K_i(\mathbf{f}; M)$. Then u can uniquely be written as $u = u_0 + u_1 \wedge e_m$ with $u_0 \in K_i(\mathbf{f}; M)$ and $u_1 \in K_{i-1}(\mathbf{f}'; M)$. We define the R-module homomorphism

$$\pi_i\ K_i(\mathbf{f}; M) \longrightarrow K_{i-1}(\mathbf{f}'; M), \quad u \mapsto u_1.$$

By (2.5), we have $\partial_i(u) = (\partial_i(u_0) + (-1)^{i-1} f_m u_1) + \partial_{i-1}(u_1) \wedge e_m$. Therefore,

$$\pi_{i-1}(\partial_i(u)) = \partial_{i-1}(u_1) = \partial'_{i-1}(\pi_i(u)).$$

Here, $\partial'_\bullet$ is the chain map of $K_\bullet(\mathbf{f}'; M)$, which is just the restriction of $\partial_\bullet$ to $K_\bullet(\mathbf{f}'; M)$. This shows that $\pi_\bullet\ K_\bullet(\mathbf{f}; M) \to K_\bullet(\mathbf{f}'; M)[-1]$ is a complex homomorphism. For each i, $\operatorname{Ker} \pi_i = K_i(\mathbf{f}'; M)$. Thus we obtain the short exact sequence of complexes

$$0 \longrightarrow K_\bullet(\mathbf{f}'; M) \longrightarrow K_\bullet(\mathbf{f}; M) \to K_\bullet(\mathbf{f}'; M)[-1] \longrightarrow 0,$$

since $\pi_\bullet$ is surjective. The resulting long exact sequence is the desired long exact sequence stated in the theorem.

For each i, the map $H_i(\mathbf{f}'; M) \to H_i(\mathbf{f}'; M)$ is a connecting homomorphism. To see that this connecting homomorphism is multiplication by $\pm f_m$, we recall the definition of connecting homomorphisms, see the proof of Theorem A.4. Consider the commutative diagram

$$\begin{array}{ccccccccc}
0 & \longrightarrow & K_{i+1}(\mathbf{f}'; M) & \longrightarrow & K_{i+1}(\mathbf{f}; M) & \xrightarrow{\pi_{i+1}} & K_i(\mathbf{f}'; M) & \longrightarrow & 0 \\
 & & \downarrow{\scriptstyle \partial'_{i+1}} & & \downarrow{\scriptstyle \partial_{i+1}} & & \downarrow{\scriptstyle \partial'_i} & & \\
0 & \longrightarrow & K_i(\mathbf{f}'; M) & \longrightarrow & K_i(\mathbf{f}; M) & \xrightarrow{\pi_i} & K_{i-1}(\mathbf{f}'; M) & \longrightarrow & 0.
\end{array}$$

Now, let $u \in K_i(\mathbf{f}'; M)$ be an i-cycle. We have to choose $v \in K_{i+1}(\mathbf{f}; M)$ with $\pi_{i+1}(v) = u$. Then the connecting homomorphism $\delta\ H_i(\mathbf{f}'; M) \to H_i(\mathbf{f}'; M)$ is given by $\delta[u] = [\partial_{i+1}(v)]$. For v we may choose $u \wedge e_m$. Then, since u is a cycle, it follows that

$$\partial_{i+1}(u \wedge e_m) = \partial_i(u) \wedge e_m + (-1)^i u f_m = (-1)^i u f_m.$$

This yields the desired conclusion. □

2.2.2 Vanishing of Koszul Homology

The following theorem gives a criterion for the acyclicity of the Koszul complex.

Theorem 2.12
(a) Let M be an R-module, and let $\mathbf{f}$ be an M-sequence in R. Then $H_i(\mathbf{f}; M) = 0$ for $i > 0$.

(b) Conversely, assume that R is a graded K-algebra with graded maximal ideal $\mathfrak{m}$, $M \in \mathcal{M}_R$ and $\mathbf{f} = f_1, \ldots, f_m$ be a sequence of homogeneous elements in R with $(\mathbf{f}) \subseteq \mathfrak{m}$ and $H_1(\mathbf{f}; M) = 0$. Then $\mathbf{f}$ is an M-sequence.

In other words, under the conditions given above, $\mathbf{f}$ is an M-sequence if and only if $H_i(\mathbf{f}; M) = 0$ for all $i > 0$.

A similar result holds in the local case.

Proof. (a) Let $\mathbf{f} = f_1, \ldots, f_m$, and assume that $\mathbf{f}$ is an M-sequence. We proceed by induction on m to show that $H_i(\mathbf{f}; M) = 0$ for $i > 0$. If $i > 1$, then $H_i(f_1; M) = 0$. By Theorem 2.11 we have the exact sequence

$$0 \longrightarrow H_1(f_1; M) \longrightarrow M \xrightarrow{f_1} M. \tag{2.8}$$

By assumption, f_1 is regular on M. Hence the multiplication map $M \xrightarrow{f_1} M$ is injective. Therefore, $H_1(f_1; M) = 0$, as well.

Now let $m > 1$, and let $\mathbf{f}' = f_1, \ldots, f_{m-1}$. By Theorem 2.11.

$$H_i(\mathbf{f}'; M) \longrightarrow H_i(\mathbf{f}; M) \longrightarrow H_{i-1}(\mathbf{f}'; M)$$

is exact for all i. Since $\mathbf{f}'$ is an M-sequence, our induction hypothesis implies that $H_i(\mathbf{f}'; M) = 0$ for $i \geq 1$. It follows that $H_i(\mathbf{f}; M) = 0$ for $i > 1$. For $i = 1$, we have the exact sequence

$$0 \longrightarrow H_1(\mathbf{f}; M) \longrightarrow M/(\mathbf{f}')M \xrightarrow{f_m} M/(\mathbf{f}')M.$$

Since $\mathbf{f}$ is an M-sequence, it follows from this sequence that $H_1(\mathbf{f}; M) = 0$, as well.

(b) We may assume that R is graded. The proof in the local case is word by word the same. Our assumptions and Nakayama's lemma imply that $M/(\mathbf{f})M \neq 0$. It remains to be shown that $M/(f_1, \ldots, f_{i-1})M \xrightarrow{f_i} M/(f_1, \ldots, f_{i-1})M$ is injective for $i = 1, \ldots, m$. For the proof we proceed by induction. For $m = 1$, the assertion follows from (2.8). Now let $m > 1$. By Theorem 2.11 we have the exact sequence

$$H_1(\mathbf{f}'; M) \xrightarrow{f_m} H_1(\mathbf{f}'; M) \longrightarrow H_1(\mathbf{f}; M) \longrightarrow M/(\mathbf{f}')M \xrightarrow{f_m} M/(\mathbf{f}')M.$$

By our assumption, $H_1(\mathbf{f}; M) = 0$. This implies that $M/(\mathbf{f}')M \xrightarrow{f_m} M/(\mathbf{f}')M$ is injective, and that $H_1(\mathbf{f}'; M)/f_m H_1(\mathbf{f}'; M) = 0$. Then $H_1(\mathbf{f}'; M) = 0$, by Nakayama's lemma. Our induction hypothesis implies that $\mathbf{f}'$ is an M-sequence, and hence $M/(f_1, \ldots, f_{i-1})M \xrightarrow{f_i} M/(f_1, \ldots, f_{i-1})M$ is injective for $i = 1, \ldots, m-1$. $\square$

Corollary 2.13
Let K be a field, let $S = K[x_1, \ldots, x_n]$ be the polynomial ring over K in the variables $x_1, \ldots, x_n$, and set $\mathfrak{m} = (x_1, \ldots, x_n)$. Then $K_\bullet(\mathbf{x}; S)$ is a free S-resolution (of length n) of $S/\mathfrak{m}$.

Proof. By Example 1.87, $\mathbf{x} = x_1, \ldots, x_n$ is an S-sequence. Thus, Theorem 2.12 implies that $K_\bullet(\mathbf{x}; S)$ is acyclic with $H_0(\mathbf{x}; S) = S/\mathfrak{m}$. □

Exercise 2.14
Let R be a Noetherian ring, and let $0 \to U \to N \to M \to 0$ be an exact sequence of R-modules. Use (2.7) and Theorem 2.12 to show that if $f_1, \ldots, f_m \in R$ is an M-sequence and $I = (f_1, \ldots, f_m)$, then $0 \to U/IU \to N/IN \to M/IM \to 0$ is exact, as well.

Exercise 2.15
Let M be an R-module and let $\mathbf{f} = f_1, \ldots, f_{m+1}$ be a sequence in R such that $f_1, \ldots, f_m$ is an M-sequence. Show that $H_i(\mathbf{f}; M) = 0$ for $i > 1$ and $H_1(\mathbf{f}; M) \cong ((f_1, \ldots, f_m) :_M f_{m+1})/(f_1, \ldots, f_m)M$.

2.3 Minimal Graded Free Resolutions

2.3.1 Graded Free Resolutions

Let K be a field and let R be a finitely generated graded K-algebra with graded maximal ideal $\mathfrak{m}$. Let $M \in \mathscr{M}_R$. By Lemma 1.12, M admits a finite homogeneous system of generators, say, $m_1, \ldots, m_r$. Let F_0 be a graded free R-module with homogeneous basis $f_1, \ldots, f_r$ with $\deg f_i = \deg m_i$ for all i, and consider the map $\varepsilon\ F_0 \longrightarrow M$ with $\varepsilon(f_i) = m_i$ for $i = 1, \ldots, r$. Then ε is a graded R-module homomorphism, that is, ε is a morphism in the category $\mathscr{M}_R$.

By Proposition 1.4, $\operatorname{Ker} \varepsilon \in \mathscr{M}_R$. For the finitely generated graded R-module $\operatorname{Ker} \varepsilon$, we choose again a free presentation as we did it for M and obtain a surjective graded R-module homomorphism $F_1 \to \operatorname{Ker} \varepsilon$. Composing the morphism $F_1 \to \operatorname{Ker} \varepsilon$ with the inclusion map $\operatorname{Ker} \varepsilon \hookrightarrow F_0$, we obtain the exact sequence $F_1 \longrightarrow F_0 \longrightarrow M \longrightarrow 0$. Proceeding in this way, we obtain an exact sequence

$$\cdots \longrightarrow F_i \longrightarrow \cdots \longrightarrow F_1 \longrightarrow F_0 \longrightarrow M \longrightarrow 0, \tag{2.9}$$

of graded modules, where all F_i are finitely generated graded free R-modules.

Replacing M by 0 in this exact sequence gives us the graded complex

$$F_\bullet \quad \cdots \longrightarrow F_i \longrightarrow \cdots \longrightarrow F_1 \longrightarrow F_0 \longrightarrow 0 \tag{2.10}$$

of graded free R-modules with $H_i(F_\bullet) = 0$ for $i > 0$ and $H_0(F_\bullet) \cong M$. A complex of free modules as in (2.10) is called a *graded free resolution* of M. The canonical morphism $F_0 \to H_0(F_\bullet) \cong M$ is called the *augmentation map* of the resolution $F_\bullet$. The exact sequence in (2.9) is called the *augmented graded free resolution* of M.

In the Examples 1.1 we notice that any finitely generated graded free R-module F is of the form $F = \bigoplus_j R(-j)^{b_j}$. Here b_j counts how many of the basis elements of F are of degree j. Thus,

$$F_\bullet \quad \cdots \longrightarrow \bigoplus_j R(-j)^{b_{ij}} \longrightarrow \cdots \longrightarrow \bigoplus_j R(-j)^{b_{1j}} \longrightarrow \bigoplus_j R(-j)^{b_{0j}} \longrightarrow 0$$

for suitable integers b_{ij}.

The concept of homotopic chain maps, which is used in the following theorem, has been introduced in Sect. A.1.

Theorem 2.16

Let $M, N \in \mathcal{M}_R$ and $\varphi\ M \to N$ be a morphism. Furthermore, let $F_\bullet$ be a graded free R-resolution of M and $G_\bullet$ be a graded free R-resolution of N. Then the following holds:

(a) There is a chain map $\varphi_\bullet\ F_\bullet \to G_\bullet$ with $H_0(\varphi_\bullet) = \varphi$.

(b) If $\psi_\bullet\ F_\bullet \to G_\bullet$ is another chain map with $H_0(\psi_\bullet) = \varphi$, then $\varphi_\bullet \sim \psi_\bullet$.

Moreover, the components φ_i and ψ_i of the chain maps and of the homotopy can all be chosen to be morphisms in $\mathcal{M}_R$.

A chain map $\varphi_\bullet$ as given in (a) is called a *lifting* of φ. Part (b) of the theorem says that any two liftings of φ are homotopic.

Proof. (of Theorem 2.16) (a) Let $\varepsilon\ F_0 \to M$ be the augmentation map of $F_\bullet$ and $\varepsilon'\ G_0 \to N$ be the augmentation map of $G_\bullet$. Then we have the diagram

$$\begin{array}{ccc} F_0 & & G_0 \\ \downarrow{\scriptstyle\varepsilon} & & \downarrow{\scriptstyle\varepsilon'} \\ M & \xrightarrow{\varphi} & N. \end{array}$$

By induction on i, we construct the chain map $\varphi_\bullet = (\varphi_i)_{i\geq 0}$. Since F_0 is a free module and ε' is surjective, the map $\varphi \circ \varepsilon$ can be lifted to φ_0 which then yields the commutative diagram

$$\begin{array}{ccc} F_0 & \xrightarrow{\varphi_0} & G_0 \\ \downarrow{\scriptstyle \varepsilon} & & \downarrow{\scriptstyle \varepsilon'} \\ M & \xrightarrow{\varphi} & N. \end{array}$$

Suppose we have already constructed the chain of maps $\varphi_0, \ldots, \varphi_{i-1}$. Let (∂_i) and (∂_i') be the differentials of $F_\bullet$ and $G_\bullet$, respectively. Since $\varphi_{i-2} \circ \partial_{i-1} = \partial_{i-1}' \circ \varphi_{i-1}$, the morphism $\varphi_{i-1}\ F_{i-1} \to G_{i-1}$ induces the morphism $\alpha\ Z_{i-1}(F_\bullet) \to Z_{i-1}(G_\bullet)$, and we obtain the commutative diagram

$$\begin{array}{ccc} F_i & & G_i \\ \downarrow & & \downarrow \\ Z_{i-1}(F_\bullet) & \xrightarrow{\alpha} & Z_{i-1}(G_\bullet) \\ \downarrow & & \downarrow \\ F_{i-1} & \xrightarrow{\varphi_{i-1}} & G_{i-1}. \end{array}$$

(with $\partial_i: F_i \to F_{i-1}$ and $\partial_i': G_i \to G_{i-1}$ as the outer curved arrows)

The map $F_i \to Z_{i-1}(F_\bullet) \xrightarrow{\alpha} Z_{i-1}(G_\bullet)$ can be lifted to $\varphi_i\ F_i \to G_i$, and it follows that $\varphi_{i-1} \circ \partial_i = \partial_i' \circ \varphi_i$. Proceeding in this way we get the chain map $\varphi_\bullet\ F_\bullet \to G_\bullet$, and it is clear that $H_0(\varphi_\bullet) = \varphi$ after identifying $H_0(F_\bullet)$ with M and $H_0(G_\bullet)$ with N.

(b) By induction on i, we construct a chain homotopy $\eta_\bullet$, that is, morphisms $\eta_i\ F_i \to G_{i+1}$ such that $\varphi_i - \psi_i = \partial_{i+1}' \circ \eta_i + \eta_{i-1} \circ \partial_i$ for all $i \geq 0$. Here, $\eta_{-1} = 0$. Let $i = 0$. Since $H_0(\varphi_\bullet) = H_0(\psi_\bullet)$, it follows that $H_0(\varphi_\bullet - \psi_\bullet) = 0$. This implies that $\mathrm{Im}(\varphi_0 - \psi_0) \subseteq Z_0(G_\bullet)$. Since F_0 is a free module, there exists a morphism η_0 to make the diagram

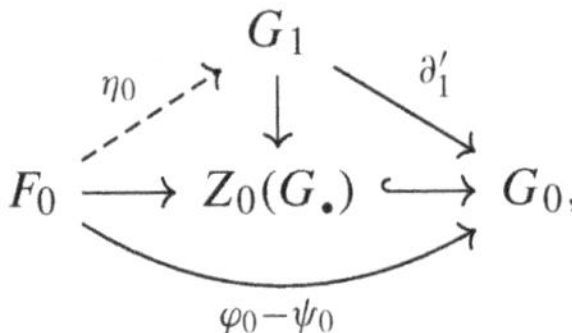

commutative. Now, let $i > 0$ and suppose η_j for $j < i$ has already been constructed. We have

$$\partial_i' \circ (\varphi_i - \psi_i) = (\varphi_{i-1} - \psi_{i-1}) \circ \partial_i = (\partial_i' \circ \eta_{i-1} + \eta_{i-2} \circ \partial_{i-1}) \circ \partial_i = \partial_i' \circ \eta_{i-1} \circ \partial_i,$$

and hence $\partial_i' \circ ((\varphi_i - \psi_i) - \eta_{i-1} \circ \partial_i) = 0$. Set $\alpha = (\varphi_i - \psi_i) - \eta_{i-1} \circ \partial_i$. Then $\mathrm{Im}\alpha \subseteq Z_i(G_\bullet)$ and by lifting α we obtain the commutative diagram

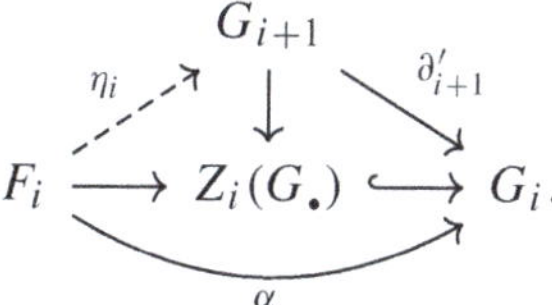

As a result we have $\varphi_i - \psi = \partial'_{i+1} \circ \eta_i + \eta_{i-1} \circ \partial_i$, as desired. □

Exercise 2.17

Let $S = K[x, y]$ be the polynomial ring over a field K, let $M = S/(x^2, xy, y^2)$, $N = S/(x, y)$ and let $\varphi : M \to N$ be the natural residue class map.

(a) Show that M has a resolution $F_\bullet$ of the form $0 \to S^2 \to S^3 \to S \to 0$ and N has a resolution $G_\bullet$ of the form $0 \to S \to S^2 \to S \to 0$.

(b) Find two distinct liftings $\varphi_\bullet : F_\bullet \to G_\bullet$ and construct a chain homotopy between the two liftings.

2.3.2 Minimal Graded Free Resolutions and Betti Numbers

We have seen in Theorem 2.16 that any two graded free R-resolutions of a module $M \in \mathcal{M}_R$ are homotopically equivalent. This result can be strengthened. Indeed, let $M \in \mathcal{M}_R$, and denote by $\mathfrak{m}$ the graded maximal ideal of R. A graded free R-resolution $(F_\bullet, \partial_\bullet)$ of M is called *minimal*, if $\partial_{i+1}(F_{i+1}) \subseteq \mathfrak{m}F_i$ for all $i \geq 0$. Why do we call a resolution satisfying this condition minimal? This is explained by the fact that in the construction of the free resolution this condition is satisfied, if and only if in each step we choose the number of the generators of F_i as small as possible.

Lemma 2.18

Let $M \in \mathcal{M}_R$, let $u_1, \ldots, u_r$ be a set of homogeneous generators of M, and let F_0 be the free R-module with basis $e_1, \ldots, e_r$. We set $\deg e_i = \deg u_i$ for $i = 1, \ldots, r$, and $\varepsilon\ F_0 \to M$ the graded R-module homomorphism with $\varepsilon(e_i) = u_i$ for $i = 1, \ldots, r$. Let $Z_0 = \operatorname{Ker} \varepsilon$. Then $Z_0 \subseteq \mathfrak{m}F_0$ if and only if $u_1, \ldots, u_r$ is a minimal set of generators of M.

Proof. The sequence $Z_0/\mathfrak{m}Z_0 \to F_0/\mathfrak{m}F_0 \to M/\mathfrak{m}M \to 0$ is an exact sequence of K-vector spaces. Note that we have $Z_0 \subseteq \mathfrak{m}F_0$ if and only if $Z_0/\mathfrak{m}Z_0 \to F_0/\mathfrak{m}F_0$ is the zero map. This map is zero if and only if the map $F_0/\mathfrak{m}F_0 \to M/\mathfrak{m}M$ is an isomorphism, i.e., $\dim_K F_0/\mathfrak{m}F_0 = \dim_K M/\mathfrak{m}M$. By Corollary 1.24 this is the case if and only if $u_1, \ldots, u_r$ is a minimal set of generators of M. □

We call a graded epimorphism $\varepsilon\ F_0 \to M$ as defined in Lemma 2.18 with $\operatorname{Ker}\varepsilon \subseteq \mathfrak{m}F_0$ a *minimal free presenting* of M. Obviously in a minimal free presenting, F_0 cannot have less generators than the minimal number of generators of M.

Proposition 2.19
Any $M \in \mathscr{M}_R$ admits a minimal graded free R-resolution.

Proof. Let $u_1, \ldots, u_r$ be a minimal set of homogeneous generators of M. Let F_0 and Z_0 be as in Lemma 2.18. Then $Z_0 \subseteq \mathfrak{m}F_0$. Suppose that the free R-resolution of M has already been constructed such that $Z_j \subseteq \mathfrak{m}F_j$ for $j = 0, \ldots, i$. We choose a minimal free presenting $F_{i+1} \to Z_i$, and set $Z_{i+1} = \operatorname{Ker}(F_{i+1} \to Z_i)$. Then, as before, $Z_{i+1} \subseteq \mathfrak{m}F_{i+1}$. Proceeding in this way we obtain the desired minimal graded free R-resolution of M. □

Similarly, one can construct minimal free resolutions for finitely generated modules over local rings.

Theorem 2.20
Let $M \in \mathscr{M}_R$, and let $F_\bullet$ and $G_\bullet$ be graded minimal free R-resolutions of M. Then $F_\bullet \cong G_\bullet$.

Proof. By Theorem 2.16 there exists a chain map $\varphi_\bullet\ F_\bullet \to G_\bullet$ with $H_0(\varphi_\bullet) = \mathrm{id}_M$. According to Proposition A.33, $\varphi_\bullet$ induces an isomorphism

$$H_i(\varphi_\bullet \underline{\otimes} R/\mathfrak{m})\ H_i(F_\bullet \underline{\otimes} R/\mathfrak{m}) \to H_i(G_\bullet \underline{\otimes} R/\mathfrak{m}).$$

Note that $F_i \underline{\otimes} R/\mathfrak{m} = F_i/\mathfrak{m}F_i$ and $G_i \underline{\otimes} R/\mathfrak{m} = G_i/\mathfrak{m}G_i$. Let $\partial_\bullet$ be the differential of $F_\bullet$. Since the resolution $F_\bullet$ is minimal, the induced differential $\partial_\bullet \underline{\otimes} R/\mathfrak{m}\ F_i/\mathfrak{m}F_i \to F_{i-1}/\mathfrak{m}F_{i-1}$ is the zero map for all i. This implies that $H_i(F_\bullet \underline{\otimes} R/\mathfrak{m}) = F_i/\mathfrak{m}F_i$ for all i. Similarly, we have $H_i(G_\bullet \underline{\otimes} R/\mathfrak{m}) = G_i/\mathfrak{m}G_i$. Thus for all i, the graded R-module homomorphism $\varphi_i\ F_i \to G_i$ induces an isomorphism $\overline{\varphi}_i\ F_i/\mathfrak{m}F_i \to G_i/\mathfrak{m}G_i$. The next lemma shows us that φ_i itself is an isomorphism. □

Lemma 2.21
Let $M, N \in \mathscr{M}_R$, and let $\varphi\ M \to N$ be a graded R-module homomorphism. Suppose that N is free and that the induced map $\overline{\varphi}\ M/\mathfrak{m}M \to N/\mathfrak{m}N$ is an isomorphism. Then φ is an isomorphism.

Proof. Let $C = \operatorname{Coker}\varphi$ and $U = \operatorname{Ker}\varphi$. Then $M \to N \to C \to 0$ is an exact sequence. Since $W\underline{\otimes}R/\mathfrak{m} = W/\mathfrak{m}W$ for any $W \in \mathcal{M}_R$ and since the tensor product is right exact, we obtain the exact sequence $M/\mathfrak{m}M \to N/\mathfrak{m}N \to C/\mathfrak{m}C \to 0$. By assumption $\overline{\varphi}$ is an isomorphism. This implies that $C/\mathfrak{m}C = 0$. Thus, Nakayama's lemma (Corollary 1.24) yields that $C = 0$. Hence we obtain the short exact sequence $0 \to U \to M \to N \to 0$. This gives us the long exact sequence $\cdots \to \underline{\operatorname{Tor}}_1^R(N, R/\mathfrak{m}) \to U/\mathfrak{m}U \to M/\mathfrak{m}M \to N/\mathfrak{m}N \to 0$. Since N is free, it follows that $\underline{\operatorname{Tor}}_1^R(N, R/\mathfrak{m}) = 0$, so that $U/\mathfrak{m}U \cong \operatorname{Ker}\overline{\varphi}$. Hence $U/\mathfrak{m}U = 0$, because $\overline{\varphi}$ is an isomorphism. Applying again Nakayama's lemma, we see that $U = 0$, and this shows that φ is an isomorphism, as desired. □

A graded complex of the form $0 \to R(-d) \xrightarrow{\text{id}} R(-d) \to 0$ is called a *short trivial complex*, and the direct sum of finitely many short trivial complexes is called a *trivial complex*.

Theorem 2.20 can be generalized as follows. For the proof of the next statement you may consult [60, Theorem 7.5].

Theorem 2.22
Let M be a finitely generated graded S-module and $F_\bullet$ be a minimal graded free resolution of M. Then for any graded free resolution $G_\bullet$ of M, there exists a trivial complex $F'_\bullet$ such that $G_\bullet = F_\bullet \oplus F'_\bullet$.

Exercise 2.23
Let $F_\bullet$ be a finite complex of finitely generated free R-modules. Show that $F_\bullet$ is a trivial complex if and only if it is exact.

Let $M \in \mathcal{M}_R$, and let $F_\bullet$ be a graded minimal free R-resolution of M. Furthermore, for $i \geq 0$, let $F_i \cong \bigoplus_j R(-j)^{\beta_{i,j}}$. Since $F_\bullet$ is unique up to isomorphism, it follows that the integers $\beta_{i,j}$ are uniquely determined by M. These integers are called the *graded Betti numbers* of M. For a fixed integer $i \geq 0$, one has $\beta_{i,j} = 0$ for all but finitely many j. We call these integers the *ith graded Betti numbers* of M. The *ith Betti number* of M is defined to be $\beta_i = \sum_j \beta_{i,j}$. Sometimes we write $\beta_{i,j}(M)$ for $\beta_{i,j}$ and $\beta_i(M)$ for β_i in order to indicate that these are the graded Betti numbers of the module M.

Proposition 2.24
Let $M \in \mathcal{M}_R$ with graded Betti numbers $\beta_{i,j}$. Then $\underline{\operatorname{Tor}}_i^R$ $(M, R/\mathfrak{m})$ is a finitely generated graded K-vector space and

$$\dim_K \underline{\operatorname{Tor}}_i^R(M, R/\mathfrak{m})_j = \beta_{i,j}$$

for all i and j.

Proof. Let $F_\bullet$ be the graded minimal free R-resolution of M. Then one has $\underline{\mathrm{Tor}}_i^R(M, R/\mathfrak{m}) = H_i(F_\bullet \underline{\otimes} R/\mathfrak{m}) = F_i/\mathfrak{m}F_i$. This shows that $\underline{\mathrm{Tor}}_i^R(M, R/\mathfrak{m})$ is a graded K-vector space. Since $F_i = \bigoplus_j R(-j)^{\beta_{i,j}}$, it follows that $F_i/\mathfrak{m}F_i = \bigoplus_j K(-j)^{\beta_{i,j}}$. This yields the desired conclusion. □

Exercise 2.25

Let $M \in \mathscr{M}_R$, and let $f \in R$ be a homogeneous element which is R-regular and M-regular. Show that

(a) $\underline{\mathrm{Tor}}_i^R(R/(f), M) = 0$ for $i > 0$.

(b) Let $F_\bullet$ be a (minimal) graded free resolution of M. Use (a) to show that $F_\bullet/fF_\bullet$ is a (minimal) graded free resolution of M/fM.

We use Corollary 2.13 to get more detailed information about the Betti numbers.

Proposition 2.26

Let S be a graded K-algebra with $\deg x_i = a_i > 0$ for $i = 1, \ldots, n$. Let $M \in \mathscr{M}_S$ with graded minimal free S-resolution $F_\bullet$, and let $\mathbf{x} = x_1, \ldots, x_n$ and $\mathfrak{m} = (\mathbf{x})$. Then for $i = 1, \ldots, n$, there exists a graded isomorphism of K-vector spaces

$$H_i(\mathbf{x}; M) \cong F_\bullet/\mathfrak{m}F_\bullet.$$

In particular, $\beta_{i,j}(M) = \dim_K(H_i(\mathbf{x}; M))_j$ for all i and j.

Proof. By Corollary 2.13, the Koszul complex $K_\bullet(\mathbf{x}; S)$ is a graded minimal free resolution of $S/\mathfrak{m}$. Thus, by using Theorem A.36, we obtain

$$H_i(\mathbf{x}; M) \cong \underline{\mathrm{Tor}}_i^S(S/\mathfrak{m}, M) \cong H_i(S/\mathfrak{m} \underline{\otimes} F_\bullet) \cong H_i(F_\bullet/\mathfrak{m}F_\bullet) \cong F_\bullet/\mathfrak{m}F_\bullet.$$

The most right isomorphism results from the fact that $F_\bullet$ is a minimal free S-resolution of M. □

The following corollary is an immediate consequence of Propositions 2.10 and 2.26.

Corollary 2.27
Let K be a field, and let $S = K[x_1, \ldots, x_n]$ be the polynomial ring with $\deg x_i = a_i > 0$ for $i = 1, \ldots, n$. Furthermore, let $M \in \mathcal{M}_S$. Then

(a) $\bigoplus_j K(-j)^{\beta_{n,j}(M)} \cong \underline{\mathrm{Tor}}_n^S(S/\mathfrak{m}, M) \cong \mathrm{Soc}(M)(-a_1 - a_2 - \cdots - a_m)$.
(b) $\beta_{n,j}(M) = \dim_K \mathrm{Soc}(M)_{j-a_1-a_2-\cdots-a_m}$ for all j.

Since for local rings minimal free resolutions can also be defined as mentioned before, one has in the local case analogue statements as in Theorem 2.20 and Proposition 2.3.2.

The following corollary follows from the graded version of Theorem A.36.

Corollary 2.28
Let $M, N \in \mathcal{M}_R$ such that $\underline{\mathrm{Tor}}_i^R(M, N) = 0$ for $i > 0$. Let $F_\bullet$ be a graded free R-resolution of M, and let $G_\bullet$ be a graded free R-resolution of N. Then $F_\bullet \underline{\otimes} G_\bullet$ is a graded free R-resolution of $M \underline{\otimes} N$. If $F_\bullet$ and $G_\bullet$ are minimal resolutions, then $F_\bullet \underline{\otimes} G_\bullet$ is also minimal.

Coming back to general commutative rings one defines the *projective dimension* of a module as the shortest length of a projective resolution of M (which may be infinite). Projective resolutions are defined like free resolutions with the free modules in the resolution replaced by projective modules. The projective dimension of a module in the category Mod_R is denoted by $\mathrm{pd}_R(M)$ and the projective dimension of a module M the category $\mathcal{M}_R$ is denoted by $\underline{\mathrm{pd}}_R(M)$. The lower index R is omitted when it is clear from the context over which ring M is defined.

Let $M \in \mathcal{M}_R$. Then the projective dimension of M is indeed the shortest length of a graded free resolution of M, as the following theorem shows.

Theorem 2.29
Let K be a field, and let R be a Noetherian local ring or a finitely generated graded K-algebra. Then any finitely generated (graded) projective R-module is free.

Proof. We assume that R is local with maximal ideal $\mathfrak{m}$. The proof in the case that R is a finitely generated graded K-algebra is the same.

Let $p_1, \ldots, p_m$ be a minimal system of (homogeneous) generators of P, and let F be the free R-module with basis $e_1, \ldots, e_m$. Since F is free, there exists an R-module homomorphism $\varphi F \to P$ with $\varphi(e_i) = p_i$ for all i. Since $p_1, \ldots, p_m$ is a system of generators of P, it follows that φ is surjective. Let $U = \operatorname{Ker} \varphi$. Then we obtain the exact sequence

$$0 \to U \to F \to P \to 0,$$

which induces the exact sequence

$$U/\mathfrak{m}U \to F/\mathfrak{m}F \to P/\mathfrak{m}P \to 0.$$

We claim that $U/\mathfrak{m}U \to F/\mathfrak{m}F$ is injective. Assuming this for the moment and noticing that $U/\mathfrak{m}U \to F/\mathfrak{m}F$ is the zero map, because $U \subseteq \mathfrak{m}F$, due to the fact that $p_1, \ldots, p_m$ be a minimal system of generators of P, we conclude that $U/\mathfrak{m}U = 0$. Now we apply Nakayama's lemma and see that $U = 0$. This implies that $P \cong F$.

It remains to prove the claim. Since P is projective, the exact sequence $0 \to U \to F \to P \to 0$ is split exact. Therefore, if $\iota U \to F$ is the inclusion map in the exact sequence, then there exists an R-module homomorphism $\pi F \to U$ such that $\pi \circ \iota = \mathrm{id}_U$. It follows that

$$(\pi \otimes R/\mathfrak{m}) \circ (\iota \otimes R/\mathfrak{m}) = (\pi \circ \iota) \otimes R/\mathfrak{m} = \mathrm{id}_U \otimes R/\mathfrak{m} = \mathrm{id}_{U/\mathfrak{m}U},$$

and this implies that $\iota \otimes R/\mathfrak{m}\ U/\mathfrak{m}U \to F/\mathfrak{m}F$ is injective, as desired. □

Whenever a minimal graded free resolution of M exists, the projective dimension of M is indeed the length of any minimal graded free resolution of M. This is shown in

Proposition 2.30
For any $M \in \mathscr{M}_R$,

$$\underline{\mathrm{pd}}(M) = \limsup\{i\ \beta_i \neq 0\}.$$

Proof. Let $G_\bullet$ be a minimal graded free resolution of M and $F_\bullet$ be an arbitrary graded free resolution of M. As we saw in the proof of Theorem 2.20, there exists a chain map $\varphi_\bullet\ F_\bullet \to G_\bullet$ with $H_0(\varphi_\bullet) = \mathrm{id}_M$ and an isomorphism

$$H_i(\varphi_\bullet \underline{\otimes} R/\mathfrak{m})\ H_i(F_\bullet \underline{\otimes} R/\mathfrak{m}) \to H_i(G_\bullet \underline{\otimes} R/\mathfrak{m}).$$

Moreover, since $G_\bullet$ is minimal, we have $H_i(G_\bullet \underline{\otimes} R/\mathfrak{m}) = G_i \underline{\otimes} R/\mathfrak{m} = G_i/\mathfrak{m}G_i$ for all i. Hence $G_i/\mathfrak{m}G_i = \mathrm{Im} H_i(\varphi_\bullet \underline{\otimes} R/\mathfrak{m}) = (\mathrm{Im}\varphi_i + \mathfrak{m}G_i)/\mathfrak{m}G_i$. Thus by Nakayama's lemma, $\mathrm{Im}\varphi_i = G_i$. This implies that the sequence $0 \longrightarrow \operatorname{Ker} \varphi_i \longrightarrow F_i \xrightarrow{\varphi_i} G_i \longrightarrow 0$ is exact. It is indeed split exact, since G_i is a free module. So $F_i \cong G_i \oplus \operatorname{Ker} \varphi_i$ for all i. Thus $G_\bullet$ has the minimum length among the graded free resolutions of M and this length is $\limsup\{i\ \beta_i \neq 0\}$. Therefore $\underline{\mathrm{pd}}(M) = \limsup\{i\ \beta_i \neq 0\}$. □

A module $M \in \mathscr{M}_R$ need not to have finite projective dimension, as the following example shows. Let $S = K[x]$ be the polynomial ring in one indeterminant, and let $R = S/(x^2)$. Then the ideal $I = (x)R$ has the following minimal graded free resolution

$$\cdots \xrightarrow{x} R(-j) \xrightarrow{x} \cdots \xrightarrow{x} R(-2) \xrightarrow{x} R(-1) \xrightarrow{x} R \to I \longrightarrow 0.$$

The more surprising is Hilbert's syzygy theorem which says that all finitely generated graded modules over the polynomial ring have finite projective dimension.

Theorem 2.31
(Hilbert's syzygy theorem) Let K be a field, let $S = K[x_1, \ldots, x_n]$ be the polynomial ring over K, and let $M \in \mathscr{M}_S$. Then

$$\underline{\mathrm{pd}}(M) \leq n.$$

Proof. Let $\mathfrak{m} = (x_1, \ldots, x_n)$. The complex $K_\bullet(\mathbf{x}; S)$, where $\mathbf{x} = x_1, \ldots, x_n$ is a minimal free S-resolution of $S/\mathfrak{m}$ of length n, see Corollary 2.13. So $\underline{\mathrm{pd}}(S/\mathfrak{m}) = n$. Thus, the graded version of Theorem A.36 implies that $\underline{\mathrm{Tor}}_i^S(M, S/\mathfrak{m}) = 0$ for $i > n$. Hence the desired conclusion follows from Proposition 2.3.2, where it is shown that $\underline{\mathrm{pd}}(M) = \max\{i \ \underline{\mathrm{Tor}}_i^S(M, S/\mathfrak{m}) \neq 0\}$. □

Corollary 2.32
Let K be a field, let $S = K[x_1, \ldots, x_n]$ be the polynomial ring over K, and let $M \in \mathscr{M}_S$. Then rank $M = \sum_{i=0}^{n}(-1)^i \beta_i(M)$.

Proof. Let $F_\bullet$ be the graded minimal free resolution of M. By Theorem 2.31, $F_i = 0$ for $i > n$, and by Exercise 1.35 we have $\beta_i(M) = \operatorname{rank} F_i$ for all i. Thus the desired result follows from Lemma 1.36. □

Exercise 2.33
Let K be a field, and let $S = K[x_1, \ldots, x_n]$ be the polynomial ring. Furthermore, let $M \in \mathscr{M}_S$. First show that $\dim_K \mathrm{Soc}(M) = \beta_n(M)$, and then conclude that $\mathrm{Soc}(M) \neq 0$ if and only if $\underline{\mathrm{pd}}(M) = n$.

Exercise 2.34
Let $M \in \mathscr{M}_R$, and let $\mathfrak{m}$ be the graded maximal ideal of R. Prove that $\underline{\mathrm{pd}}_R(M) = \mathrm{pd}_{R_\mathfrak{m}}(M_\mathfrak{m})$.

2.4 Injective Hulls and Graded Injective Resolutions

2.4.1 Injective Resolutions

Let R be a ring. By definition, a module E is *injective* if $\operatorname{Hom}_R(-, E)$ is an exact functor. Since $\operatorname{Hom}_R(-, W)$ is left exact for any R-module W, it follows that E is injective if and only if for every injective R-module homomorphism $\iota : N \to M$ the induced R-module homomorphism $\operatorname{Hom}_R(M, E) \to \operatorname{Hom}_R(N, E)$ is surjective. This is the case, if and only if for any R-module homomorphism $\varphi\ N \to E$ there exists an R-module homomorphism $\psi\ M \to E$ such that $\psi \circ \iota = \varphi$. That is, the R-module homomorphism $\varphi\, N \to E$ can be extended to the R-module homomorphism $\psi\ M \to E$.

$$\begin{array}{ccc} M & & \\ {\scriptstyle\iota}\uparrow & \searrow^{\psi} & \\ N & \xrightarrow{\varphi} & E \end{array}$$

Exercise 2.35
Let M be an R-module, and let $E \subseteq M$ be an injective submodule of M. Then E is a direct summand of M.

The *homomorphism extension property* mentioned before, only needs to be tested for ideals. This is shown in the next theorem for which proof we will have to apply *Zorn's lemma*. It says that a partially ordered set containing upper bounds for every totally ordered subset, contains at least one maximal element.

Theorem 2.36
(Baer) The R-module E is injective if and only if for any ideal $J \subseteq R$, any R-module homomorphism $J \to E$ can be extended to R.

Proof. The 'only if' part is clear from the definition of an injective module. Given an injective R-module homomorphism $N \to M$ and an R-module homomorphism $\varphi\ N \to E$. We need to show that φ can be extended to M. Identifying N with its image in M, we may assume that $N \subseteq M$.

Consider the set of pairs (U, ψ), where U is a submodule of M with $N \subseteq U$ and where ψ extends φ. We define a partial order on this set by letting $(U_1, \psi_1) \leq (U_2, \psi_2)$ if $U_1 \subseteq U_2$ and $\psi_1 = \psi_2|_{U_1}$. Zorn's lemma implies that there exists a maximal element(U', ψ') in this set. Suppose $U' \neq M$. Then there exists an R-submodule W of M with $U' \subsetneq W$ such that $W/U' \cong R/J$ for some proper ideal J of R. Thus we obtain the exact sequence $0 \to U' \to M \to R/J \to 0$, which, by

Theorem A.40, induces the exact sequence

$$\mathrm{Hom}_R(M, E) \to \mathrm{Hom}_R(U', E) \to \mathrm{Ext}^1_R(R/J, E).$$

Similarly, the exact sequence $0 \to J \to R \to R/J \to 0$ induces the exact sequence

$$\mathrm{Hom}_R(R, E) \to \mathrm{Hom}_R(J, E) \to \mathrm{Ext}^1_R(R/J, E) \to \mathrm{Ext}^1_R(R, E).$$

Since by assumption, any R-module homomorphism $J \to R$ can be extended to R, it follows that $\mathrm{Hom}_R(R, E) \to \mathrm{Hom}_R(J, E)$ is surjective, and since R is free, it follows that $\mathrm{Ext}^1_R(R, E) = 0$. This implies that $\mathrm{Ext}^1_R(R/J, E) = 0$, so that $\mathrm{Hom}_R(M, E) \to \mathrm{Hom}_R(U', E)$ is surjective. Consequently, any R-module homomorphism $U' \to E$ can be lifted to M, contradicting the choice of U'. □

Use Theorem 2.36 to solve

Exercise 2.37
Let R be a Noetherian ring. Show that a direct sum of R-modules is injective if and only if each of its direct summands is injective.

A module M is called *divisible* if for each $m \in M$ and each regular element $r \in M$ there exists $m' \in M$ such that $m = rm'$.

Corollary 2.38
(a) Let E be an injective R-module. Then E is divisible.
(b) A $\mathbb{Z}$-module is divisible, if and only if it is injective.

Proof. (a) and (b) follow from Theorem 2.36. For the proof of (a) consider $m \in E$ and a regular element $r \in R$. Set $J = (r)$. Since r is regular, it follows that J is a free R-module. Therefore there exists an R-module homomorphism $\varphi\ J \to E$ with $\varphi(r) = m$. Let $\psi\ R \to E$ be an extension of φ, and let $\psi(1) = m'$. Then $m = \varphi(r) = \psi(r) = r\psi(1) = rm'$.

For the proof of (b) we observe that all ideals in $\mathbb{Z}$ are principal. □

Let M be a $\mathbb{Z}$-module (which is nothing but an abelian group). Then $M \cong \mathbb{Z}^{\Lambda}/U$ for some set Λ and subgroup U of $\mathbb{Z}^{\Lambda}$. Since $\mathbb{Q}$ is a divisible, it follows that $E = \mathbb{Q}^{\Lambda}/U$ is divisible, as well (why?). Thus by Corollary 2.38, E is an injective $\mathbb{Z}$-module. Composing the isomorphism $M \cong \mathbb{Z}^{\Lambda}/U$ with the natural inclusion map $\mathbb{Z}^{\Lambda}/U \to \mathbb{Q}^{\Lambda}/U$ yields an injective $\mathbb{Z}$-module homomorphism $M \to E$. The next result shows that a similar result holds for any module over an arbitrary ring.

Theorem 2.39
Every R-module can be embedded into an injective R-module.

Proof. For $R = \mathbb{Z}$ we know the result already. Now let R be arbitrary. The map $\varphi\ M \to \mathrm{Hom}_{\mathbb{Z}}(R, M)$ with $\varphi(x)(r) = rx$ for $x \in M$ and $r \in R$ is an injective R-module homomorphism. As a $\mathbb{Z}$-module, M can be embedded into an injective $\mathbb{Z}$-module E. Let $\psi\ M \to E$ be the embedding. Then we obtain the injective R-module homomorphism $\mathrm{Hom}_{\mathbb{Z}}(R, \psi)\ \mathrm{Hom}_{\mathbb{Z}}(R, M) \to \mathrm{Hom}_{\mathbb{Z}}(R, E)$, which gives us the embedding

$$\mathrm{Hom}_{\mathbb{Z}}(R, \psi) \circ \varphi\ M \to \mathrm{Hom}_{\mathbb{Z}}(R, E).$$

The desired result follows once we have shown that $\mathrm{Hom}_{\mathbb{Z}}(R, E)$ is an injective R-module. In order to see this we notice that there is an isomorphism of the functor $\mathrm{Hom}_R(-, \mathrm{Hom}_{\mathbb{Z}}(R, E))\ \mathrm{Mod}_R \to \mathrm{Mod}_R$ with the functor $\mathrm{Hom}_{\mathbb{Z}}(-, E)\ \mathrm{Mod}_R \to \mathrm{Mod}_R$. It follows that the functor $\mathrm{Hom}_R(-, \mathrm{Hom}_{\mathbb{Z}}(R, E))$ is right exact, since this is the case for $\mathrm{Hom}_{\mathbb{Z}}(-, E)$, as E is an injective $\mathbb{Z}$-module. This shows that $\mathrm{Hom}_{\mathbb{Z}}(R, E)$ is an injective R-module, as desired. □

Exercise 2.40
Let W be an R-module. Then W is injective if and only if Ext^1_R $(R/J, W) = 0$ for all ideals $J \subseteq R$.

We use the characterization of injective modules given in this exercise to prove

Proposition 2.41
Let R be a Noetherian ring, let $S \subseteq R$ be a multiplicatively closed set, and let E be an injective R-module. Then $S^{-1}E$ is an injective $S^{-1}R$-module.

Proof. Let J be an ideal in R. Since R is Noetherian, it follows from Proposition A.39 that $\mathrm{Ext}^1_{S^{-1}R}(S^{-1}R/JS^{-1}R, S^{-1}E) \cong S^{-1}\,\mathrm{Ext}^1_R(R/J, E) = 0$. Since any ideal in $S^{-1}R$ is of the form $JS^{-1}R$ for some ideal $J \subseteq R$, it follows from Exercise 2.40 that $S^{-1}E$ is an injective $S^{-1}R$-module. □

Let M be an R-module. A complex $E^{\bullet}$ of injective modules

$$E^{\bullet} : 0 \to E^0 \to E^1 \to \cdots \to E^j \to \cdots$$

is called an *injective resolution* of M if $H^j(E^\bullet) = 0$ for $j > 0$ and $H^0(E^\bullet) \cong M$. It follows that $E^\bullet$ is an injective resolution of M, if and only if there exists a monomorphism $M \to E^0$ such that $0 \to M \to E^0 \to E^1 \to \cdots$ is exact.

Thanks to Theorem 2.39 we can show that any R-module admits an injective resolution. Indeed, let M be an R-module. By Theorem 2.39, there exists an injective R-module E^0 and a monomorphism $M \to E^0$. Let M_1 be the cokernel of $M \to E^0$. If $M_1 = 0$, then the injective resolution of M is $0 \to E^0 \to 0 \to 0 \cdots$. Otherwise, $M_1 \neq 0$ and there exists an injective R-module and a monomorphism $M_1 \to E^1$. Composing the epimorphism $E^0 \to M_1$ with the monomorphism $M_1 \to E^1$, we obtain the exact sequence $0 \to M \to E^0 \to E^1$. Proceeding in this way we obviously obtain the desired injective resolution of M.

Exercise 2.42

Let M and N be R-modules, and $\varphi\ M \to N$ be an R-module homomorphism. Furthermore, let $E^\bullet$ be an injective resolution of M and $I^\bullet$ be an injective resolution of N. Show that there exists a chain map $\varphi^\bullet\ E^\bullet \to I^\bullet$ with $H^0(\varphi^\bullet) = \varphi$.

The injective resolution of a module is by no means unique. But as for free resolutions, for injective resolutions there is also the concept of minimal resolutions, which we will now discuss. To this end, we first define essential extensions and injective hulls.

Let N be an R-module, and let M be a submodule of N. The inclusion $M \subseteq N$ is said to be an *essential extension*, if $U \cap M \neq 0$ for any nonzero submodule U of N. The module N is called an essential extension of M, if $M \subseteq N$ is an essential extension. An essential extension $M \subseteq N$ is called *proper*, if $N \neq M$. More generally, we call an injective R-module homomorphism $\varphi\ M \to N$ an essential extension, if $\mathrm{Im}\varphi \subseteq N$ is an essential extension.

Injective modules are characterized by the following interesting property.

Theorem 2.43

Let M be an R-module. Then M is injective if and only if M has no proper essential extensions.

Proof. Suppose M is injective and M is a proper submodule of N. By Exercise 2.35, M is a direct summand of N. Thus $N = M + W$ for an R-module W with $W \neq 0$ and $M \cap W = 0$. Thus the extension is not essential.

Conversely, assume that M has no proper essential extensions. We have to show that if N is an R-module and $U \subseteq N$ is a submodule of N, then any R-module homomorphism $\varphi\ U \to M$ can be extended to an R-module homomorphism $\varphi'\ N \to M$.

Let $W = (N \oplus M)/V$, where V is the submodule of $N \oplus M$ whose elements are $(u, -\varphi(u))$ with $u \in U$. Then we obtain the commutative diagram

$$\begin{array}{ccc} N & \longrightarrow & W \\ \uparrow & & \uparrow \psi \\ U & \xrightarrow{\varphi} & M, \end{array}$$

where $N \to W$ assigns to each $n \in N$ the element $(n, 0) + V$, and ψ $M \to W$ is defined by setting $\psi(m) = (0, m) + V$ for all $m \in M$. Thus, the diagram is in fact commutative, because for all $u \in U$ we have $\psi(\varphi(u)) = (0, \varphi(u)) + V = (u, 0) + V$. The last equation holds, since $(u, 0) - (0, \varphi(u)) = (u, -\varphi(u)) \in V$.

Note that ψ is injective. Indeed, suppose that $\psi(m) = 0$. Then there exists $u \in U$ with $(0, m) = (u, \varphi(u))$. Then $u = 0$ and $m = \varphi(u) = \varphi(0) = 0$. Now we may identify M with its image in W, and hence may assume that M is a submodule of W. By Zorn's lemma, there exists a submodule $D \subsetneq W$ with $D \cap M = 0$, and for any submodule D' of W which properly contains D we have $D' \cap M \neq 0$. Then M may be viewed as a submodule of W/D, and W/D is an essential extension of M. Therefore, our assumption on M implies that $M = W/D$. Hence the composition of the map $N \to W$ with the canonical epimorphism $W \to W/D = M$ yields the desired extension of φ. □

Exercise 2.44

Let $U \subseteq M \subseteq N$ be R-modules. Show that $U \subseteq N$ is an essential extension if and only if $U \subseteq M$ and $M \subseteq N$ are essential extensions.

2.4.2 Injective Hulls and Minimal Injective Resolutions

We will see that any R-module M can be embedded into a 'minimal' injective R-module.

Theorem 2.45

Let M be an R-module. Then

(a) There exists an injective R-module E such that $M \subseteq E$ is an essential extension.
(b) Let E be an injective R-module such that $M \subseteq E$ is an essential extension. Then for any monomorphism $M \to E'$ such that E' is an injective R-module, there exists a monomorphism φ $E \to E'$ such that the diagram

$$\begin{array}{ccc} M & \hookrightarrow & E \\ \downarrow & \swarrow \varphi & \\ E' & & \end{array}$$

is commutative.

Proof. (a) By Theorem 2.39, M can be embedded into an injective R-module $\widetilde{E}$. Let $\mathscr{A}$ be the set of essential extensions N of M with $N \subseteq \widetilde{E}$. By Zorn's lemma, $\mathscr{A}$ admits a maximal element E with respect to inclusion. We claim that E has no proper essential extension. Then Theorem 2.43 implies that E is injective.

Assume to the contrary that E has a proper essential extension E'. Since $\widetilde{E}$ is injective, the inclusion map $E \to \widetilde{E}$ can be extended to an R-module homomorphism $\varphi\ E' \to \widetilde{E}$. Since $\operatorname{Ker}\varphi \cap E = 0$, and since the extension $E \to E'$ is essential, it follows that $\operatorname{Ker}\varphi = 0$. Then $\operatorname{Im}\varphi$ is a proper essential extension of E in $\widetilde{E}$. So by Exercise 2.44, $\operatorname{Im}\varphi$ is a proper essential extension of M in $\widetilde{E}$, with $E \subsetneq \operatorname{Im}\varphi$, contradicting the choice of E.

(b) Since E' is injective, the inclusion map $M \to E'$ can be extended to an R-module homomorphism $\alpha\ E \to E'$. As in the proof of part (a), it can be seen that $\operatorname{Ker}\alpha = 0$. Hence α is a monomorphism, as desired. □

Corollary 2.46
Let E and E' be injective modules such that $M \subseteq E$ and $M \subseteq E'$ are essential extensions. Then there exists an isomorphism $\varphi\ E \to E'$ such that the diagram

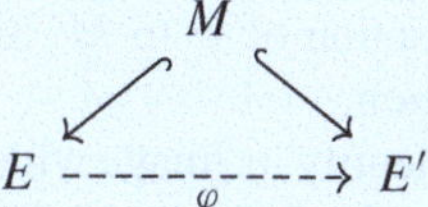

is commutative.

Theorems 2.45 and 2.43 imply that for R-modules M and E the following statements are equivalent:

(i) E is an injective module and an essential extension of M.
(ii) E is a maximal essential extension of M.
(iii) E is the smallest injective R-module in which M can be embedded.

An R-module E satisfying one of the conditions above is called an *injective hull* of M. Corollary 2.46 tells us that an injective hull of M is uniquely determined up to isomorphism. We denote it by $E_R(M)$ or simply $E(M)$.

Exercise 2.47
Let R be a Noetherian domain with the quotient field Q. Show that $E(R) = Q$.

Corollary 2.46 can be slightly generalized as stated in

Lemma 2.48
Let M and N be R-modules, and let $M \to E(M)$ and $N \to E(N)$ be the inclusion maps into the injective hulls of M and N, respectively. Furthermore, let $\varphi M \to N$ be an isomorphism. Then there exists an R-module homomorphism $\psi : E(M) \to E(N)$ such that the diagram

$$\begin{array}{ccc} M & \longrightarrow & E(M) \\ \downarrow{\scriptstyle\varphi} & & \downarrow{\scriptstyle\psi} \\ N & \longrightarrow & E(N). \end{array}$$

is commutative, and any such ψ is an isomorphism.

Proof. The existence of an R-module homomorphism $\psi : E(M) \to E(N)$ which makes the diagram commutative, follows easily from Theorem 2.45(b).

Now let $\psi : E(M) \to E(N)$ be any R-module homomorphism making the diagram commutative. Suppose ψ is not injective. Then $\operatorname{Ker} \psi \neq 0$. Since the extension $M \subseteq E(M)$ is essential, it follows that $M \cap \operatorname{Ker} \psi \neq 0$. Let $x \in M \cap \operatorname{Ker} \psi$. Then $\varphi(x) = 0$, because φ is the restriction of ψ to M. Since φ is an isomorphism, we conclude that $x = 0$, a contradiction.

Since ψ is injective, we may identify its image with $E(M)$ and hence may assume that $E(M) \subseteq E(N)$. Since the extensions $M \subseteq E(N)$ is essential, the extension $E(M) \subseteq E(N)$ is essential, as well, see Exercise 2.44. Hence Theorem 2.43 implies that $E(M) = E(N)$. It follows that ψ is surjective. □

Injective hulls localize.

Proposition 2.49
Let R be a Noetherian ring, $M \in \operatorname{Mod}_R$ and $S \subset R$ be a multiplicatively closed set. Then $S^{-1}E_R(M) \cong E_{S^{-1}R}(S^{-1}M)$.

Proof. By Proposition 2.41, $S^{-1}E_R(M)$ is an injective module, and it remains to be shown that $S^{-1}M \to S^{-1}E_R(M)$ is an essential extension.

Let U be an $S^{-1}R$-submodule of $S^{-1}E_R(M)$ such that $U \neq 0$. We have to show that $U \cap S^{-1}M \neq 0$. It suffices to show that $S^{-1}Ru \cap S^{-1}M \neq 0$ for some $u \in U$. To find a suitable u we argue as follows. Because $U \neq 0$, we may choose $0 \neq v \in U$. Since $v \in S^{-1}E_R(M)$, there exists $x \in E_R(M)$ such that $S^{-1}Rv = S^{-1}(Rx)$, and hence we may assume that $v \in E_R(M)$. Now we consider the set of ideals

$\mathscr{A} = \{\text{Ann}(sv) s \in S\}$. Since R is Noetherian, there exists (with respect to inclusion) a maximal element $\text{Ann}(s'v)$ in $\mathscr{A}$. Since $S^{-1}R(s'v) = S^{-1}Rv$, we may replace v by $u = s'v$, and hence we may assume that $\text{Ann}(u) = \text{Ann}(tu)$ for any $t \in S$.

With this choice of u we show that $Ru \cap S^{-1}M \neq 0$. Note that $Ru \cap M = Iu$ for some ideal $I \subseteq R$. Since the extension $M \subset E_R(M)$ is essential, it follows that $Iu \neq 0$. Assume that $S^{-1}Ru \cap S^{-1}M = 0$. Since $S^{-1}Iu = S^{-1}Ru \cap S^{-1}M$, we see that $S^{-1}Iu = 0$. Therefore, there exists $t \in S$ such that $tIu = 0$. Hence, $I \subseteq \text{Ann}(tu) = \text{Ann}(u)$, and so $Iu = 0$, a contradiction. $\square$

We construct the *minimal injective resolution*

$$0 \to E^0(M) \xrightarrow{\partial^0} E^1(M) \xrightarrow{\partial^1} \cdots \xrightarrow{\partial^{i-1}} E^i(M) \xrightarrow{\partial^i} \cdots$$

of M in the obvious way as follows: we let $E^0(M) = E(M)$, and let $E^1(M)$ be the injective hull of the cokernel C of $M \to E^0(M)$. The differential ∂^0 is defined to be the composition of the maps $E^0(M) \to C$ and the inclusion map $C \to E^1(M)$. Next consider Coker ∂^0 and its injective hull, which we denote by $E^2(M)$, and proceed in this way. Corollary 2.46 implies that the minimal injective resolution $(E^\bullet(M), \partial^\bullet)$ is unique up to isomorphism.

We want to better understand the structure of the injective modules $E^i(M)$ in the minimal injective resolution of M. This needs some preparation.

An R-module M is called *indecomposable*, if $M \neq 0$ and the only direct summands of M are 0 and M itself. Otherwise, M is called *decomposable*.

Theorem 2.50
Let R be a ring.

(a) For each $P \in \text{Spec}(R)$, the module $E(R/P)$ is indecomposable.
(b) Let $E \neq 0$ be an injective R-module and $P \in \text{Ass}(E)$. Then $E(R/P)$ is a direct summand of E.

Proof. (a) Suppose $E(R/P)$ is decomposable. Then there exist nonzero submodules $U_1, U_2 \subsetneq E(R/P)$ such that $U_1 \cap U_2 = 0$. Then $(U_1 \cap R/P) \cap (U_2 \cap R/P) = 0$. Since $R/P \subseteq E(R/P)$ is an essential extension, it follows that $U_i \cap R/P \neq 0$ for $i = 1, 2$. This is a contradiction, because R/P is a domain. Indeed in a domain, the intersection of two nonzero ideals can never be the zero ideal.

(b) Since $P \in \text{Ass}(E)$, it follows that R/P may be considered as a submodule of E. By Theorem 2.45, there exists a monomorphism $E(R/P) \to E$. Thus we may assume that $E(R/P) \subseteq E$. The desired result follows now from Exercise 2.35. $\square$

Exercise 2.51
Let $P \in \operatorname{Spec}(R)$, and let E be an indecomposable injective R-module with $R/P \subseteq E$. Then $E \cong E(R/P)$.

For a prime ideal P we denote by $K(P)$ the residue class field R_P/PR_P.

Proposition 2.52
Let R be a Noetherian ring, and let $P, Q \in \operatorname{Spec}(R)$. Then

$$\operatorname{Hom}_{R_Q}(K(Q), E(R/P)_Q) \cong \begin{cases} K(P), & \text{if } P = Q, \\ 0, & \text{otherwise.} \end{cases}$$

In particular, $E(R/P) \cong E(R/Q)$ if and only if $P = Q$.

Proof. By Propositions A.39 and 2.49 we have

$$\begin{aligned} \operatorname{Hom}_R(R/Q, E(R/P))_Q &\cong \operatorname{Hom}_{R_Q}(K(Q), E(R/P)_Q) \\ &\cong \operatorname{Hom}_{R_Q}(K(Q), E_{R_Q}((R/P)_Q)). \end{aligned}$$

If $P \not\subseteq Q$, then $(R/P)_Q = 0$, and so $\operatorname{Hom}_{R_Q}(K(Q), E_{R_Q}((R/P)_Q)) = 0$. Now let $P \subseteq Q$. If $\operatorname{Hom}_{R_Q}(K(Q), E(R/P)_Q) \neq 0$, then $\operatorname{Hom}_R(R/Q, E(R/P)) \neq 0$. Hence there exists a nonzero element $x \in E(R/P)$ with $Qx = 0$. Since the extension $R/P \to E(R/P)$ is essential, we have that $Rx \cap R/P \neq 0$. Hence $Q \subseteq \operatorname{Ann}(Rx \cap (R/P)) = P$. It follows that $P = Q$, if $\operatorname{Hom}_{R_Q}(K(Q), E(R/P)_Q) \neq 0$.

Now, suppose that $P = Q$. Then

$$\operatorname{Hom}_{R_Q}(K(Q), E(R/P))_Q = \operatorname{Hom}_{R_P}(K(P), E_{R_P}(K(P))).$$

Replacing R by R_P, we may assume that R is local with maximal ideal $\mathfrak{m}$, and we have to show that $\operatorname{Hom}_R(R/\mathfrak{m}, E(R/\mathfrak{m})) \cong R/\mathfrak{m}$.

Observe that $\operatorname{Hom}_R(R/\mathfrak{m}, E(R/\mathfrak{m}))$ can be identified with the submodule $U = 0 :_{E(R/\mathfrak{m})} \mathfrak{m}$ of $E(R/\mathfrak{m})$. Certainly, $R/\mathfrak{m} \subseteq U$. Note that U is an $R/\mathfrak{m}$-vector space. Therefore, $U = R/\mathfrak{m} \oplus W$. Suppose that $U \neq R/\mathfrak{m}$. Then $W \neq 0$ but $W \cap R/\mathfrak{m} = 0$, contradicting the fact that the extension $R/\mathfrak{m} \subseteq E(R/\mathfrak{m})$ is essential. □

Each injective R-module decomposes into a direct sum of indecomposables.

Theorem 2.53
Let R be a Noetherian ring, and let E be an injective R-module. Then

$$E = \bigoplus_i E(R/P_i).$$

where the P_i are prime ideals of R.

Proof. By Zorn's Lemma, there exists a maximal family $\{E_j\}_{j\in\mathscr{J}}$ of injective submodules of E such that $E_j \cong E(R/P_j)$ with $P_j \in \operatorname{Spec}(R)$ for all j, and such that the sum E' of the E_j in E is a direct sum. By Exercise 2.37, E' is an injective module, and by Exercise 2.35, E' is a direct summand of E. Hence there exists a submodule E'' of E such that $E' \oplus E'' = E$. Suppose $E'' \neq 0$. Then we choose a nonzero element $x \in E''$ and $P \in \operatorname{Ass}(Rx)$. Then $R/P \subseteq E''$. It follows from Theorem 2.50 and Exercise 2.37 that $E(R/P)$ is a direct summand of E''. This contradicts the maximality of the family $\{E_j\}_{j\in\mathscr{J}}$. □

Let R be a Noetherian ring, $P \in \operatorname{Spec}(R)$ and let M be a finitely generated R-module. The finite number $\mu_i(P, M) = \dim_{K(P)} \operatorname{Ext}^i_{R_P}(K(P), M_P)$ is called the *ith Bass number* of M with respect to P.

Now we come to the main result about minimal injective resolutions.

Theorem 2.54
Let R be a Noetherian ring, and let M be a finitely generated R-module. Then

$$E^i(M) = \bigoplus_{P\in\operatorname{Spec}(R)} E(R/P)^{\mu_i(P,M)}.$$

Proof. Let $0 \to E^0(M) \to E^1(M) \to \cdots$ be the minimal injective resolution of M. It follows from Theorem 2.53 that $E^i(M) = \bigoplus_{P\in\operatorname{Spec}(R)} E(R/P)^{\rho_i(P,M)}$ for integers $\rho_i(P, M)$ which depend on M, P and i. We need to show that $\rho_i(P, M) = \mu_i(P, M)$ for all i.

Let $P \in \operatorname{Spec}(R)$. Since localization is an exact functor, it follows from Proposition 2.49 that $0 \to E^0(M)_P \to E^1(M)_P \to \cdots$ is the minimal injective R_P-resolution of M_P. Therefore, $\operatorname{Ext}^i_{R_P}(K(P), M_P) = H^i(C^\bullet)$, where

$$C^\bullet\ 0 \to \operatorname{Hom}_{R_P}(K(P), E^0(M)_P) \to \operatorname{Hom}_{R_P}(K(P), E^1(M)_P) \to \cdots.$$

It follows that $H^i(C^\bullet) \cong K(P)^{\mu_i(P,M)}$.

Proposition 2.52 implies that $\mathrm{Hom}_{R_P}(K(P), E^i(M)_P) \cong K(P)^{\rho_i(P,M)}$, so that

$$C^\bullet\, 0 \to K(P)^{\rho_0(P,M)} \to K(P)^{\rho_1(P,M)} \to \cdots .$$

We claim that the maps in this complex are all the zero maps. For $\varphi \in \mathrm{Hom}_{R_P}(K(P), E^i(M)_P)$ we need to show that for all $i > 0$ the composition

$$K(P) \xrightarrow{\varphi} E^i(M)_P \xrightarrow{\partial} E^{i+1}(M)_P$$

is the zero map. Indeed, we may assume that $\varphi(x) \neq 0$ for some $x \in K(P)$. Let $U \subset E^i(M)_P$ be the image of $E^{i-1}(M)_P \to E^i(M)_P$. Since $E^\bullet(M)_P$ is a minimal injective resolution of M_P, it follows that $U \subset E^i(M)_P$ is an essential extension. Therefore, there exists $a \in R_P$ such that $a\varphi(x)$ is a nonzero element in U. Note that $a \notin PR_P$, because $Px = 0$ for $x \in K(P)$. It follows that a is a unit in R_P, and hence $\varphi(x) \in U = \mathrm{Ker}\,\partial$. This argument applies to all $x \in K(P)$. Thus $\partial \circ \varphi = 0$, and the claim is proved.

The claim implies that $H^i(C^\bullet) \cong K(P)^{\rho_i(P,M)}$. Therefore, the $K(P)$-vector spaces $K(P)^{\rho_i(P,M)}$ and $K(P)^{\mu_i(P,M)}$ are isomorphic. Thus $\rho_i(P, M) = \mu_i(P, M)$ for all $P \in \mathrm{Spec}(R)$ and all i, as desired. □

Let R be a ring, let $M \in \mathrm{Mod}_R$ and let $E^\bullet$ be an injective resolution of M. The *injective dimension* of M is the shortest length (which may be infinite) of all injective resolutions of M. We denote the injective dimension of M by $\mathrm{id}(M)$.

We do not have to consider all injective resolutions of M to determine its injective dimension. Indeed, we have

Proposition 2.55

Let R be a ring, and let $M \in \mathrm{Mod}_R$. Then $\mathrm{id}(M)$ is the length of the minimal injective resolution of M.

Proof. Let $E^\bullet$ be an injective resolution of M. By Exercise 2.42 there exists a chain map $\varphi^\bullet\ E^\bullet(M) \to E^\bullet$ with $H^0(\varphi^\bullet) = \mathrm{id}_M$ and a chain map $\psi^\bullet\ E^\bullet \to E^\bullet(M)$ with $H^0(\psi^\bullet) = \mathrm{id}_M$. Let $\gamma^\bullet = \psi^\bullet \circ \varphi^\bullet$. Then $\gamma^\bullet\ E(M)^\bullet \to E(M)^\bullet$ is chain map with $H^0(\gamma^\bullet) = \mathrm{id}_M$.

We claim that γ^i is an isomorphism for all i. The claim implies that φ^i is an injective map for all i, which in turn yields the desired conclusion.

For $i \geq 0$, let $\partial^i : E^i(M) \to E^{i+1}(M)$ be the differential maps of $E^\bullet(M)$. The proof of the claim follows by repeated application of Lemma 2.48. Indeed this lemma implies that γ_0 is an isomorphism. The isomorphism γ_0 induces an isomorphim $\alpha\ \mathrm{Im}(\partial_0) \to \mathrm{Im}(\partial_0)$, and we have a commutative diagram

$$\begin{array}{ccc} \operatorname{Im}(\partial_0) & \longrightarrow & E^1(M) \\ \downarrow{\alpha} & & \downarrow{\gamma_1} \\ \operatorname{Im}(\partial_0) & \longrightarrow & E^1(M). \end{array}$$

Lemma 2.48 applied again, shows that γ_1 is an isomorphism. This isomorphism induces an isomorphism $\operatorname{Im}(\partial_1) \to \operatorname{Im}(\partial_1)$. Proceeding in this way, we see that all γ_i are isomorphisms. □

Proposition 2.56
Let R be a ring, and let $M \in \operatorname{Mod}_R$. Then the following conditions are equivalent:

(i) $\operatorname{id}(M) \le n$.
(ii) $\operatorname{Ext}_R^{n+1}(N, M) = 0$ for all $N \in \operatorname{Mod}_R$.
(iii) $\operatorname{Ext}_R^{n+1}(R/I, M) = 0$ for all ideals $I \subseteq R$.

Proof. (i) $\Rightarrow$ (ii): By assumption M admits an injective resolution $E^\bullet$ of length $\le n$. Since $\operatorname{Ext}_R^{n+1}(N, M) \cong H^{n+1}(\operatorname{Hom}_R(N, E^\bullet))$ and since $E^{n+1} = 0$, we see that $\operatorname{Ext}_R^{n+1}(N, M) = 0$.

(ii) $\Rightarrow$ (iii) is trivial.

(iii) $\Rightarrow$ (i): Let $(E^\bullet, \partial^\bullet)$ be an injective resolution of M, and let W be the image of ∂^{n-1}. Then we obtain the exact sequence $0 \to M \to E^0 \to \cdots \to E^{n-1} \to W \to 0$. Let $I \subseteq R$ be an ideal. By assumption, $\operatorname{Ext}_R^{n+1}(R/I, M) = 0$, and from the definition of Ext, we obtain $\operatorname{Ext}_R^{n+1}(R/I, M) \cong \operatorname{Ext}_R^1(R/I, W)$. Therefore, $\operatorname{Ext}_R^1(R/I, W) = 0$.

Now the exact sequence $0 \to I \to R \to R/I \to 0$ yields the exact sequence $\operatorname{Hom}_R(R, W) \to \operatorname{Hom}_R(I, W) \to \operatorname{Ext}_R^1(R/I, W) = 0$. This implies that any R-module homomorphism $I \to W$ can be extended to an R-module homomorphism $R \to W$. This together with Theorem 2.36 shows that W is injective, and it implies that $\operatorname{id}(M) \le n$. □

2.4.3 Graded Injective Hulls and Graded Injective Resolutions

Injective modules and essential extensions within the category GMod_R are formally defined as we defined them in the category of all R-modules. So, a module $E \in \operatorname{GMod}_R$ is called *graded injective*, if the functor

$$\underline{\operatorname{Hom}}_R(-, E)\ \operatorname{GMod}_R \to \operatorname{GMod}_R$$

is exact. Furthermore, an extension $N \subseteq M$ of graded R-modules is called *graded essential* in GMod_R, if for any graded submodule $0 \neq U \subseteq M$ one has that $U \cap N \neq 0$. The graded essential extension is called *proper* if $M \neq N$.

Similarly, as in Theorem 2.43 one shows that a module in GMod_R is graded injective if and only if it has no proper graded essential extensions. If $M \in \mathrm{GMod}_R$, then $E \in \mathrm{GMod}_R$ is called a graded injective hull of M if E is graded injective and a graded essential extension of M. We denote the graded injective hull of M by $\underline{E}(M)$. Next we will show that graded injective hulls of graded modules exist.

Proposition 2.57
Any graded R-module M admits a graded injective hull which is unique up to isomorphism.

We sketch the steps of the proof. By Theorem 2.39 there exists an embedding $M \subseteq L$, where L is an injective R-module, and by using Zorn's lemma there exists a maximal graded essential extension $E \subseteq L$ of M. It remains to be shown that E is graded injective. Suppose this is not the case. Then E has a proper graded essential extension $E \subsetneq E'$. Since L is injective, there exists an R-module homomorphism $\varphi\ E' \to L$ for which $\varphi_{|E}$ is the inclusion map $E \hookrightarrow L$. One checks that φ is an injective map. Let E'' be the image of φ. Then, since φ is an injective map, E'' inherits a natural graded structure from E' and $E \subsetneq E''$ is a proper graded essential extension of E in L. This is a contradiction.

Example 2.58
Let $S = K[x]$ be the polynomial ring over the field K with standard grading. Then the graded injective hull of S is $S_x = K[x, x^{-1}]$. Since the extension S_x of S is obviously graded essential, it remains to be shown that S_x is graded injective. Indeed, let $M \in \mathrm{GMod}(S)$, and consider the graded natural S_x-module homomorphism $(M_x)_0 \underline{\otimes}_K S_x \to M_x$. For each $a \in \mathbb{Z}$, we have

$$((M_x)_0 \underline{\otimes}_K S_x)_a \cong (M_x)_0 \underline{\otimes}_K K x^a \cong x^a (M_x)_0 = (M_x)_a.$$

This shows that $(M_x)_0 \underline{\otimes}_K S_x \to M_x$ is a graded isomorphism.

By using this isomorphism and the universal property of localization we obtain

$$\begin{aligned}\underline{\mathrm{Hom}}_S(M, S_x) &\cong \underline{\mathrm{Hom}}_{S_x}(M_x, S_x) \cong \underline{\mathrm{Hom}}_{S_x}((M_x)_0 \underline{\otimes}_K S_x, S_x)\\ &\cong \underline{\mathrm{Hom}}_K((M_x)_0, K) \underline{\otimes}_K S_x.\end{aligned}$$

The last isomorphism is given by the map $\varphi\ :\ \underline{\mathrm{Hom}}_K((M_x)_0, K) \underline{\otimes}_K S_x \to \underline{\otimes}_K S_x, S_x)$ with $\varphi(\lambda \otimes x^i)(v \otimes x^j) = \lambda(v) x^{i+j}$. Note that $\underline{\mathrm{Hom}}_K((M_x)_0, K) \underline{\otimes}_K S_x$ is a composition of exact functors applied to M, and hence is an exact functor. This proves that S_x is a graded injective S-module.

An R-module M is called *graded indecomposable*, if it is indecomposable in GMod_R. The graded indecomposable injective R-modules are precisely the modules $\underline{E}(R/P)(a)$, where P is a graded prime ideal. Moreover, $\mathrm{Ass}(\underline{E}(R/P)(a)) = \{P\}$.

As in the nongraded case one then obtains

Theorem 2.59
Let R be a finitely generated graded K-algebra, and let $M \in \mathscr{M}_R$. Then there exists an acyclic graded complex

$$0 \to \underline{E}^0(M) \to \underline{E}^1(M) \to \cdots \to \underline{E}^i(M) \to \cdots$$

of graded injective modules such that *(i)* $M \cong H^0(\underline{E}^{\bullet}(M))$, *(ii)* $\underline{E}^0(M) = \underline{E}(M)$, *(iii)* $\underline{E}^1(M)$ is the graded injective hull of $\mathrm{Coker}(M \to \underline{E}^0(M))$, and for each $i \geq 2$, $\underline{E}^i(M)$ is the graded injective hull of $\mathrm{Coker}(\underline{E}^{i-2}(M) \to \underline{E}^{i-1}(M))$.

Each $\underline{E}^i(M)$ is a direct sum of modules of the form $\underline{E}(R/P)(a)$, where P is a graded prime ideal. For each graded prime ideal P the module $\underline{E}^i(M)$ contains only finitely many summands of the form $\underline{E}(R/P)$ with suitable shifts. The number of these summands coincides with the Bass number $\mu_i(P, M)$.

The complex $\underline{E}^{\bullet}(M)$ in Theorem 2.59 is called the *minimal graded injection resolution* of M.

Exercise 2.60
Let R be a graded K-algebra, and let P a graded prime ideal of R. Then the graded injective hull $\underline{E}(R/P)$ of R/P is characterized by the property that (i) $R/P \subseteq \underline{E}(R/P)$, (ii) $\underline{E}(R/P)$ is graded indecomposable and (iii) $\underline{E}(R/P)$ is graded injective.

Let R be a finitely graded K-algebra, and let $M \in \mathrm{GMod}_R$. A *graded injective resolution* of M is an injective resolution of M in the category GMod_R. Such resolutions exist because each $M \in \mathrm{GMod}_R$ admits a graded injective hull.

The *graded injective dimension* of M in GMod_R is the shortest length of all graded injective resolutions of M. and is denoted by $\underline{\mathrm{id}}(M)$.

In analogy to Proposition 2.56 and with essentially the same proof we have

Proposition 2.61
Let R be a graded K-algebra, and let $M \in \mathrm{GMod}_R$. Then the following conditions are equivalent:

(i) $\underline{\mathrm{id}}(M) \leq n$.
(ii) $\underline{\mathrm{Ext}}_R^{n+1}(N, M) = 0$ for all $N \in \mathrm{GMod}_R$.
(iii) $\underline{\mathrm{Ext}}_R^{n+1}(R/I, M) = 0$ for all graded ideals $I \subseteq R$.

If R is a finitely generated K-algebra and $M \in \mathscr{M}_R$, then we may replace $\underline{\mathrm{Ext}}_R^{n+1}(R/I, M)$ in *(iii)* by $\mathrm{Ext}_R^{n+1}(R/I, M)$.

Cohen-Macaulay Modules and the Canonical Module

3

Cohen-Macaulay rings and modules play a central role in modern commutative algebra. This class of modules is distinguished in many ways, which will become apparent in this chapter. Polynomial rings over a field, complete intersection rings and Artinian rings are examples of Cohen-Macaulay rings. We review the main properties of Cohen-Macaulay modules and the canonical module of a finitely generated graded K-algebras.

3.1 Cohen-Macaulay Modules

3.1.1 Cohen-Macaulay Rings and Modules

In order to introduce the concept of a Cohen-Macaulay module, we first need to define and study the depth of a module and related homological invariants.

Let R be a ring and M be an R-module. Furthermore, let I be an ideal of R. An M-sequence $\mathbf{f} = f_1, \ldots, f_m$ of elements of I is called a *maximal* M-sequence in I, if for any $f \in I$, the new sequence $f_1, \ldots, f_m, f$ is not an M-sequence anymore, which means that $I \subseteq Z(M/(f_1, \ldots, f_m)M)$. When R is a Noetherian ring, such maximal M-sequences exist as we see in

J. Herzog et al., *Numerical Semigroups*, Compact Textbooks in Mathematics,
https://doi.org/10.1007/978-3-032-05424-1_3

Lemma 3.1
Let R be a Noetherian ring and M be an R-module. Then

(a) Any M-sequence in R has finite length.
(b) An M-sequence in an ideal $I \subseteq R$ can be extended to a maximal M-sequence.

Proof. (a) Suppose that there exists an infinite M-sequence $f_1, f_2, \ldots$ in R. Then one obtains a strictly increasing chain of ideals $(f_1) \subsetneq (f_1, f_2) \subsetneq \cdots$, which is not possible, since R is Noetherian.

(b) follows from (a). □

Moreover, we have

Theorem 3.2
(Rees) Let R be a Noetherian ring, M be a finitely generated R-module, and let I be an ideal of R with $IM \neq M$. Furthermore, let $\mathbf{f} = f_1, \ldots, f_n$ be a maximal M-sequence in I. Then $\operatorname{Ext}^i_R(R/I, M) = 0$ for $i < n$ and $\operatorname{Ext}^n_R(R/I, M) \neq 0$.

In particular, the length of all maximal M-sequences in I is equal to

$$\min\{i : \operatorname{Ext}^i_R(R/I, M) \neq 0\}.$$

Proof. By Proposition A.43, for any $i \leq n$ we have

$$\operatorname{Ext}^i_R(R/I, M) \cong \operatorname{Hom}_R(R/I, M/(f_1, \ldots, f_i)M).$$

For $i < n$, the element f_{i+1} is regular on $M/(f_1, \ldots, f_i)M$. So by Exercise A.42, $\operatorname{Hom}_R(R/I, M/(f_1, \ldots, f_i)M) = 0$. Moreover, $\operatorname{Ext}^n_R(R/I, M) \cong \operatorname{Hom}_R(R/I, M')$, where $M' = M/(f_1, \ldots, f_n)M$. Hence we have to show that $\operatorname{Hom}_R(R/I, M') \neq 0$. Note that $M' \neq 0$, since $M \neq IM$. Since $\mathbf{f}$ is a maximal M-sequence in I, we have $I \subseteq Z(M')$. So by Lemmas 1.47 and 1.67, $I \subseteq P$ for some $P \in \operatorname{Ass}(M')$. Then there is a natural epimorphism $\varphi : R/I \to R/P$, and since $P \in \operatorname{Ass}(M')$, there exists a natural inclusion $\psi : R/P \hookrightarrow M'$. Therefore, $\psi \circ \varphi$ is a nonzero map in $\operatorname{Hom}_R(R/I, M')$. □

Keeping the assumptions of Theorem 3.2, the common length of the maximal M-sequences in I is called the *grade of I with respect to M* or the *M-grade* of I, and is denoted by grade(I, M). When $M = R$, we denote grade(I, R) simply by grade(I). It follows from Theorem 3.2 that

$$\operatorname{grade}(I, M) = \min\{i : \operatorname{Ext}^i_R(R/I, M) \neq 0\}.$$

Example 3.3
Let $S = K[x, y]$ be the polynomial ring over a field K, and let $I = (x - y, x^2 + y^2)$. Then $S/(x - y) \cong K[x]$ and $x^2 + y^2 \equiv 2x^2(\mod x - y)$. Hence $x^2 + y^2$ is a regular element on $S/(x - y)$ if and only if char$K \neq 2$. In particular, grade$(I) = 2$ if char$K \neq 2$ and grade$(I) = 1$, if char$K = 2$.

Exercise 3.4
Let R be a Noetherian ring, M be a finitely generated R-module and I be an ideal of R. Show that $IM = M$ if and only if $\operatorname{Ext}^i_R(R/I, M) = 0$ for all i.

Theorem 3.2 and Exercise 3.4 suggest to define grade$(I, M) = \infty$ when $IM = M$.

Let $(R, \mathfrak{m})$ be a Noetherian local ring and M be a finitely generated R-module. The *depth* of M is defined to be the maximal length of an M-sequence. We denote the depth of M by depth$_R M$ or briefly depth$M$. Since $M/\mathbf{f}M \neq 0$ for any M-sequence $\mathbf{f}$, it follows that $(\mathbf{f}) \subseteq \mathfrak{m}$. This shows that

$$\operatorname{depth} M = \operatorname{grade}(\mathfrak{m}, M).$$

Therefore, Theorem 3.2 implies that depth$M = \min\{i : \operatorname{Ext}^i_R(R/\mathfrak{m}, M) \neq 0\}$, and in particular, all maximal M-sequence have the same length.

The following simple lemma shows the behaviour of grade and depth under reduction modulo regular sequences and under localization.

Lemma 3.5
Let R be a Noetherian ring, let M be a finitely generated R-module, let I be an ideal of R with $IM \neq M$, and let $\mathbf{f} = f_1, \ldots, f_i$ be an M-sequence in I. Then

(a) grade$(I, M/(f_1, \ldots, f_i)M) = $ grade$(I, M) - i$,
(b) If $(R, \mathfrak{m})$ is local, then depth$M/(f_1, \ldots, f_i)M = $ depth$M - i$.
(c) If S is a multiplicatively closed subset of R with $I \cap S = \emptyset$, then grade$(I, M) \leq$ grade$(S^{-1}I, S^{-1}M)$.

Proof. (a) Let $g = \operatorname{grade}(I, M)$. By Lemma 3.1, the M-sequence $x_1 \ldots, x_i$ can be extended to a maximal M-sequence, which by Theorem 3.2 has length g. Say, $x_1, \ldots, x_g$ in I is this sequence. Then $x_{i+1}, \ldots, x_g$ is a maximal $M/(x_1, \ldots, x_i)M$-sequence in I. This shows that $\operatorname{grade}(I, M/(f_1, \ldots, f_i)M) = g - i = \operatorname{grade}(I, M) - i$.

(b) follows from (a), if we replace I by $\mathfrak{m}$.

(c) Let $g = \operatorname{grade}(I, M)$, and let $f_1, \ldots, f_g$ be a maximal M-sequence in I. Since $S \cap I = \emptyset$, it follows that $S^{-1}I \neq S^{-1}R$ and $f_1/1, \ldots, f_g/1 \in S^{-1}I$. We show that $f_1/1, \ldots, f_g/1$ is an $S^{-1}M$-sequence. By assumption, $f_1 \in I$ is a non-zerodivisor on M. This is equivalent to saying that $0 \longrightarrow M \xrightarrow{f_1} M \longrightarrow M/(f_1)M \longrightarrow 0$ is exact. Since localization is an exact functor, we obtain the exact sequence

$$0 \longrightarrow S^{-1}M \xrightarrow{f_1/1} S^{-1}M \longrightarrow S^{-1}M/(f_1/1)S^{-1}M \to 0.$$

This shows that $f_1/1$ is a non-zerodivisor on $S^{-1}M$. An easy induction argument completes the proof. □

The following proposition gives us an upper bound for the depth of a module.

Proposition 3.6

Let $(R, \mathfrak{m})$ be a Noetherian local ring and M be a finitely generated R-module. Then $\operatorname{depth} M \leq \dim R/P$ for any $P \in \operatorname{Ass}(M)$. In particular, $\operatorname{depth} M \leq \dim M$.

Proof. We prove the inequality by induction on $n = \operatorname{depth} M$. If $n = 0$, there is nothing to prove. Suppose $n > 0$ and $f_1, \ldots, f_n$ be a maximal M-sequence in $\mathfrak{m}$. Set $L = M/f_1M$. Then $f_2, \ldots, f_n$ is a maximal L-sequence. Hence $\operatorname{depth} L = n-1$. Let $P \in \operatorname{Ass}(M)$. Then R/P is isomorphic to a submodule of M, which implies that $\operatorname{Hom}_R(R/P, M) \neq 0$. The short exact sequence $0 \to M \xrightarrow{f_1} M \to L \to 0$ induces the exact sequence

$$0 \to \operatorname{Hom}_R(R/P, M) \xrightarrow{f_1} \operatorname{Hom}_R(R/P, M) \to \operatorname{Hom}_R(R/P, L).$$

If $\operatorname{Hom}_R(R/P, L) = 0$, then $\operatorname{Hom}_R(R/P, M) = f_1\operatorname{Hom}_R(R/P, M)$. Then by Nakayama's lemma we have $\operatorname{Hom}_R(R/P, M) = 0$, a contradiction. Thus $\operatorname{Hom}_R(R/P, L) \neq 0$ and hence $\operatorname{grade}(P, L) = 0$. This means that $P \subseteq Z(L)$. So by Lemmas 1.47 and 1.67, there exists $Q \in \operatorname{Ass}(L)$ such that $P \subseteq Q$. On the other hand $f_1 L = 0$. Thus $f_1 \in Q$. Because f_1 is a non-zerodivisor of M, we have $f_1 \notin P$. Therefore $P \subsetneq Q$. This implies that $\dim R/Q < \dim R/P$. By induction hypothesis $\operatorname{depth} L \leq \dim R/Q$. Therefore, $\operatorname{depth} M = \operatorname{depth} L + 1 \leq \dim R/Q + 1 \leq \dim R/P$. □

Exercise 3.7
Let $(R, \mathfrak{m})$ be a Noetherian local ring and M be a finitely generated R-module. Show that

(a) $\operatorname{depth} M = 0$ if and only if $\mathfrak{m} \in \operatorname{Ass}(M)$.
(b) $\operatorname{depth} M \geq 1$ if and only if $\operatorname{Soc}(M) = 0$.

Let $(R, \mathfrak{m})$ be a Noetherian local ring and M be a finitely generated R-module. In Proposition 3.6 we have seen that $\operatorname{depth} M \leq \dim M$. The module M is called *Cohen-Macaulay*, if $\operatorname{depth} M = \dim M$. Moreover, R is called a *Cohen-Macaulay ring*, if R is Cohen-Macaulay as an R-module.

For a prime ideal $P \in \operatorname{Supp}(M)$, the *$M$-height* of P, which we denote by $\operatorname{height}_M P$, is defined to be $\dim_{R_P} M_P$. This number is indeed the supremum of lengths of chains of prime ideals in $\operatorname{Supp}(M)$ which end at P. Moreover, for an ideal I of R with $IM \neq M$, we set

$$\operatorname{height}_M I = \min\{\operatorname{height}_M P : P \in \operatorname{Supp}(M) \cap V(I)\},$$

and call $\operatorname{height}_M I$ the *M-height* of I.

Note that $\operatorname{Supp}(M) \cap V(I) = \operatorname{Supp}(M/IM)$. Therefore, $\operatorname{Supp}(M) \cap V(I) \neq \emptyset$, because our assumption implies that $M/IM \neq 0$. Hence, $\operatorname{height}_M I = \min \{\operatorname{height}_M P \ P \in \operatorname{Min}(M/IM)\}$. Therefore this minimum is taken over a finite set, see Proposition 1.61. When $M = R$, then the M-height is just the height of I.

The next proposition shows how $\operatorname{grade}(I, M)$ and $\operatorname{height}_M I$ are related.

Proposition 3.8
Let R be a Noetherian ring, let M be a finitely generated R-module, and let I be an ideal of R such that $IM \neq M$. Then

(a) $\operatorname{grade}(I, M) \leq \operatorname{height}_M I$.
(b) If I is generated by an M-sequence of length m, then $\operatorname{grade}(I, M) = \operatorname{height}_M I = m$.

Proof. We begin with the proof of (b). Let I be generated by the M-sequence $\mathbf{f} = f_1, \ldots, f_m$. The equality $\operatorname{grade}(I, M) = m$ follows from Lemma 3.1. We show $\operatorname{height}_M I = m$ by induction on m. If $m = 0$, then there is nothing to prove. So we may assume that $m > 0$. Set $J = (f_1, \ldots, f_{m-1})$. Since $f_1, \ldots, f_{m-1}$ is again an M-sequence, by induction hypothesis $\operatorname{height}_M J = m - 1$. Also from $J \subseteq I$ we have $m - 1 = \operatorname{height}_M J \leq \operatorname{height}_M I$. Let $P \in \operatorname{Supp}(M) \cap V(I)$ be such that $\operatorname{height}_M I = \operatorname{height}_M P$. Then P is a minimal prime ideal of M/IM. Thus

$$\dim_{R_P} M_P/(f_1/1,\ldots,f_m/1)M_P = 0.$$

So by Theorem 1.84(a), $m \geq \dim_{R_P} M_P = \operatorname{height}_M P = \operatorname{height}_M I$. Therefore $m-1 \leq \operatorname{height}_M I \leq m$. Assume that $\operatorname{height}_M I = m-1$. Then $\operatorname{height}_M J = m-1 = \operatorname{height}_M P$. Together with the fact that $J \subseteq P$, we conclude that P is a minimal prime ideal of M/JM. Thus by Proposition 1.61(a), $P \in \operatorname{Ass}(M/JM)$. Lemma 1.47 implies that $P \subseteq Z(M/JM)$. Hence $f_m \in Z(M/JM)$, a contradiction. So $\operatorname{height}_M I = m$, as desired.

(a) Let $\operatorname{grade}(I, M) = m$. Then we may choose an M-sequence $\mathbf{f} = f_1,\ldots,f_m$ in I. As shown in part (b), $\operatorname{height}_M(\mathbf{f}) = m$. The inclusion $(\mathbf{f}) \subseteq I$ implies that $\operatorname{height}_M I \geq m$. □

The following result shows that the Cohen-Macaulay property behaves well under localization.

Proposition 3.9
Let $(R,\mathfrak{m})$ be a Noetherian local ring, and let M be a finitely generated R-module. Then M is Cohen-Macaulay if and only if M_P is a Cohen-Macaulay R_P-module for all $P \in \operatorname{Supp}(M)$.

Proof. Suppose that M_P is Cohen-Macaulay for all $P \in \operatorname{Supp}(M)$. Then M is Cohen–Macaulay, since $M_{\mathfrak{m}} \cong M$.

Conversely, assume that M is Cohen-Macaulay, and let $P \in \operatorname{Supp}(M)$. We prove that M_P is Cohen-Macaulay. Let $\operatorname{grade}(P, M) = t$ and let $\mathbf{f} = f_1,\ldots,f_t$ be a maximal M-sequence in P. Let $n = \dim M$. Then one can extend $\mathbf{f}$ to a maximal M-sequence $f_1,\ldots,f_n$ in $\mathfrak{m}$. We set $L = M/(f_1,\ldots,f_t)M$. We have $P \subseteq Z(L) = \bigcup_{Q\in\operatorname{Ass}(L)} Q$ and by Exercise 1.57, $\operatorname{Ass}(L)$ is a finite set. Thus by Lemma 1.67, $P \subseteq Q$ for some $Q \in \operatorname{Ass}(L)$. Hence $Q \in \operatorname{Supp}(M)$, and therefore $\operatorname{height}_M Q + \dim R/Q \leq \dim M = n$. By using Proposition 3.6 we get

$$\operatorname{height}_M P \leq \operatorname{height}_M Q \leq n - \dim R/Q \leq n - \operatorname{depth} L.$$

By Lemma 3.5(b) we have $\operatorname{depth} L = n - t$. Hence $\operatorname{height}_M P \leq t = \operatorname{grade}(P, M)$. Lemma 3.5(c) together with Proposition 3.8 implies that $\operatorname{depth} M_P \geq \operatorname{grade}(P, M) \geq \operatorname{height}_M P = \dim M_P$. The other side of inequality holds by Proposition 3.6. □

In general, if R is Noetherian but not local, and M is a finitely generated R-module, then M is called a *Cohen-Macaulay* module if M_P is a Cohen-Macaulay R_P-module for any prime ideal $P \in \operatorname{Supp}(M)$.

As an immediate consequence of Proposition 3.9 we have

Corollary 3.10

An R-module M is Cohen-Macaulay if and only if for any maximal ideal $\mathfrak{m}$ of R, $M_{\mathfrak{m}}$ is Cohen-Macaulay.

Proof. Let $P \in \operatorname{Supp}(M)$. Then there exists a maximal ideal $\mathfrak{m}$ with $P \subseteq \mathfrak{m}$. Note that $PR_{\mathfrak{m}}$ is a prime ideal in the local ring $R_{\mathfrak{m}}$. If we assume that the $R_{\mathfrak{m}}$-module $M_{\mathfrak{m}}$ is Cohen Macaulay, then it follows from Proposition 3.9 that $(M_{\mathfrak{m}})_{PR_{\mathfrak{m}}}$ is Cohen-Macaulay. Since $M_P \cong (M_{\mathfrak{m}})_{PR_{\mathfrak{m}}}$, it follows that M_P is Cohen-Macaulay. This shows that M is Cohen-Macaulay. The converse direction is obvious. □

Trivial examples of Cohen-Macaulay rings are zero-dimensional local rings and 1-dimensional domains. However, not all 1-dimensional rings are Cohen-Macaulay. Consider the ring $R = K[x, y]/(x^2, xy)$, where K is a field. Then for the graded maximal ideal $\mathfrak{m} = (x, y)/(x^2, xy)$, we have $\dim R_{\mathfrak{m}} = 1$, while $\operatorname{depth} R_{\mathfrak{m}} = 0$. Indeed, $\mathfrak{m} \in \operatorname{Ass}(M)$ since $x\mathfrak{m} = 0$.

Exercise 3.11

Let R be a Noetherian local ring or a graded K-algebra, let M be a finitely generated (graded) R-module, and let $\mathbf{f} = f_1, \ldots, f_r$ be a (homogeneous) M-sequence in R. Show that M is Cohen-Macaulay if and only if $M/(\mathbf{f})M$ is Cohen-Macaulay.

Hint: use Proposition 1.89 and Lemma 3.5(b).

The following theorem gives an equivalent condition for a module to be Cohen-Macaulay.

Theorem 3.12

Let R be a Noetherian ring and M be a finitely generated R-module. The following conditions are equivalent:

(i) M is Cohen-Macaulay.
(ii) For any ideal I of R with $IM \neq M$, $\operatorname{grade}(I, M) = \operatorname{height}_M I$.

Proof. (i) ⇒(ii): We prove the result by induction on $m = \operatorname{grade}(I, M)$. First assume that $m = 0$. Then $I \subseteq Z(M)$. By Lemmas 1.47 and 1.67, there exists $P \in \operatorname{Ass}(M)$ such that $I \subseteq P$. Then $PR_P \in \operatorname{Ass}(M_P)$. Hence by Proposition 3.6, $\operatorname{depth} M_P \leq \dim R_P/PR_P = 0$. Since we assume that M_P is a Cohen-Macaulay R_P-module, it follows $\dim M_P = 0$. This implies that PR_P is a minimal prime ideal of M_P. Hence $\operatorname{height}_{M_P} PR_P = 0$. Note that $\operatorname{height}_{M_P} PR_P = \operatorname{height}_M P$. This implies that $\operatorname{height}_M P = 0$. Since $P \in \operatorname{Supp}(M) \cap V(I)$, we obtain $\operatorname{height}_M I = 0$, as desired.

Now, assume that $m > 0$ and choose an M-regular element $f \in I$. Set $M' = M/fM$, which by Exercise 3.11 is a Cohen-Macaulay module. We have $\operatorname{grade}(I, M') = \operatorname{grade}(I, M) - 1$, see Lemma 3.5. We claim that $\operatorname{height}_M I \leq \operatorname{height}_{M'} I + 1$. Assuming this inequality holds, Proposition 3.8(a) together with our induction hypothesis, we obtain $\operatorname{grade}(I, M) = \operatorname{grade}(I, M') + 1 = \operatorname{height}_{M'} I + 1 \geq \operatorname{height}_M I \geq \operatorname{grade}(I, M)$, which implies $\operatorname{grade}(I, M) = \operatorname{height}_M I$.

In order to prove the claim, let $P \in \operatorname{Supp}(M') \cap V(I)$ be such that $\operatorname{height}_{M'} I = \operatorname{height}_{M'} P$. We have

$$\operatorname{height}_{M'} P = \dim M'_P = \dim M_P/(f/1)M_P = \dim M_P - 1 = \operatorname{height}_M P - 1.$$

Since $I \subseteq P$, we obtain $\operatorname{height}_M I \leq \operatorname{height}_M P = \operatorname{height}_{M'} P + 1 = \operatorname{height}_{M'} I + 1$.

(ii) $\Rightarrow$(i): Let $P \in \operatorname{Supp}(M)$. By Lemma 3.5 we have $\operatorname{grade}(PR_P, M_P) \geq \operatorname{grade}(P, M)$. Therefore, $\operatorname{depth} M_P \geq \operatorname{grade}(P, M) = \operatorname{height}_M P = \dim M_P$. Hence $\operatorname{depth} M_P = \dim M_P$ and M_P is Cohen-Macaulay. □

3.1.2 Graded Cohen-Macaulay Modules Over Graded K-Algebras

Recall from Lemma 1.77 that for any prime ideal P in the graded ring R, the ideal P^* which is generated by all homogeneous elements in P, is again a prime ideal.

Proposition 3.13

Let R be a graded K-algebra with the graded maximal ideal $\mathfrak{m}$, M be a finitely generated graded R-module, and let P be a nongraded prime ideal in $\operatorname{Supp}(M)$. Then

(a) $\dim M_P = \dim M_{P^*} + 1$.
(b) $\operatorname{depth} M_P = \operatorname{depth} M_{P^*} + 1$.
(c) M_P is Cohen-Macaulay if and only if M_{P^*} is Cohen-Macaulay.

Proof. (a) Note that $\dim M_P = \operatorname{height}_M P$. By Exercise 1.78, $P \in \operatorname{Supp}(M)$ if and only if $P^* \in \operatorname{Supp}(M)$, and with the same arguments as in the proof of Theorem 1.79(c) one can see that $\operatorname{height}_M P = \operatorname{height}_M P^* + 1$.

(b) Let $S = \{r \in R \; r \text{ is homogeneous}, r \notin P\}$. The set S is multiplicatively closed. Set $R_{(P)} = S^{-1}R$ and $M_{(P)} = S^{-1}M$. Since R_P and R_{P^*} are localizations of $R_{(P)}$, there are natural ring homomorphisms $R_{(P)} \to R_P$ and $R_{(P)} \to R_{P^*}$. Hence M_P and M_{P^*} are $R_{(P)}$-modules and both may be viewed as localizations of $M_{(P)}$. Set $R' = R_{(P)}$ and $M' = M_{(P)}$. Observe that any nonzero homogeneous element of R'/P^*R' is invertible. Therefore by Lemma 1.80, we have $R'/P^*R' \cong K[t, t^{-1}]$ as graded K-algebras, for some homogeneous element t of positive degree which is transcendental over K. Note that $R'/P^*R' \neq K$, since $P^*R' \subsetneq PR'$. Using the fact that $K[t, t^{-1}]$ is a localization of $K[t]$, it follows that $K[t, t^{-1}]$ is a principal ideal domain. Thus PR'/P^*R' is a principal ideal. So there exists $a \in PR' \setminus P^*R'$ be such that $PR' = R'a + P^*R'$. Then we have the short exact sequence

$$0 \to R'/P^*R' \xrightarrow{a} R'/P^*R' \to R'/PR' \to 0,$$

where the first map is multiplication by a. This sequence induces the long exact sequence

$$\operatorname{Ext}^i_{R'}(R'/P^*R', M') \xrightarrow{a} \operatorname{Ext}^i_{R'}(R'/P^*R', M') \to \operatorname{Ext}^{i+1}_{R'}(R'/PR', M') \to \cdots.$$

Note that $\operatorname{Ext}^i_{R'}(R'/P^*R', M')$ is a graded R'/P^*R'-module. Since $R'/P^*R' = K[t, t^{-1}]$, by Exercise 3.14, $\operatorname{Ext}^i_{R'}(R'/P^*R', M')$ is a free R'/P^*R'-module. Using the facts that P^*R' is a prime ideal and $a \notin P^*R'$, we conclude that the homomorphisms $\operatorname{Ext}^i_{R'}(R'/P^*R', M') \xrightarrow{a} \operatorname{Ext}^i_{R'}(R'/P^*R', M')$ are injective. We apply this fact to the above long exact sequence to obtain

$$\operatorname{Ext}^i_{R'}(R'/P^*R', M')/a\operatorname{Ext}^i_{R'}(R'/P^*R', M') \cong \operatorname{Ext}^{i+1}_{R'}(R'/PR', M')$$

for all i. This means that

$$\operatorname{Ext}^{i+1}_{R'}(R'/PR', M') \cong (R'/R'a) \otimes_{R'} \operatorname{Ext}^i_{R'}(R'/P^*R', M').$$

Let $\operatorname{Ext}^i_{R'}(R'/P^*R', M') = \bigoplus R'/P^*R'$. Then

$$(R'/R'a) \otimes_{R'} (\bigoplus R'/P^*R') \cong \bigoplus R'/(P^*R' + R'a) = \bigoplus R'/PR'.$$

Therefore $\operatorname{Ext}^{i+1}_{R'}(R'/PR', M')$ is a free R'/PR'-module which has the same rank as $\operatorname{Ext}^i_{R'}(R'/P^*R', M')$. Thus we may conclude that

$$\operatorname{Ext}^{i+1}_{R'_P}(R'_{PR'}/PR'_{PR'}, M'_{PR'}) = 0 \quad \Leftrightarrow \quad \operatorname{Ext}^i_{R'_{P^*R'}}(R'_{P^*R'}/P^*R'_{P^*R'}, M'_{P^*R'}) = 0.$$

This implies that $\operatorname{depth} M'_{PR'} = \operatorname{depth} M'_{P^*R'} + 1$. Now, since $M'_{PR'} \cong M_P$ and $M'_{P^*R'} \cong M_{P^*}$, the assertion holds.

Part (c) is clear from (a) and (b). □

Exercise 3.14
Let $R = K[t, t^{-1}]$ be a graded ring, where K is a field and t is a homogeneous element of positive degree in R which is transcendental over K. Then every graded R-module M is free.

Let R be a finitely generated graded K-algebra with the graded maximal ideal $\mathfrak{m}$, and let M be a finitely generated graded R-module. Since R is Noetherian, it is already defined what it means that M is Cohen-Macaulay. In Corollary 3.10 it is shown that M is Cohen-Macaulay if and only if $M_{\mathfrak{m}'}$ is Cohen-Macaulay for all maximal ideal $\mathfrak{m}'$ in R. Here, since R is graded, we have a unique graded maximal ideal $\mathfrak{m}$. The surprising fact is that in order to prove that M is Cohen-Macaulay, it is enough to check that $M_\mathfrak{m}$ is Cohen-Macaulay, as we will see below. Due to this fact we define in the graded case $\operatorname{depth} M = \operatorname{grade}(\mathfrak{m}, M)$. By Theorem 1.72 and Proposition 3.8, $\operatorname{depth} M = \operatorname{grade}(\mathfrak{m}, M) \le \operatorname{height}_M(\mathfrak{m}) = \dim M$.

Proposition 3.15
Let R be a graded K-algebra with the graded maximal ideal $\mathfrak{m}$ and M be a finitely generated graded R-module. Then M is Cohen-Macaulay if and only if $\operatorname{depth} M = \dim M$.

Proof. If M is Cohen-Macaulay, it follows from Theorem 3.12 that $\operatorname{depth} M = \dim M$. Conversely, assume that $\operatorname{depth} M = \dim M$. Then $\operatorname{depth} M_\mathfrak{m} \ge \operatorname{depth} M = \dim M \ge \dim M_\mathfrak{m}$. The first inequality follows from Lemma 3.5(c). Hence $M_\mathfrak{m}$ is Cohen-Macaulay. Let P be an arbitrary prime ideal in $\operatorname{Supp}(M)$. If P is a graded ideal, then $P \subseteq \mathfrak{m}$ and $M_P \cong (M_\mathfrak{m})_{PR_\mathfrak{m}}$ is Cohen-Macaulay. Now, assume that P is not graded. Then $P^* \ne P$ and M_{P^*} is Cohen-Macaulay. Thus by Proposition 3.13, M_P is Cohen-Macaulay. □

Exercise 3.16
Let R be a graded K-algebra, let I be a graded ideal of R, and let M be a finitely generated graded R-module with $\operatorname{grade}(I, M) = g$. Show that there exists a homogeneous M-sequence of length g in I.

Exercise 3.17
Let R be a standard graded K-algebra, and let $M \in \mathcal{M}_R$ with $\operatorname{depth} M = g > 0$. Suppose that K is infinite. Show that there exists an M-regular sequence $f_1, \dots, f_g$ with $f_i \in R_1$ for all i.

Hint: look at the proof of Theorem 1.84.

A graded K-algebra R is called a *complete intersection*, if there exists an isomorphism $R \cong S/(f_1, \dots, f_m)$ of graded K-algebras, where S is the polynomial ring and $f_1, \dots, f_m$ is a homogeneous S-sequence.

Exercise 3.18
Let R be a graded K-algebra with a minimal presentation S/I. Show that R is a complete intersection if and only if I is generated by a regular sequence.

Exercise 3.11 implies that complete intersections are Cohen-Macaulay. In general it is not so easy to find out whether a given graded K-algebra is a complete intersection. For example, the K-algebra $K[t^3, t^4, t^5]$ is not a complete intersection, while $K[t^4, t^5, t^6]$ is a complete intersection, see Corollary 4.37.

Exercise 3.19
Formulate and prove the graded version of Proposition 3.6.

Theorem 3.20
Let R be a Noetherian local ring or a graded K-algebra, let M be a finitely generated (graded) R-module, and let I be a (graded) ideal of R with $IM \neq M$. Then

$$\operatorname{height}_M I + \dim M/IM \leq \dim M,$$

and equality holds if M is Cohen-Macaulay.

Proof. Let $\operatorname{height}_M I = h$ and $\dim M/IM = r$. Then there exists a chain of prime ideals $P_0 \subsetneq P_1 \subsetneq \cdots \subsetneq P_r$ in $\operatorname{Supp}(M/IM)$. Since $I \subsetneq P_0$, we have $\operatorname{height}_M P_0 \geq h$. This means that the above chain of prime ideals can be extended to a chain of prime ideals of length at least $r + h$ in $\operatorname{Supp}(M)$. Hence $\dim M \geq r + h$.

Now, in addition assume that M is Cohen-Macaulay. Then by Theorem 3.12 there exists an M-sequence $f_1, \ldots, f_h$ in I such that $I \subseteq Z(M')$, where $M' = M/(f_1, \ldots, f_h)M$. In the graded case we may assume that $f_1, \ldots, f_h$ are homogeneous (see Exercise 3.16). Hence $I \subseteq P$ for some $P \in \operatorname{Ass}(M')$. By the graded version of Proposition 3.6 we have $\operatorname{depth} M' \leq \dim R/P$. Note that by Lemma 3.5, $\operatorname{depth} M = \operatorname{depth} M' + h$. Moreover, since $P \in V(I) \cap \operatorname{Supp}(M) = \operatorname{Supp}(M/IM)$, we have $\dim M/IM \geq \dim R/P$. Therefore, $\dim M = \operatorname{depth} M = \operatorname{depth} M' + h \leq \dim R/P + h \leq \dim M/IM + \operatorname{height}_M I$. Hence equality holds. □

Let R be a Noetherian ring and I be an ideal of R with $h = \operatorname{height} I$ which is generated by m elements. By Corollary 1.71 we have $h \leq m$. The ideal I is called an ideal of *the principal class,* if I is generated by h elements.

Proposition 3.21
Let R be a local ring or a graded K-algebra with the (graded) maximal ideal $\mathfrak{m}$. Moreover, let R be Cohen-Macaulay, and let $I \subsetneq R$ be a (graded) ideal. Then I is of the principal class of height h and generated by h (homogeneous) elements if and only if I is generated by an R-sequence.

Proof. Suppose that I is of the principal class of height h and generated by homogeneous elements $f_1, \ldots, f_h$. By induction on h we show that $f_1, \ldots, f_h$ is an R-sequence. Let $d = \dim R$. We claim that $(*)$ $\dim R/(f_1, \ldots, f_i) = d - i$ for any $1 \leq i \leq h$. Indeed, by Corollary 1.86, $\dim R/(f_1, \ldots, f_i) \geq d - i$. On the other hand, since R is Cohen-Macaulay, by Theorem 3.20, $d - h = \dim R/I = \dim \overline{R}/(f_{i+1}, \ldots, f_h)\overline{R} \geq \dim \overline{R} - h + i$, where $\overline{R} = R/(f_1, \ldots, f_i)$. Hence

$\dim \overline{R} \leq d-i$. So we have $\dim \overline{R} = d-i$. In particular, $\dim R/(f_1) = d-1$. Suppose in contrary that $f_1 \in Z(R)$. Then there exists $P \in \operatorname{Ass}(R)$ such that $f_1 \in P$. Since R is Cohen-Macaulay, we have $d = \dim R/P \leq \dim R/(f_1)$, which is a contradiction. By induction suppose that $f_1, \ldots, f_i$ is an R-sequence. Set $\overline{R} = R/(f_1, \ldots, f_i)$. Then it follows from Proposition 1.89 that $\overline{R}$ is a Cohen-Macaulay ring. By $(*)$ we have $\dim \overline{R}/(f_{i+1})\overline{R} = \dim \overline{R} - 1$. As before, we obtain that f_{i+1} is regular on $\overline{R}$, as desired.

The other direction of the statement follows from Proposition 3.8. □

An R-module M over a Noetherian ring R is called *maximal Cohen-Macaulay* if M is Cohen-Macaulay and $\dim M = \dim R$. We observe the following useful fact about maximal Cohen-Macaulay modules.

Proposition 3.22
Let R be a Noetherian local ring or a graded K-algebra, and let M be a (graded) maximal Cohen-Macaulay R-module. Then any (homogeneous) R-sequence is an M-sequence.

Proof. Let $\mathbf{x} = x_1, \ldots, x_r$ be a (homogeneous) R-sequence. Since $\operatorname{depth} M = \dim M = \dim R$, it follows from Proposition 3.6 and Exercise 3.19 that $\dim R/P = \dim R$ for any $P \in \operatorname{Ass}(M)$. This means that $\operatorname{Ass}(M) \subseteq \operatorname{Ass}(R)$, which yields $Z(M) \subseteq Z(R)$. Hence x_1 is regular on M, as well. Suppose inductively that $x_1, \ldots, x_{r-1}$ is an M-sequence. Then it follows from the graded version of Proposition 1.89 that $\overline{M} = M/(x_1, \ldots, x_{r-1})M$ is a maximal Cohen-Macaulay $\overline{R}$-module, where $\overline{R} = R/(x_1, \ldots, x_{r-1})R$. Then as before it follows that $Z(\overline{M}) \subseteq Z(\overline{R})$, and then x_r is regular on $\overline{M}$. □

3.1.3 Auslander-Buchsbaum and Hilbert-Burch Theorems

The following important theorem relates the depth of a module to its projective dimension, provided the module has finite projective dimension.

Theorem 3.23
*(Auslander-Buchsbaum)*Let R be a Noetherian local ring or a graded K-algebra, and let M be a nonzero finitely generated (graded) R-module with $\operatorname{pd}(M) < \infty$. Then

$$\operatorname{pd}(M) + \operatorname{depth} M = \operatorname{depth} R.$$

Proof. Let $\mathfrak{m}$ be the (graded) maximal ideal of R, and let $F_\bullet$ be a minimal (graded) free resolution of M. Then

$$0 \longrightarrow F_p \xrightarrow{\varphi_p} F_{p-1} \xrightarrow{\varphi_{p-1}} \cdots \longrightarrow F_1 \xrightarrow{\varphi_1} F_0 \xrightarrow{\varphi_0} M \longrightarrow 0$$

is an exact sequence, where $p = \mathrm{pd}(M)$. We prove the assertion by induction on $\operatorname{depth} R$.

Let $\operatorname{depth} R = 0$. Then $\mathfrak{m} \in \mathrm{Ass}(R)$. So there is an injective map $\delta : R/\mathfrak{m} \longrightarrow R$. Thus the map $\delta \otimes F_p$ is injective, as well. Similarly, $R \otimes \varphi_p$ is an injective map. If $p > 0$, then considering the commutative diagram

$$\begin{array}{ccc} R/\mathfrak{m} \otimes F_p & \xrightarrow{R/\mathfrak{m}\otimes\varphi_p} & R/\mathfrak{m} \otimes F_{p-1} \\ \big\downarrow{\scriptstyle \delta\otimes F_p} & & \big\downarrow{\scriptstyle \delta\otimes F_{p-1}} \\ R \otimes F_p & \xrightarrow{R\otimes\varphi_p} & R \otimes F_{p-1} \end{array}$$

we conclude that $\overline{\varphi}_p = R/\mathfrak{m} \otimes \varphi_p$ is injective.

The short exact sequence

$$0 \to F_p \to F_{p-1} \to \mathrm{Im}(\varphi_{p-1}) \to 0$$

induces the long exact homology sequence

$$\begin{aligned} \cdots \to\ & \mathrm{Tor}_1^R(R/\mathfrak{m}, F_{p-1}) \to \mathrm{Tor}_1^R(R/\mathfrak{m}, \mathrm{Im}(\varphi_{p-1})) \\ \to\ & R/\mathfrak{m} \otimes F_p \to R/\mathfrak{m} \otimes F_{p-1} \to R/\mathfrak{m} \otimes \mathrm{Im}(\varphi_{p-1}) \to 0. \end{aligned}$$

Since $\mathrm{Tor}_i^R(R/\mathfrak{m}, F_{p-1}) = \mathrm{Tor}_i^R(R/\mathfrak{m}, F_p) = 0$ for all $i > 0$, and since $R/\mathfrak{m} \otimes F_p \to R/\mathfrak{m} \otimes F_{p-1}$ is injective, it follows that $\mathrm{Tor}_i^R(R/\mathfrak{m}, \mathrm{Im}(\varphi_{p-1})) = 0$ for all $i > 0$. Hence the local version of Proposition 2.24 implies that $\beta_i(\mathrm{Im}(\varphi_{p-1})) = 0$ for $i > 0$. Consequently, by the local version of Proposition 2.30 we have $\mathrm{pd}(\mathrm{Im}(\varphi_{p-1})) = 0$. This implies that $\mathrm{Im}(\varphi_{p-1})$ is free. Therefore,

$$0 \longrightarrow \mathrm{Im}(\varphi_{p-1}) \longrightarrow F_{p-2} \xrightarrow{\varphi_{p-2}} \cdots \longrightarrow F_1 \xrightarrow{\varphi_1} F_0 \longrightarrow 0$$

is a free resolution of M of length $p-1$, which is a contradiction. So we must have $p = 0$ and then M is a free R-module. Therefore $\operatorname{depth} M = \operatorname{depth} R = 0$, and in this case we are done.

Now, assume that $\operatorname{depth} R > 0$ and $\operatorname{depth} M > 0$. Then we may choose an element x of R, which is a non zerodivisor on R and M. Then we obtain the short exact sequence $0 \to R \xrightarrow{x} R \to R/xR \to 0$, which induces the long exact sequence

$$\cdots \to \mathrm{Tor}_1^R(R, M) \to \mathrm{Tor}_1^R(R/xR, M) \to M \xrightarrow{x} M \to M/xM \to 0.$$

Since $\mathrm{Tor}_i^R(R, M) = 0$ for $i > 0$, and since $M \xrightarrow{x} M$ is injective, we conclude that $\mathrm{Tor}_i^R(R/xR, M) = 0$ for $i > 0$, and x is a non-zerodivisor on M.

Let $F_\bullet$ be a minimal free resolution of M. Then $H_i(R/xR \otimes_R F_\bullet) = \mathrm{Tor}_i^R(R/xR, M) = 0$ for $i > 0$. It follows that $F_\bullet/xF_\bullet$ is a minimal free R/xR-resolution of M/xM. Thus $\mathrm{pd}_{R/xR}(M/xM) = p$, as required.

It remains to settle the case $\operatorname{depth} R > 0$ and $\operatorname{depth} M = 0$. Since $\operatorname{depth} M \neq \operatorname{depth} R$, we know that M is not a free R-module and hence $p > 0$. Let $Z = \operatorname{Im}\varphi_1$ be the first syzygy module of M. Then $\operatorname{pd}(Z) = p - 1$. Also applying the long exact Ext-sequence to the short exact sequence $0 \longrightarrow Z \longrightarrow F_0 \longrightarrow M \longrightarrow 0$ we obtain $\operatorname{depth} Z = 1$. Then by the previous case $\operatorname{depth} Z + \operatorname{pd}(Z) = \operatorname{depth} R$. Thus $\operatorname{depth} R = \operatorname{depth} M + \operatorname{pd}(M)$. □

Let R be a Noetherian ring, and let $I \subsetneq R$ be an ideal. Then $\operatorname{grade}(I, R) \leq \operatorname{pd}(R/I)$. This inequality can be seen by computing $\operatorname{Ext}^i_R(R/I, R)$ via a projective resolution of R/I. The ideal I is called *perfect*, if $\operatorname{grade}(I, R) = \operatorname{pd}(R/I)$.

Here is a simple example of a perfect ideal of grade 2. Let K be a field, and let $K[x, y]$ be the polynomial ring in 2 variables. The ideal $I = (x^2, xy, y^2) \subsetneq S$ has the resolution $0 \to S^2 \xrightarrow{\varphi} S^3 \to I \to 0$, where φ is given by the matrix

$$\begin{bmatrix} y & 0 \\ -x & y \\ 0 & -x \end{bmatrix}.$$

It is clear that I is perfect of grade 2, because I contains the regular sequence x^2, y^2. Note that the three 2-minors of the matrix are the generators of I. This is not coincidentally, as we will see in the next theorem which is known as the Hilbert-Burch theorem. We first prove the following

Lemma 3.24

Let R be a Noetherian ring, let I be a proper ideal of R and let M be a finitely generated R-module. Consider the R-module homomorphism $f : M \to \operatorname{Hom}(I, M)$ with $f(m)(a) = am$ for $m \in M$ and $a \in I$. Then $\operatorname{grade}(I, M) \geq 2$ if and only if f is an isomorphism.

Proof. By Theorem A.40 we have the exact sequence

$$\begin{aligned} 0 \to \operatorname{Hom}_R(R/I, M) &\to \operatorname{Hom}_R(R, M) \xrightarrow{\varphi} \operatorname{Hom}_R(I, M) \\ \to \operatorname{Ext}^1_R(R/I, M) &\to \operatorname{Ext}^1_R(R, M) = 0. \end{aligned}$$

Note that $f = \varphi \circ \psi$, where $\psi : M \to \operatorname{Hom}_R(R, M)$ is the natural isomorphism with $\psi(m)(r) = rm$. Hence f is an isomorphism if and only if φ is an isomorphism.

From the above exact sequence we deduce that φ is an isomorphism if and only if $\operatorname{Hom}_R(R/I, M) = \operatorname{Ext}^1_R(R/I, M) = 0$. By Theorem 3.2 this is equivalent to say that $\operatorname{grade}(I, M) \geq 2$. □

Exercise 3.25
Let $I \subsetneq R$ be an ideal with $\text{grade}(I) \geq 2$, and let $\varphi\ I \to R$ be an R-module homomorphism. Use Lemma 3.24 to show that there exists $a \in R$ such that $\varphi(b) = ab$ for all $b \in I$.

The Hilbert-Burch theorem identifies certain ideals as ideals of minors of a matrix. Before stating this theorem we need some preparation.

Let $A = [a_{ij}]$ be an $m \times n$ matrix with entries in R, and let $1 \leq s \leq \min(m, n)$ be an integer. We denote by $I_s(A)$ the ideal generated by all s-minors of A. One can define the ideal of minors attached to an R-module homomorphism $F \to G$ of free R-modules. Let $F = R^n$ and $G = R^m$ be free R-modules, and let $\varphi\ F \to G$ be an R-module homomorphism. Let $f_1, \dots, f_n$ be a basis of F. Then the wedge products $f_{i_1} \wedge f_{i_2} \wedge \cdots \wedge f_{i_s}$ with $i_1 < i_2 < \cdots < i_s$ form a basis of $\bigwedge^s F$. We define an R-module homomorphism $\bigwedge^s \varphi\ \bigwedge^s F \to \bigwedge^s G$ by setting

$$(\bigwedge^s \varphi)(f_{i_1} \wedge f_{i_2} \wedge \cdots \wedge f_{i_s}) = \varphi(f_{i_1}) \wedge \varphi(f_{i_2}) \wedge \cdots \wedge \varphi(f_{i_s})$$

for all $1 \leq i_1 < i_2 < \cdots < i_s \leq n$.

Lemma 3.26
With the notation introduced we have

(a) For any sequence of elements $g_1, \dots, g_s \in F$, one has

$$(\bigwedge^s \varphi)(g_1 \wedge g_2 \wedge \cdots \wedge g_s) = \varphi(g_1) \wedge \varphi(g_2) \wedge \cdots \wedge \varphi(g_{g_s}).$$

(b) The definition of $\bigwedge^s \varphi$ does not depend on the particular choice of the basis $f_1, \dots, f_n$.

Proof. (a) Let $g_j = \sum_{i=1}^n a_{ij} f_i$. Then

$$g_1 \wedge \cdots \wedge g_s = (\sum_{i=1}^n a_{i1} f_i) \wedge \cdots \wedge (\sum_{i=1}^n a_{is} f_i).$$

Expanding the right hand side of this equation, we find that

$$g_1 \wedge \cdots \wedge g_s = \sum_{1 \leq i_1 < \cdots < i_s \leq n} [i_1, \dots, i_s] f_{i_1} \wedge \cdots \wedge f_{i_s}, \tag{3.1}$$

where $[i_1, \dots, i_s]$ is the maximal minor of the $n \times s$-matrix $A = (a_{ij})$, whose rows are $i_1, \dots, i_s$.

It follows that

$$
\begin{aligned}
(\bigwedge^{s}\varphi)(g_1\wedge\cdots\wedge g_s) &= \sum_{1\le i_1<\cdots<i_s\le n}[i_1,\ldots,i_s](\bigwedge^{s}\varphi)(f_{i_1}\wedge\cdots\wedge f_{i_s})\\
&= \sum_{1\le i_1<\cdots<i_s\le n}[i_1,\ldots,i_s]\varphi(f_{i_1})\wedge\cdots\wedge\varphi(f_{i_s}).
\end{aligned}
$$

On the other hand, we have

$$
\begin{aligned}
\varphi(g_1)\wedge\cdots\wedge\varphi(g_s) &= \varphi(\sum_{i=1}^{n}a_{i1}f_i)\wedge\cdots\wedge\varphi(\sum_{i=1}^{n}a_{is}f_i)\\
&= (\sum_{i=1}^{n}a_{i1}\varphi(f_i))\wedge\cdots\wedge(\sum_{i=1}^{n}a_{is}\varphi(f_i))\\
&= \sum_{1\le i_1<\cdots<i_s\le n}[i_1,\ldots,i_s]\varphi(f_{i_1})\wedge\cdots\wedge\varphi(f_{i_s}).
\end{aligned}
$$

This proves the desired equation.

(b) is an immediate consequence of (a). □

Use Lemma 3.26 to solve

Exercise 3.27

Let F, G and H be finitely generated free R-modules, and let $\varphi\ F\to G$ and $\psi: G\to H$ be R-module homomorphisms. Show that

$$
\bigwedge^{s}(\psi\circ\varphi)=(\bigwedge^{s}\psi)\circ(\bigwedge^{s}\varphi)\quad\text{for all } s.
$$

Corollary 3.28

Let $\varphi\ F\to G$ be an isomorphism. Then $\bigwedge^s\varphi\ \bigwedge^s F\to\bigwedge^s G$ is an isomorphism for all s.

Proof. Let φ^{-1} be the inverse map of φ. Then $(\bigwedge^s\varphi^{-1})\circ(\bigwedge^s\varphi)=\bigwedge^s(\varphi^{-1}\circ\varphi)=\bigwedge^s\mathrm{id}_F=\mathrm{id}_{\bigwedge^s F}$, and similarly $(\bigwedge^s\varphi)\circ(\bigwedge^s\varphi^{-1})=\mathrm{id}_{\bigwedge^s G}$. This shows that $\bigwedge^s\varphi$ is an isomorphism. □

Let $f_1,\ldots,f_n$ be a basis of F and $g_1,\ldots,g_m$ be a basis of G, and let $A=(a_{ij})$ be the $m\times n$-matrix defined by $\varphi(f_j)=\sum_{i=1}^{m}a_{ij}g_i$ for $j=1,\ldots,n$. We apply

(3.1), where $g_1, \dots, g_s$ is replaced by $\varphi(f_{j_1}) \wedge \cdots \wedge \varphi(f_{j_s})$ and $f_{i_1} \wedge \cdots \wedge f_{i_s}$ is replaced by $g_{i_1} \wedge \cdots \wedge g_{i_s}$ and obtain

$$\begin{aligned}(\bigwedge^s \varphi)(f_{j_1} \wedge \cdots \wedge f_{j_s}) &= \varphi(f_{j_1}) \wedge \cdots \wedge \varphi(f_{j_s}) \\ &= \sum_{1 \le i_1 < \cdots i_s \le m} [i_1, \dots, i_s | j_1, \dots, j_s] g_{i_1} \wedge \cdots \wedge g_{i_s},\end{aligned}$$

where $[i_1, \dots, i_s | j_1, \dots, j_s]$ denotes the minor of A with rows $i_1, \dots, i_s$ and columns $j_1, \dots, j_s$.

Thus, if $A^{(s)}$ denotes the matrix representing $\bigwedge^s \varphi$, then

$$A^{(s)} = ([i_1, \dots, i_s | j_1, \dots, j_s])_{\substack{1 \le i_1 < \cdots < i_s \le m \\ 1 \le j_1 < \cdots < j_s \le n}}. \tag{3.2}$$

In particular, we have

$$I_s(A) = I_1(A^{(s)}) \quad \text{for all} \quad s. \tag{3.3}$$

Exercise 3.29

Show that the ideal $I_s(A)$ only depends on φ (and not on the choice of the bases which gives us the matrix A).

We set $I_s(\varphi) = I_s(A)$.

Theorem 3.30

*(Hilbert-Burch)*Let R be a Noetherian ring, and let I be an ideal of R with $\operatorname{grade}(I) > 1$ and with an augmented free resolution

$$0 \longrightarrow R^s \xrightarrow{\varphi} R^{s+1} \longrightarrow I \longrightarrow 0.$$

Then $I_s(\varphi) = I$.

Proof. Let $A = [a_{ij}]$ be an $(s+1) \times s$ matrix which represents φ with respect to bases $e_1, \dots, e_{s+1}$ and $f_1, \dots, f_s$ of R^{s+1} and R^s, respectively. For any $1 \le i \le s+1$, let δ_i be the minor of A which is obtained by removing its ith row. Then $I_s(\varphi) = (\delta_1, \dots, \delta_{s+1})$. Consider the R-module homomorphism $\sigma : R^{s+1} \to I_s(\varphi)$ which maps each basis element e_i to $(-1)^i \delta_i$, for $1 \le i \le s+1$. Note that for any basis element $f_i \in R^s$, we have $(\sigma \circ \varphi)(f_i) = \sum_{i=1}^{s} (-1)^i a_{ij} \delta_i = 0$. Indeed, let A' be $(s+1) \times (s+1)$ matrix obtained by adding the jth column of A to A. Then $\det A' = 0$, since one column is repeated. Expanding the determinant A' with respect to the new column, we get $0 = \det A' = \sum_{i=1}^{s} (-1)^i a_{ij} \delta_i$. Thus, if we denote by $U \subseteq R^{s+1}$ the image of φ (which is also the kernel of $R^{s+1} \to I$), and if we denote by V the kernel of $R^{s+1} \to I_s(\varphi)$, then it follows that $U \subseteq V$, and we have a natural epimorphism $\epsilon\ I \to I_s(\varphi)$ with kernel $N = U/V$. This gives us the exact sequence

$$0 \to N \to I \to I_s(\varphi) \to 0. \qquad (3.4)$$

In the next step, first we show that $I \cong I_s(\varphi)$. This will be the case if $N = 0$. Suppose $N \neq 0$. Then there exists $P \in \operatorname{Ass}(N)$, and we have $N_p \neq 0$ and $\operatorname{depth} N_P = 0$.

It follows from the exact sequence (3.4) that $P \in \operatorname{Ass}(I)$, which implies that $\operatorname{depth} IR_P = 0$. Suppose that $I \subseteq P$. Then $IR_P \subseteq PR_P$ and $1 < \operatorname{grade}(I) \leq \operatorname{grade}(IR_P) \leq \operatorname{grade}(PR_P) = \operatorname{depth} R_P$. On the other hand, from the exact sequence $0 \to R^s \to R^{s+1} \to I \to 0$ we obtain the exact sequence

$$0 \to R_P^s \to R_P^{s+1} \to IR_P \to 0. \qquad (3.5)$$

Since $\operatorname{depth} IR_P = 0$, it follows that $\operatorname{depth} R_P \leq 1$, see Theorem 3.23. This yields a contradiction. Therefore $I \nsubseteq P$, which implies that $IR_P = R_P$, and together (3.4) we obtain the exact sequence

$$0 \to N_P \to R_P \to I_s(\varphi)R_P \to 0. \qquad (3.6)$$

Again we use that $IR_P = R_P$, and obtain from (3.5) the exact sequence

$$0 \to R_P^s \to R_P^{s+1} \to R_P \to 0, \qquad (3.7)$$

which is indeed split exact. Therefore, we may choose the basis $e_1, \dots, e_{s+1}$ such that $\varphi(f_i) = e_i$ for $i = 1, \dots, s$. With respect to this new basis, the matrix representing φ is of the form

$$\begin{bmatrix} 1 & 0 & \cdots & 0 \\ 0 & 1 & \cdots & 0 \\ \vdots & \vdots & \ddots & \vdots \\ 0 & 0 & \cdots & 1 \\ 0 & 0 & \cdots & 0 \end{bmatrix}.$$

By Exercise 3.29, this implies that $I_s(\varphi)R_P \cong R_P$. Hence the sequence (3.6) becomes the exact sequence $0 \to N_P \to R_P \to R_P \to 0$. It follows that $N_P = 0$. This is impossible, since by our assumption $N \neq 0$ and $P \in Ass(N)$. Hence we must have that $N = 0$. So we conclude that $I \cong I_s(\varphi)$, and that the following sequence

$$0 \longrightarrow R^s \xrightarrow{\varphi} R^{s+1} \xrightarrow{\sigma} I_s(\varphi) \longrightarrow 0. \qquad (3.8)$$

is exact.

In the final step of our proof we show that $I = I_s(\varphi)$.

Since $\operatorname{grade}(I) \geq 2$, by Exercise 3.25, the composition map $I \xrightarrow{\epsilon} I_s(\varphi) \hookrightarrow R$ is a multiplication map by an element $a \in R$. Hence $I_s(\varphi) = aI$.

Now, we show that a is a unit in R. Let P be an arbitrary prime ideal with $\operatorname{height} P \leq 1$. Then $I \nsubseteq P$, since by Proposition 3.8, $\operatorname{height} I \geq \operatorname{grade}(I) \geq 2$. Hence $IR_P = R_P$, and we obtain the split exact sequence $0 \longrightarrow R_P^s \xrightarrow{\varphi_P} R_P^{s+1} \longrightarrow$

$R_P \longrightarrow 0$. Therefore, as above, we deduce that $I_s(\varphi)R_P = R_P$. This means that $I_s(\varphi) \nsubseteq P$ for any prime ideal P with height$P \le 1$. So height$I_s(\varphi) \ge 2$.

Suppose a is not a unit. Then, since $I_s(\varphi) = aI \subseteq (a)$, Krull's Principal Ideal Theorem (Theorem 1.66) implies that height$I_s(\varphi) \le 1$, a contradiction. □

It is worthwhile to mention that when R is a local ring or a graded K-algebra, then the assumptions of Theorem 3.30 on the ideal I are equivalent to saying that I is a perfect ideal of grade 2.

3.2 The Canonical Module

In this section we review the main properties of the canonical module for finitely generated graded K-algebras. Its definition and its properties will be derived from the graded version of Grothendieck's duality theorem. Throughout this section R is a finitely generated positively graded K-algebra over a field K with the graded maximal ideal $\mathfrak{m}$.

3.2.1 Matlis Duality and Local Cohomology of Graded Modules

Consider the category $\mathscr{K}$ of $\mathbb{Z}$-graded K-vector spaces $V = \bigoplus_{i\in\mathbb{Z}} V_i$ for which each V_i is a finitely generated K-vector space and whose morphisms are the $\mathbb{Z}$-graded K-vector space homomorphisms. We define a functor $(-)^\vee\ \mathscr{K} \to \mathscr{K}$, $V \mapsto V^\vee$, where

$$(V^\vee)_i = \operatorname{Hom}_K(V_{-i}, K) \quad \text{for all} \quad i.$$

Exercise 3.31

Let $W_1, \ldots, W_n$ be a family of K-vector spaces in $\mathscr{K}$. Show that $(\bigoplus_{i=1}^n W_i)^\vee \cong \bigoplus_{i=1}^n W_i^\vee$.

Note that $(-)^\vee$ is an exact and contravariant functor with $(V^\vee)^\vee = V$ for all $V \in \mathscr{K}$.

For us it is important to study the subcategory $\mathscr{F}_R$ of $\mathscr{K}$, whose objects are the graded R-modules M with $\dim_K M_i < \infty$ for all $i \in \mathbb{Z}$ and whose morphisms are the graded R-module homomorphisms of degree 0. For example, if $R = K[x]$ is the polynomial ring, then $K[x, x^{-1}] \in \mathscr{F}_R$.

For an R-module $M = \bigoplus_{i\in\mathbb{Z}} M_i$ in $\mathscr{F}_R$, the K-vector space $\bigoplus_{i\in\mathbb{Z}}(M^\vee)_i$ represents the graded structure of $M^\vee$ as an R-module. The action of R on $M^\vee$ is as follows: for any homogeneous element $r \in R$ of degree i and any $\varphi \in (M^\vee)_j$ let $r\varphi : M_{-i-j} \to K$ be the K-vector space homomorphism with $(r\varphi)(x) = \varphi(rx)$ for all x. Then $r\varphi \in (M^\vee)_{i+j}$. It can be easily verified that $M^\vee$ is an R-module with this action. Hence the functor $(-)^\vee$ maps $\mathscr{F}_R$ to $\mathscr{F}_R$.

We consider two important subcategories of $\mathscr{F}_R$. We denote by $\mathscr{A}_R$ the category of Artinian graded R-modules. If $M \in \mathscr{A}_R$, then $\dim_K M_i < \infty$ for all i, see Exercise 3.32. Thus, the category $\mathscr{A}_R$ as well as the category $\mathscr{M}_R$ of finitely generated graded R-modules are subcategories of $\mathscr{F}_R$, which itself is a subcategory of the category GMod_R of graded R-modules.

Exercise 3.32
Let M be a Artinian graded R-module. Show that $\dim_K M_i < \infty$ for all i.

Exercise 3.33
Let $S = K[x_1, \ldots, x_n]$ be the polynomial ring over K. Show that $S^\vee = K[x_1^{-1}, \ldots, x_n^{-1}]$ with the following S-module structure: for any $u = x_1^{a_1} \cdots x_n^{a_n} \in S$ and $v = x_1^{-b_1} \cdots x_n^{-b_n} \in K[x_1^{-1}, \ldots, x_n^{-1}]$, set

$$uv = \begin{cases} x_1^{a_1-b_1} \cdots x_n^{a_n-b_n}, & \text{if } a_i < b_i \text{ for all } i, \\ 0, & \text{otherwise.} \end{cases} \tag{3.9}$$

Note that $S^\vee$ in Exercise 3.33 is a module in the category $\mathscr{A}_R$. The following result shows that this is not surprising.

Proposition 3.34
(Matlis duality for graded modules) Let $M \in \mathscr{F}_R$. Then the following holds:

(a) $M \in \mathscr{M}_R$ if and only if $M^\vee \in \mathscr{A}_R$.
(b) $N \in \mathscr{A}_R$ if and only if $N^\vee \in \mathscr{M}_R$.

Thus, the functor $(-)^\vee\ \mathscr{M}_R \to \mathscr{A}_R$ establishes an anti-equivalence between the categories $\mathscr{M}_R$ and $\mathscr{A}_R$.

Proof. (a) Let $M_1 \subset M_2 \subset \cdots \subset M$ be an ascending chain of graded submodules of M. Then by using the fact that $(-)^\vee$ is exact and contravariant, we obtain a descending sequence $M^\vee \supseteq (M/M_1)^\vee \supseteq (M/M_2)^\vee \supseteq \cdots$ of graded submodules of $M^\vee$. If we assume that $M^\vee \in \mathscr{A}_R$, then this sequence terminates, that is, there exists an integer i_0 such that $(M/M_{i_0})^\vee = (M/M_i)^\vee$ for all $i \geq i_0$. Then $M/M_{i_0} = ((M/M_{i_0})^\vee)^\vee = ((M/M_i)^\vee)^\vee = M/M_i$ for all i, which implies that $M_i = M_{i_0}$ for all $i \geq i_0$. This together with Proposition 1.20 shows that M is a Noetherian module, and hence finitely generated, and so $M \in \mathscr{M}_R$. The converse is proved in the same way.

(b) Since $N = (N^\vee)^\vee$, if follows from (a) that $N^\vee \in \mathscr{M}_R$. The converse is proved in the same way. □

Exercise 3.35
Let $M \in \mathscr{M}_R$ and $N \in \mathscr{A}_R$.

(a) Show that $M_i = 0$ for $i \ll 0$ and $N_i = 0$ for $i \gg 0$.
(b) Show that any graded submodule and any graded quotient module of M (resp. of N) belongs to $\mathscr{M}_R$ (resp. to $\mathscr{A}_R$).
(c) Show that if $\ell(M) < \infty$, then $\ell(M^\vee) < \infty$ and $\ell(M) = \ell(M^\vee)$.

In GMod_R the injective hull $\underline{E}(R/\mathfrak{m})$ of $K = R/\mathfrak{m}$ is characterized by the property that (1) $R/\mathfrak{m} \subseteq \underline{E}(R/\mathfrak{m})$, (2) $\underline{E}(R/\mathfrak{m})$ is indecomposable, and (3) $\underline{E}(R/\mathfrak{m})$ is an injective object in GMod_R, see Exercise 2.51.

We show

Proposition 3.36
$R^\vee = \underline{E}(R/\mathfrak{m})$.

Proof. The epimorphism $R \to R/\mathfrak{m} = K$ induces the monomorphism $R/\mathfrak{m} = (R/\mathfrak{m})^\vee \to R^\vee$. Furthermore, R is indecomposable in GMod_R, since any proper graded ideal of R is contained in $\mathfrak{m}$. Now, using this fact together with the equality $(R^\vee)^\vee = R$ and Exercise 3.31, we conclude that $R^\vee$ is indecomposable. It remains to be shown that $R^\vee$ is injective. Given a monomorphism $\iota\ N \to M$ and a morphism $f\ N \to R^\vee$, we obtain the epimorphism $\iota^\vee\ M^\vee \to N^\vee$ and the morphism $f^\vee\ R \to N^\vee$. Let $x = f^\vee(1)$. Since $\iota^\vee$ is surjective, there exists a homogeneous element $y \in M^\vee$ with $\iota^\vee(y) = x$. Let $g\ R \to M^\vee$ be the R-module homomorphism with $g(1) = y$. Then $g\ R \to M^\vee$ is a morphism in GMod_R and $\iota^\vee \circ g = f^\vee$. This implies that $f = g^\vee \circ \iota$ with $g^\vee\ M \to R^\vee$, and shows that $R^\vee$ is injective. □

Let $M \in \mathscr{M}_R$. By Nakayama's lemma, the number $\mu(M) = \dim_K(M/\mathfrak{m}M)$ is the minimal number of generators of M. When R is zero-dimensional, this number coincides with the so-called type of $M^\vee$, as we see in the next proposition.

The *type* of M is defined as $r(M) = \dim_K \underline{\mathrm{Ext}}^t_R(R/\mathfrak{m}, M)$, where $\mathrm{depth}(M) = t$. If M happens to be Cohen–Macaulay, the number $r(M)$ is also called the *Cohen–Macaulay type* of M. In the case that $t = 0$, we have $\underline{\mathrm{Ext}}^t_R(R/\mathfrak{m}, M) = \underline{\mathrm{Hom}}_R(R/\mathfrak{m}, M) \cong 0 :_M \mathfrak{m}$, where $0 :_M \mathfrak{m}$ is the socle of M.

▶ **Remark 3.37** Let $M \in \mathscr{M}_R$, and let $\mathbf{x} = x_1, \ldots, x_n$ be an M-sequence. Then it follows from Proposition A.43 that $r(M/\mathbf{x}M) = r(M)$.

Proposition 3.38
Let $\dim R = 0$, and let $M \in \mathcal{M}_R$. Then

(a) $\underline{\mathrm{Hom}}_R(M, R^\vee) \cong M^\vee$.
(b) $M \cong \underline{\mathrm{Hom}}_R(\underline{\mathrm{Hom}}_R(M, R^\vee), R^\vee)$.
(c) $\mu(M) = r(M^\vee)$ and $r(M) = \mu(M^\vee)$.

Proof. (a) Since $\dim(M) = 0$, by Theorem 1.83, M and R have only finitely many nonzero components, $R^\vee = \mathrm{Hom}_K(R, K)$ and $M^\vee = \mathrm{Hom}_K(M, K)$. Therefore, $\underline{\mathrm{Hom}}_R(M, R^\vee) = \underline{\mathrm{Hom}}_R(M, \underline{\mathrm{Hom}}_K(R, K)) \cong \underline{\mathrm{Hom}}_K(M, K) = M^\vee$.

(b) By using (a) we get

$$\underline{\mathrm{Hom}}_R(\underline{\mathrm{Hom}}_R(M, R^\vee), R^\vee) \cong (M^\vee)^\vee \cong M.$$

(c) The exact sequence $0 \to \mathfrak{m}M \to M \to M/\mathfrak{m}M \to 0$ induces the exact sequence

$$0 \to (M/\mathfrak{m}M)^\vee \to M^\vee \to (\mathfrak{m}M)^\vee \to 0.$$

Thus, $\varphi \in M^\vee$ belongs to $\varphi \in (M/\mathfrak{m}M)^\vee$ if and only if $\mathfrak{m}\varphi(M) = \varphi(\mathfrak{m}M) = 0$. This shows that $(M/\mathfrak{m}M)^\vee = 0 :_{M^\vee} \mathfrak{m}$. It follows that

$$\mu(M) = \dim_K M/\mathfrak{m}M = \dim_K(M/\mathfrak{m}M)^\vee = \dim_K(0 :_{M^\vee} \mathfrak{m}) = r(M^\vee).$$

Replacing M by $M^\vee$ and by using that $(M^\vee)^\vee = M$, the other equation follows. □

Exercise 3.39
In Proposition 3.38(a) we used that

$$\underline{\mathrm{Hom}}_R(M, \mathrm{Hom}_K(R, K)) \cong \mathrm{Hom}_K(M, K).$$

Prove this isomorphism.

Hint: for $\varphi \in \underline{\mathrm{Hom}}_R(M, \mathrm{Hom}_K(R, K))$, define $\alpha(\varphi) \in \mathrm{Hom}_K(M, K)$ by setting $\alpha(\varphi)(x) = \varphi(x)(1)$ for all $x \in M$, and show that α is an isomorphism.

Matlis duality plays an important role in Grothendieck's local duality theorem which relates local cohomology modules to Ext-modules. The story begins with the *section functor* $\underline{\Gamma}_\mathfrak{m}$ with support in the graded maximal ideal $\mathfrak{m}$ of R. Let $M \in \mathrm{GMod}_R$. We set

$$\underline{\Gamma}_\mathfrak{m}(M) = \{x \in M \ \ \mathfrak{m}^k x = 0 \text{ for some } k > 0\}.$$

In other words, $\underline{\Gamma}_\mathfrak{m}(M) = \bigcup_{k=1}^\infty (0 :_M \mathfrak{m}^k)$.

Any morphism $\varphi M \to N$ in GMod_R induces by restriction to $\underline{\Gamma}_{\mathfrak{m}}(M)$ a morphism $\underline{\Gamma}_{\mathfrak{m}}(\varphi)\, \underline{\Gamma}_{\mathfrak{m}}(M) \to \underline{\Gamma}_{\mathfrak{m}}(N)$. Hence $\underline{\Gamma}_{\mathfrak{m}}(-)$ is a functor from GMod_R to GMod_R.

Exercise 3.40
Let $M \in \mathscr{A}_R$. Show that $\underline{\Gamma}_{\mathfrak{m}}(M) = M$.

Lemma 3.41
$\underline{\Gamma}_{\mathfrak{m}}$ is a covariant and left exact functor.

Proof. Let $0 \to A \xrightarrow{\varphi} B \xrightarrow{\psi} C \to 0$ be an exact sequence in the category GMod_R. The map $\underline{\Gamma}_{\mathfrak{m}}(\varphi)$ is just the restriction of φ to $\underline{\Gamma}_{\mathfrak{m}}(A)$. Therefore $\underline{\Gamma}_{\mathfrak{m}}(\varphi)$ is injective. Let $b \in \text{Ker}(\underline{\Gamma}_{\mathfrak{m}}(\psi))$. Then $b \in \text{Ker}(\psi)$, which implies that there exists $a \in A$ with $\varphi(a) = b$. Since $\mathfrak{m}^k b = 0$ for some $k > 0$, we see that $\varphi(\mathfrak{m}^k a) = \mathfrak{m}^k \varphi(a) = \mathfrak{m}^k b = 0$. Since φ is injective, this implies that $a \in \underline{\Gamma}_{\mathfrak{m}}(A)$. Hence $\underline{\Gamma}_{\mathfrak{m}}(\varphi)(a) = b$, as desired. □

Exercise 3.42
Show that $\underline{\Gamma}_{\mathfrak{m}}$ is an R-linear functor.

Exercise 3.43
Let $M \in \mathscr{M}_R$. Show that

(a) $\underline{\Gamma}_{\mathfrak{m}}(M) \neq 0$ if and only if $\operatorname{depth} M = 0$.
(b) $\operatorname{depth} M/\underline{\Gamma}_{\mathfrak{m}}(M) > 0$.
(c) Let $U \subseteq \underline{\Gamma}_{\mathfrak{m}}(M)$. Then $U = \underline{\Gamma}_{\mathfrak{m}}(M)$ if and only if $\operatorname{depth} M/U > 0$.

Exercise 3.44
(a) Let $S = K[x, y]$ be the polynomial ring over K, let $I = (x^4, x^3y, x^2y^2, xy^3)$, and let $R = S/I$. Compute $\underline{\Gamma}_{\mathfrak{m}}(R)$.

(b) Let $R = K[x_1, \dots, x_n]/I$, where $I = J \cap (x_1, \dots .x_n)^k$ and $J = (x_ix_j\ 1 \leq i < j \leq n)$. Compute $\underline{\Gamma}_{\mathfrak{m}}(R)$.

Exercise 3.45
Give an example which shows that $\underline{\Gamma}_{\mathfrak{m}}$ is not an exact functor.

For any $M \in \text{GMod}_R$ a graded injective resolution exists. Therefore, the right derived functors of $\underline{\Gamma}_{\mathfrak{m}}$ exist, and we set $\underline{H}^i_{\mathfrak{m}} = R^i\underline{\Gamma}_{\mathfrak{m}}$ to be the right derived functor of $\underline{\Gamma}_{\mathfrak{m}}$. If $E^\bullet$ is an injective resolution of M in the category GMod_R, i.e., if it is a graded injective resolution, then

$$\underline{H}^i_{\mathfrak{m}}(M) \cong H^i(\underline{\Gamma}_{\mathfrak{m}}(E^\bullet)).$$

The functor $\underline{H}^i_{\mathfrak{m}}$ $\operatorname{GMod}_R \to \operatorname{GMod}_R$ is called the *ith local cohomology functor*.

The general properties of right derived functors imply

Lemma 3.46

(a) $\underline{H}^0_{\mathfrak{m}}(M) = \underline{\Gamma}_{\mathfrak{m}}(M)$ for all $M \in \operatorname{GMod}_R$.
(b) $\underline{H}^i_{\mathfrak{m}}(E) = 0$, if E is an injective module in GMod_R and $i > 0$.
(c) Each short exact sequence $0 \to U \to M \to N \to 0$ in GMod_R induces a long exact sequence

$$0 \to \underline{H}^0_{\mathfrak{m}}(U) \to \underline{H}^0_{\mathfrak{m}}(M) \to \underline{H}^0_{\mathfrak{m}}(N) \to \underline{H}^1_{\mathfrak{m}}(U) \to \underline{H}^1_{\mathfrak{m}}(M) \to \cdots.$$

The R-module $\underline{\Gamma}_{\mathfrak{m}}(M)$ may also be viewed as the direct limit of the direct system $\{(0 :_M \mathfrak{m}^k)\}_{k=1}^{\infty}$, with inclusion maps $(0 :_M \mathfrak{m}^i) \hookrightarrow (0 :_M \mathfrak{m}^j)$ for $i < j$.

Lemma 3.47
Let J be an $\mathfrak{m}$-primary graded ideal of R. We denote by

$$\underline{\Gamma}_J \operatorname{GMod}_R \to \operatorname{GMod}_R$$

the functor with $\underline{\Gamma}_J(M) = \{x \in M \ \ J^k x = 0\}$ for some $k \geq 0$, and by $\underline{H}^i_J$ the right derived functors of $\underline{\Gamma}_J$. Then $\underline{H}^i_{\mathfrak{m}} = \underline{H}^i_J$ for all i.

Proof. Since $\mathfrak{m}^r \subseteq J \subseteq \mathfrak{m}$ for some r, we see that $\underline{\Gamma}_J = \underline{\Gamma}_{\mathfrak{m}}$. □

This lemma entails

Corollary 3.48
Let $S \to R$ be a homomorphism of finitely generated graded K-algebras, and assume that R, viewed as an S-module, is finitely generated. Let $\mathfrak{n}$ be the graded maximal ideal of S and let $M \in \mathscr{M}_R$. Then $\underline{H}^i_{\mathfrak{n}}(M) \cong \underline{H}^i_{\mathfrak{m}}(M)$ for all i.

Proof. Since R is a finitely generated S-module, Nakayama's lemma implies that $R/\mathfrak{n}R$ is a finitely generated K-vector space. It follows from Theorem 1.83 that $\mathfrak{m}^k \subseteq \mathfrak{n}R$ for some k. Furthermore, the module M, viewed as S-module is finitely generated. Since $\underline{H}^i_{\mathfrak{n}}(M) = \underline{H}^i_{\mathfrak{n}R}(M)$, the desired result follows from Lemma 3.47. □

Proposition 3.49
Let $M \in \mathcal{M}_R$, and let $E^\bullet$ be the minimal graded injective resolution of M. Then there exist integers a_{ij} and integers $\mu_i(\mathfrak{m}, M)$ such that the cocomplex $\underline{\Gamma}_{\mathfrak{m}}(E^\bullet)$ is of the form

$$0 \longrightarrow \bigoplus_{j=1}^{\mu_0(\mathfrak{m},M)} R^\vee(a_{0j}) \longrightarrow \bigoplus_{j=1}^{\mu_1(\mathfrak{m},M)} R^\vee(a_{1j}) \longrightarrow \cdots . \qquad (3.10)$$

Proof. By Theorem 2.59 each E^i is a direct sum of modules of the form $\underline{E}(R/P)(a)$ for some prime ideals P, and for each P the number of such summands is equal to $\mu_i(P, M)$. Since $\operatorname{Ass}(\underline{E}(R/P)) = \{P\}$, for any $x \in \underline{E}(R/P)$ we have $0 :_R x \subseteq P$. Hence $\underline{\Gamma}_{\mathfrak{m}}(\underline{E}(R/P)) = 0$ if $P \neq \mathfrak{m}$, and by Propositions 3.34, 3.36 and Exercise 3.40 we obtain $\underline{\Gamma}_{\mathfrak{m}}(\underline{E}(R/\mathfrak{m})) = \underline{\Gamma}_{\mathfrak{m}}(R^\vee) = R^\vee$. Since direct sums commute with $\underline{\Gamma}_{\mathfrak{m}}$, the desired result follows. □

The integer $\mu_i(\mathfrak{m}, M)$ is called the *ith Bass number of M with respect to* $\mathfrak{m}$.

Corollary 3.50
Let $M \in \mathcal{M}_R$. Then $\underline{H}^i_{\mathfrak{m}}(M) \in \mathcal{A}_R$ for all $i \geq 0$.

Proof. By Proposition 3.49, $\underline{\Gamma}_{\mathfrak{m}}(E^\bullet)$ is a cocomplex of Artinan modules. Therefore, the ith cohomology modules $H^i_{\mathfrak{m}}(M)$ of this complex are Artinian, as well. □

Before we proceed to Grothendieck's theorem on the vanishing and non-vanishing of local cohomology, we need another description of local cohomology based on the modified Čech complex. Let R be a ring, and let M an R-module and $x \in R$. Then $S = \{1, x, x^2, \ldots\}$ is a multiplicatively closed set. It is common to write M_x for $S^{-1}M$. Note that if R is a graded K-algebra, $M \in \mathcal{M}_R$, and x is homogeneous, then R_x is a graded ring and M_x is a graded R_x-module. The ith graded component of R_x is defined to be the set of elements $\{r/x^k \; r \text{ is homogeneous} \deg r - k \deg x = i\}$ of

R_x. Similarly, the grading of M_x is defined. Observe that in general M_x has positive and negative components.

Now let R be a graded K-algebra, and let $\mathbf{x} = x_1, \ldots, x_n$ be a homogeneous sequence in R such that $(\mathbf{x})$ is $\mathfrak{m}$-primary. The *modified Čech complex*, attached to the sequence $\mathbf{x}$ is the graded cocomplex

$$C^\bullet\, 0 \to C^0 \to C^1 \to \cdots \to C^n \to 0,$$

where

$$C^i = \bigoplus_{1\le j_1<j_2<\cdots<j_i\le n} R_{x_{j_1}x_{j_2}\cdots x_{j_i}}, \quad \text{for } i > 0,$$

$C^0 = R$ and the differential $d^i : C^i \to C^{i+1}$ is defined on the components $R_{x_{j_1}x_{j_2}\cdots x_{j_i}} \to R_{x_{k_1}x_{k_2}\cdots x_{k_{i+1}}}$ as follows: it is the zero map, if $\{j_1, j_2, \ldots, j_i\} \not\subseteq \{k_1, k_2, \ldots, k_{i+1}\}$. On the other hand, if $\{j_1, j_2, \ldots, j_i\} \subseteq \{k_1, k_2, \ldots, k_{i+1}\}$ and $k_t \in \{k_1, k_2, \ldots, k_{i+1}\} \setminus \{j_1, j_2, \ldots, j_i\}$, then it is the map

$$(-1)^{t-1}\iota : R_{x_{j_1}x_{j_2}\cdots x_{j_i}} \to (R_{x_{j_1}x_{j_2}\cdots x_{j_i}})_{x_{k_t}} = R_{x_{k_1}x_{k_2}\cdots x_{k_{i+1}}},$$

where ι is the natural map from $R_{x_{j_1}x_{j_2}\cdots x_{j_i}}$ to its localization.

The grading on $R_{x_{j_1}x_{j_2}\cdots x_{j_i}}$ is the natural grading, as follows. For $r \in R$ and $a_1, \ldots, a_n \in \mathbb{Z}_+$ set $\deg(r/x_1^{a_1}\cdots x_n^{a_n}) = \deg r - \sum_{i=1}^n a_i \deg x_i$. A simple calculation shows that $C^\bullet$ is indeed a graded cocomplex.

Now we obtain

Theorem 3.51
$H^i(M\underline{\otimes}_R C^\bullet) \cong \underline{H}^i_{\mathfrak{m}}(M)$ for all i.

Proof. We prove the required isomorphisms by showing that the functors $H^i(-\underline{\otimes}_R C^\bullet)$ are the right derived functors of $\underline{\Gamma}_{\mathfrak{m}}(-)$.

Let $M \in \mathcal{M}_R$. Note that $C^\bullet : 0 \to R \to \bigoplus_{j=1}^n R_{x_j} \to \cdots$. Hence if we use that $M\underline{\otimes}_R R_{x_j} \cong M_{x_j}$, it follows that $H^0(M\underline{\otimes}_R C^\bullet) = \operatorname{Ker}(M \to \bigoplus_{j=1}^n M_{x_j})$. Let $u \in M$. Then u belongs to this kernel if and only if $u/1 = 0$ in M_{x_j} for $j = 1, \ldots, n$, and this is the case if and only if for all j, $x_j^{a_j} u = 0$ for suitable integers $a_i \ge 1$. This means that $(x_1^{a_1}, \ldots, x_n^{a_n})u = 0$. When this happens, we have $\mathfrak{m}^t \subseteq (x_1^{a_1}, \ldots, x_n^{a_n})$ for some positive integer t, since $x_1, \ldots, x_n$ is a system of parameters of R.

Hence we have shown that $H^0(M\underline{\otimes}_R C^\bullet) = \underline{\Gamma}_{\mathfrak{m}}(M)$.

Next let $0 \to U \to M \to N \to 0$ be a short exact sequence in $\mathcal{M}_R$. By using the fact that localization is an exact functor, it follows that

$$0 \longrightarrow U\underline{\otimes}_R C^\bullet \longrightarrow M\underline{\otimes}_R C^\bullet \longrightarrow N\underline{\otimes}_R C^\bullet \longrightarrow 0$$

is an exact sequence of cocomplexes. This yields the long exact cohomology sequence

$$0 \to H^0(U\underline{\otimes}_R C^\bullet) \to H^0(M\underline{\otimes}_R C^\bullet) \to H^0(N\underline{\otimes}_R C^\bullet) \to H^1(U\underline{\otimes}_R C^\bullet) \to \cdots .$$

Thus it remains to be shown that if I is a graded injective R-module, then $H^i(I\underline{\otimes}_R C^\bullet) = 0$ for all $i > 0$. Note that I is a direct sum of modules of the form $\underline{E}(R/P)$ with some shifts. Hence, we may assume that $I = \underline{E}(R/P)$, where P is a graded prime ideal of R, since tensor and homology commute with direct sums.

We consider two cases: first assume that $I = \underline{E}(R/\mathfrak{m})$. Let $a \in \underline{E}(R/\mathfrak{m})$. Since $\underline{E}(R/\mathfrak{m}) = R^\vee$ is Artinian, by Theorem 1.83 there exists for each j an integer $c_j \geq 1$ such that $x_j^{c_j} a = 0$. This implies that $a/1 = 0$ in $I_{x_{j_1}\cdots x_{j_i}}$ for all $i > 0$ and $j_1, \ldots, j_i \in [n]$. Hence $I\underline{\otimes}_R C^i = 0$ for $i > 0$.

Next assume that $I = \underline{E}(R/P)$, where $P \neq \mathfrak{m}$ is a graded prime ideal of R. Then $x_j \notin P$ for some j. Since $\operatorname{Ass}(\underline{E}(R/P)) = \{P\}$, it follows that the multiplication map $I \to I$ by x_j is injective. It is also surjective, because the image of the multiplication map is the injective submodule $x_j I \subseteq I$ which is necessarily a direct summand of I. However, since I is indecomposable, we must have $x_j I = I$.

We will show that $H^i(I\underline{\otimes}_R C^\bullet) = 0$ for all i. Note that $I\underline{\otimes} C^\bullet$ can be identified with

$$D^\bullet\, 0 \to D^0 \to D^1 \to \cdots \to D^n \to 0,$$

where

$$D^i = \bigoplus_{1\leq j_1<j_2<\cdots<j_i\leq n} I_{x_{j_1}x_{j_2}\cdots x_{j_i}}.$$

We may assume that $x_1 I = I$. The complex $D^\bullet$ can be defined for any partial sequence $x_1, \ldots, x_k$, as well, and we show by induction on k that the corresponding complexes are all exact.

For $k = 1$ we have to show that $0 \to I \to I_{x_1} \to 0$ is exact. Since $x_1 \notin P$, it follows that x_1 is a non-zerodivisor of I. Therefore, the map $I \to I_{x_1}$ is injective. Let $b \in I_{x_1}$. Then $b = a/x_1^k$ for some $a \in I$ and some $k \geq 1$. Since $I = x_1 I$, there exists $c \in I$ such that $a = x_1^k c$. It follows that $b = (x_1^k c)/x_1^k = c/1$. This shows that $I \to I_{x_1}$ is also surjective.

Now, let $k > 1$, and let $(D'')^\bullet$ be the complex for the sequence $x_1, \ldots, x_{k-1}$ and $(D')^\bullet$ be the complex for the sequence $x_1, \ldots, x_k$. Then $(D'')^\bullet_{x_k}[-1]$ is a subcomplex of $(D')^\bullet$ and $(D')^\bullet/(D'')^\bullet_{x_k}[-1] \cong (D'')^\bullet$. Thus we obtain a short exact sequence of cocomplexes

$$0 \to (D'')^\bullet_{x_k}[-1] \to (D')^\bullet \to (D'')^\bullet \to 0.$$

For example, for $k = 3$ we have the complex

$$(D')^\bullet\, 0 \to I \to I_{x_1} \oplus I_{x_2} \oplus \underline{I_{x_3}} \to I_{x_1x_2} \oplus \underline{I_{x_1x_3}} \oplus \underline{I_{x_2x_3}} \to \underline{I_{x_1x_2x_3}},$$

whose underlined part is a subcomplex of $(D')^\bullet$ which can be identified with $(D'')^\bullet_{x_3}[-1]$, where $(D')^\bullet\, 0 \to I \to I_{x_1} \oplus I_{x_2} \to I_{x_1x_2} \to 0$.

By induction hypothesis, $H^i((D'')^\bullet) = 0$ for all i. Therefore we also have that $H^i((D'')^\bullet_{x_k}[-1]) = H^{i-1}((D'')^\bullet)_{x_k} = 0$ for all i. Hence the long exact sequence arising from the above short exact sequence shows us that $H^i((D')^\bullet) = 0$ for all i. □

3.2.2 The Canonical Module and Local Duality

Local cohomology and the canonical module were introduced by Grothendieck. Here we present the graded versions of Grothendieck's local cohomology theorems.

Theorem 3.52
(Vanishing and non-vanishing of local cohomology) Let $M \in \mathscr{M}_R$ with $t = \operatorname{depth} M$ and $d = \dim M$. Then

(a) $\underline{H}^i_{\mathfrak{m}}(M) = 0$ for $i < t$ and $\underline{H}^t_{\mathfrak{m}}(M) \neq 0$.
(b) $\underline{H}^i_{\mathfrak{m}}(M) = 0$ for $i > d$ and $\underline{H}^d_{\mathfrak{m}}(M) \neq 0$.

Proof. (a) Let $E^\bullet$ be a graded injective resolution of M. For any $i < t$ we have

$$\mu_i(\mathfrak{m}, M) = \dim_K \underline{\operatorname{Ext}}^i_R(R/\mathfrak{m}, M) = 0.$$

Hence $\underline{\Gamma}_{\mathfrak{m}}(E^i) = 0$ for $i < t$, as can be seen from Proposition 3.49. It follows that $\underline{H}^i_{\mathfrak{m}}(M) = 0$ for $i < t$.

We use induction on t to show that $\underline{H}^t_{\mathfrak{m}}(M) \neq 0$. If $t = 0$, then $0 :_M \mathfrak{m} \neq 0$, and so $\underline{H}^0_{\mathfrak{m}}(M) = \underline{\Gamma}_{\mathfrak{m}}(M) \neq 0$. Suppose now that $t > 0$. We choose a homogeneous element $x \in R$, say of degree a, which is regular on M. The short exact sequence $0 \to M(-a) \xrightarrow{x} M \to M/xM \to 0$ gives rise to the long exact cohomology sequence

$$\cdots \longrightarrow \underline{H}^{t-1}_{\mathfrak{m}}(M) \longrightarrow \underline{H}^{t-1}_{\mathfrak{m}}(M/xM) \longrightarrow \underline{H}^t_{\mathfrak{m}}(M(-a)) \longrightarrow \cdots .$$

We know already that $\underline{H}^{t-1}_{\mathfrak{m}}(M) = 0$, and by our induction hypothesis we have $\underline{H}^{t-1}_{\mathfrak{m}}(M/xM) \neq 0$. Therefore, $\underline{H}^t_{\mathfrak{m}}(M(-a)) \neq 0$. Since $\underline{H}^t_{\mathfrak{m}}(M(-a)) = \underline{H}^t_{\mathfrak{m}}(M)(-a)$, we see that $\underline{H}^t_{\mathfrak{m}}(M) \neq 0$.

(b) Let $\mathfrak{n}$ be the graded maximal ideal of $R/\operatorname{Ann}(M)$. Then $\underline{H}^i_{\mathfrak{m}}(M) \cong \underline{H}^i_{\mathfrak{n}}(M)$, see Corollary 3.48. Thus we may replaces R by $R/\operatorname{Ann}(M)$, and hence may assume from the very beginning that $\dim R = \dim M$. Let $\mathbf{x} = x_1, \ldots x_d$ be a homogeneous system of parameters of R. By Theorem 3.51, $\underline{H}^i_{\mathfrak{m}}(M) \cong H^i(M \underline{\otimes}_R C^\bullet)$, where $C^\bullet$ is the modified Čech complex attached to $\mathbf{x}$. This implies that $\underline{H}^i_{\mathfrak{m}}(M) = 0$ for $i > d$.

It remains to be shown that $\underline{H}^d_{\mathfrak{m}}(M) \neq 0$. Let $S = K[x_1, \dots, x_d]$, where $x_1, \dots, x_d$ is the system of parameters of R, which we have chosen before. Then by Noether's Normalization Theorem 1.90, S is a polynomial ring and R is a finitely generated graded S-module. It follows that M is a finitely generated graded S-module, as well. Let $\mathfrak{n}$ be the graded maximal ideal of S. Applying again Corollary 3.48, we see that $\underline{H}^i_{\mathfrak{n}}(M) \cong \underline{H}^i_{\mathfrak{m}}(M)$ for all i. Hence we may replace R by S and may assume that M is an S-module.

Let $r = \operatorname{rank} M$. By Proposition 1.37 there exists an exact sequence $M \to S^r \to W \to 0$ of graded S-modules with $\dim W < d$. Let N be the kernel of $S^r \to W$, and let U be the kernel of $M \to S^r$. Then we obtain the graded exact sequences $0 \to U \to M \to N \to 0$ and $0 \to N \to S^r \to W \to 0$. The first exact sequence yields the exact sequence

$$\underline{H}^d_{\mathfrak{m}}(M) \longrightarrow \underline{H}^d_{\mathfrak{m}}(N) \longrightarrow \underline{H}^{d+1}_{\mathfrak{m}}(U).$$

Since $\underline{H}^{d+1}_{\mathfrak{m}}(U) = 0$, it follows $\underline{H}^d_{\mathfrak{m}}(M) \neq 0$ if $\underline{H}^d_{\mathfrak{m}}(N) \neq 0$. From the second short exact sequence we obtain the exact sequence

$$\underline{H}^d_{\mathfrak{m}}(N) \longrightarrow \underline{H}^d_{\mathfrak{m}}(S^r) \longrightarrow \underline{H}^d_{\mathfrak{m}}(W).$$

Since $\operatorname{depth} S^r = d$, it follows from (a) that $\underline{H}^d_{\mathfrak{m}}(S^r) \neq 0$, and since $\dim W < d$, it follows that $\underline{H}^d_{\mathfrak{m}}(W) = 0$. This implies that $\underline{H}^d_{\mathfrak{m}}(N) \neq 0$, and consequently $\underline{H}^d_{\mathfrak{m}}(M) \neq 0$, as desired. □

Let R be Cohen–Macaulay of dimension d. Corollary 3.50 together with Theorem 3.34 implies that $\underline{H}^d_{\mathfrak{m}}(R)^\vee \in \mathcal{M}_R$. We denote this module by ω_R and call it the *canonical module* of R. The study of ω_R is the main purpose of this section.

Theorem 3.53

(The local duality theorem) Let R be Cohen-Macaulay of dimension d, and let $M \in \mathcal{M}_R$. Then for all $i \in \mathbb{Z}$, there are isomorphisms

$$\underline{\operatorname{Ext}}^i_R(M, \omega_R) \cong \underline{H}^{d-i}_{\mathfrak{m}}(M)^\vee$$

of graded R-modules.

Proof. We claim that the functor $F(-) = \underline{H}^d_{\mathfrak{m}}(-)^\vee$ is representable. By Proposition A.49 we have to show: (1) F is R-linear, (2) F is left exact, (3) F commutes with finite direct summands and (4) $F(R(a)) = F(R)(-a)$ for all $a \in \mathbb{Z}$.

Proof of (1): By Exercise 3.42, $\underline{\Gamma}_{\mathfrak{m}}$ is an R-linear functor. Therefore, Proposition A.48 implies that its right derived functors $\underline{H}^i_{\mathfrak{m}}$ are also R-linear. Since $(-)^\vee$ is an R-linear functor as well, and since compositions of R-linear functors are again R-linear, we see that $\underline{H}^i_{\mathfrak{m}}(-)^\vee$ is R-linear for all i. In particular, $\underline{H}^d_{\mathfrak{m}}(-)^\vee$ is R-linear.

Proof of (2): Given the exact sequence $0 \to M \to N \to W \to 0$ in $\mathcal{M}_R$, we get the exact sequence $\underline{H}^d_{\mathfrak{m}}(M) \to \underline{H}^d_{\mathfrak{m}}(N) \to \underline{H}^d_{\mathfrak{m}}(W) \to \underline{H}^{d+1}_{\mathfrak{m}}(M)$, see Lemma 3.46(c). The vanishing theorem implies that $\underline{H}^{d+1}_{\mathfrak{m}}(M) = 0$. Thus, by applying the contravariant exact functor $(-)^\vee$ to this sequence, we obtain the exact sequence $0 \to \underline{H}^d_{\mathfrak{m}}(W)^\vee \to \underline{H}^d_{\mathfrak{m}}(N)^\vee \to \underline{H}^d_{\mathfrak{m}}(M)^\vee$.

Proof of (3): It is enough to show that $F(M \oplus N) \cong F(M) \oplus F(N)$. Let $E^\bullet$ and $E'^\bullet$ be graded injective resolutions of M and N, respectively. Then $E^\bullet \oplus E'^\bullet$ is a graded injective resolution of $M \oplus N$, and we obtain $\underline{\Gamma}_{\mathfrak{m}}(E^\bullet \oplus E'^\bullet) = \underline{\Gamma}_{\mathfrak{m}}(E^\bullet) \oplus \underline{\Gamma}_{\mathfrak{m}}(E'^\bullet)$. This implies that $\underline{H}^i_{\mathfrak{m}}(M \oplus N) \cong \underline{H}^i_{\mathfrak{m}}(M) \oplus \underline{H}^i_{\mathfrak{m}}(N)$ for all i. Note that $(-)^\vee$ commutes with finite direct summands. Therefore,

$$\underline{H}^i_{\mathfrak{m}}(M \oplus N)^\vee \cong \underline{H}^i_{\mathfrak{m}}(M)^\vee \oplus \underline{H}^i_{\mathfrak{m}}(N)^\vee$$

for all i. For $i = d$, this is what we wanted to show.

Proof of (4): Let $E^\bullet$ be a graded injective resolution of M, and let $a \in \mathbb{Z}$. Then $E^\bullet(a)$ is a graded injective resolution of $M(a)$, where $E^\bullet(a)^i = E^i(a)$ for all i. It follows that $\underline{\Gamma}_{\mathfrak{m}}(E^\bullet(a)) = \underline{\Gamma}_{\mathfrak{m}}(E^\bullet)(a)$, and therefore

$$\underline{H}^i_{\mathfrak{m}}(M(a)) = \underline{H}^i(\underline{\Gamma}_{\mathfrak{m}}(E^\bullet(a))) = \underline{H}^i(\underline{\Gamma}_{\mathfrak{m}}(E^\bullet))(a) = \underline{H}^i_{\mathfrak{m}}(M)(a).$$

Since $(W(a))^\vee = W^\vee(-a)$ for any $W \in \mathrm{GMod}_R$, the desired conclusion follows.

Now, since we know that $\underline{H}^d_{\mathfrak{m}}(-)^\vee$ is representable, we get

$$\underline{H}^d_{\mathfrak{m}}(-)^\vee \cong \underline{\mathrm{Hom}}_R(-, \underline{H}^d_{\mathfrak{m}}(R)^\vee) = \underline{\mathrm{Hom}}_R(-, \omega_R).$$

The functors $\underline{\mathrm{Ext}}^i_R(-, \omega_R)$ are the right derived functor of $\underline{\mathrm{Hom}}_R(-, \omega_R)$. Let $G(-) = \underline{H}^d_{\mathfrak{m}}(-)^\vee \cong \underline{\mathrm{Hom}}_R(-, \omega_R)$. We claim that the right derived functors $R^i G$ of G are the functors $\underline{H}^{d-i}_{\mathfrak{m}}(-)^\vee$. This will then imply that $\underline{H}^{d-i}_{\mathfrak{m}}(M)^\vee \cong \underline{\mathrm{Ext}}^i_R(M, \omega_R)$ for all $M \in \mathcal{M}_R$. In order to prove the claim we check the conditions of the right derived functors, analogue to the conditions (1)–(4) for left derived functors. We notice that for a free module $F \in \mathcal{M}_R$ we have $\underline{H}^{d-i}_{\mathfrak{m}}(F)^\vee = 0$ for $i > 0$. This follows from the vanishing theorem and the fact that $\operatorname{depth} F = d$, since R is Cohen–Macaulay. Thus condition (2) is satisfied. The other conditions (1), (3) and (4) follow from the fact that the functors $\underline{H}^i_{\mathfrak{m}}(-)$ are the right derived functors of $\underline{H}^0_{\mathfrak{m}}(-)$. □

Theorem 3.54

Let R be a Cohen-Macaulay ring of dimension d. Then the following hold:

(a) ω_R is Cohen–Macaulay of dimension d.
(b) $\underline{\mathrm{id}}(\omega_R) = d$.
(c) If $d = 0$, then $\omega_R \cong R^\vee$.
(d) let $\mathbf{x} = x_1, \ldots, x_r$ be a homogeneous regular R-sequence with $\deg(x_i) = a_i$ for $i = 1 \ldots, r$. Then $\omega_{R/(\mathbf{x})R} \cong (\omega_R/(\mathbf{x})\omega_R)(\sum_{i=1}^r a_i)$.

Proof. (a) By Theorem 3.53, $\underline{\mathrm{Ext}}^i_R(R/\mathfrak{m}, \omega_R) = \underline{H}^{d-i}_{\mathfrak{m}}(R/\mathfrak{m})^\vee$, and by Theorem 3.52, $\underline{H}^{d-i}_{\mathfrak{m}}(R/\mathfrak{m})^\vee = 0$ for $i \neq d$. This show that $\operatorname{depth}\omega_R = d$. Since $\operatorname{depth}\omega_R \leq \dim \omega_R \leq d$, the assertion follows.

(b) By Theorem 3.53, $\underline{\mathrm{Ext}}^{d+i}_R(M, \omega_R) \cong \underline{H}^{-i}_{\mathfrak{m}}(M)^\vee = 0$ for all $M \in \mathscr{M}_R$ and all $i > 0$. Hence $\underline{\mathrm{id}}(\omega_R) \leq d$, see Proposition 2.56 and its graded version Proposition 2.61. On the other hand, $\underline{\mathrm{Ext}}^d_R(R/\mathfrak{m}, \omega_R) \cong \underline{H}^0_{\mathfrak{m}}(R/\mathfrak{m})^\vee = R/\mathfrak{m} \neq 0$. Therefore, $\underline{\mathrm{id}}(\omega_R) \geq d$.

(c) If $d = 0$, then by Theorem 1.83 we have $\underline{\Gamma}_{\mathfrak{m}}(R) = R$. Thus $\omega_R = \underline{H}^0_{\mathfrak{m}}(R)^\vee = \underline{\Gamma}_{\mathfrak{m}}(R)^\vee = R^\vee$.

(d) The exact sequence $0 \to R(-a_1) \to R \to R/x_1R \to 0$ induces the exact sequence

$$\underline{H}^{d-1}_{\mathfrak{m}}(R) \to \underline{H}^{d-1}_{\mathfrak{m}}(R/x_1R) \to \underline{H}^d_{\mathfrak{m}}(R)(-a_1) \xrightarrow{x_1} \underline{H}^d_{\mathfrak{m}}(R) \to \underline{H}^d_{\mathfrak{m}}(R/x_1R).$$

By Theorem 3.52, $\underline{H}^{d-1}_{\mathfrak{m}}(R) = \underline{H}^d_{\mathfrak{m}}(R/x_1R) = 0$. This yields the short exact sequence

$$0 \to \underline{H}^{d-1}_{\mathfrak{m}}(R/x_1R) \to \underline{H}^d_{\mathfrak{m}}(R)(-a_1) \xrightarrow{x_1} \underline{H}^d_{\mathfrak{m}}(R) \to 0.$$

Note that by Corollary 3.48, $\underline{H}^{d-1}_{\mathfrak{m}}(R/x_1R) \cong \underline{H}^{d-1}_{\mathfrak{m}/x_1R}(R/x_1R)$. Then applying the functor $(-)^\vee$ to the above short exact sequence gives us the short exact sequence

$$0 \to \omega_R \xrightarrow{x_1} \omega_R(a_1) \to \omega_{R/x_1R} \to 0,$$

which yields the isomorphism $(\omega_R/x_1\omega_R)(a_1) \cong \omega_{R/x_1R}$. Induction on r completes the proof. □

Example 3.55
Let $S = K[x_1, \ldots, n]$ be the polynomial ring. Then $\omega_S \cong S(-n)$. To prove this we apply Theorem 3.54. It follows from (d) that $\omega_S/\mathfrak{m}\omega_S \cong \omega_{S/\mathfrak{m}}(-n)$. By (c), we have $\omega_{S/\mathfrak{m}} = S/\mathfrak{m}$. Thus, Nakayama's lemma implies that ω_S is generated by one element of degree n. By (a), ω_S has Krull dimension n. Then ω_S must be isomorphic to $S(-n)$.

3.2.3 Duality Properties of the Canonical Module and the Cohen-Macaulay Type

Further important properties of the canonical module will be given in the next theorems. For their proofs the following technical results are needed. The first of them, which is the next lemma, is of similar nature as Lemma 2.21.

Lemma 3.56
Let $M, N \in \mathscr{M}_R$ and let x be a homogeneous element in R which is regular on R and N. Furthermore, let $\varphi\ M \to N$ be a homogeneous map with the property that the induced map $\overline{\varphi}\ M/xM \to N/xN$ is an isomorphism. Then φ is an isomorphism.

Proof. Let $U = \operatorname{Ker}(\varphi)$ and $W = \operatorname{Coker}(\varphi)$. We have an exact sequence $M/xM \to N/xN \to W/xW \to 0$. Our assumption implies that $W/xW = 0$. Hence by Nakayama's lemma, $W = 0$. Thus we obtain the exact sequence $0 \to U \to M \to N \to 0$, which induces the long exact sequence

$$\cdots \to \underline{\operatorname{Tor}}_1^R(R/xR, N) \to U/xU \to M/xM \to N/xN \to 0.$$

Since x is regular on R, we obtain the exact sequence $0 \to R(-a) \xrightarrow{x} R \to R/xR \to 0$ with $a = \deg x$, which induces the long exact sequence

$$\cdots \to \underline{\operatorname{Tor}}_1^R(R, N) \to \underline{\operatorname{Tor}}_1^R(R/xR, N) \to N(-a) \xrightarrow{x} N \to N/xN \to 0.$$

We will show that $\underline{\operatorname{Tor}}_1^R(R/xR, N) = 0$. Note that $\operatorname{Tor}_1^R(R, N) = 0$, and $N(-a) \xrightarrow{x} N$ is injective because x is regular on N. It follows that $\underline{\operatorname{Tor}}_1^R(R/xR, N) = 0$.

We conclude that $U/xU = \operatorname{Ker}(\overline{\varphi})$. Thus our assumption implies that $U/xU = 0$. Nakayama's lemma then shows that $U = 0$. □

Lemma 3.57
Let $M, N \in \mathscr{M}_R$ be Cohen-Macaulay modules with $\dim M = \dim N = \dim R$, and let $\mathbf{x} = x_1, \ldots, x_r$ be a homogeneous R-sequence. Suppose that $\underline{\operatorname{Ext}}_R^i(M, N) = 0$ for $i > 0$. Then $\mathbf{x}$ is a regular sequence on $\underline{\operatorname{Hom}}_R(M, N)$ and

$$\underline{\operatorname{Hom}}_R(M, N)/\mathbf{x}\underline{\operatorname{Hom}}_R(M, N) \cong \underline{\operatorname{Hom}}_{R/\mathbf{x}R}(M/\mathbf{x}M, N/\mathbf{x}N).$$

Proof. Since $\mathbf{x}$ is an R-sequence and M and N are maximal Cohen-Macaulay modules, by Proposition 3.22 the sequence $\mathbf{x}$ is also regular on M and N.

We prove the lemma by induction on r. Let $r = 1$. Since x_1 is regular on M, we obtain the exact sequence $0 \to M(-a) \xrightarrow{x_1} M \to M/x_1M \to 0$, where $a = \deg(x_1)$. Since x_1 annihilates M/x_1M and is regular on N, by Lemma A.42 we obtain $\underline{\operatorname{Hom}}_R(M/x_1M, N) = 0$. Applying this together with the assumption that $\underline{\operatorname{Ext}}_R^i(M, N) = 0$ for $i > 0$ to the long exact sequence corresponding to the sequence $0 \to M(-a) \xrightarrow{x_1} M \to M/x_1M \to 0$, we get the exact sequence

$$0 \to \underline{\mathrm{Hom}}_R(M,N) \xrightarrow{x_1} \underline{\mathrm{Hom}}_R(M,N)(a) \to \underline{\mathrm{Ext}}^1_R(M/x_1M,N) \to 0.$$

Moreover, we obtain $\underline{\mathrm{Ext}}^i_R(M/x_1M,N) = 0$ for $i > 1$. In particular,

$$(\underline{\mathrm{Hom}}_R(M,N)/x_1\underline{\mathrm{Hom}}_R(M,N))(a) \cong \underline{\mathrm{Ext}}^1_R(M/x_1M,N)$$

and x_1 is regular on $\underline{\mathrm{Hom}}_R(M,N)$.

By Theorem A.44 we have $\underline{\mathrm{Ext}}^{i+1}_R(M/x_1M,N) \cong \underline{\mathrm{Ext}}^i_{R/x_1R}(M/x_1M,N/x_1N)$ (a) for all i. Hence, $\underline{\mathrm{Ext}}^1_R(M/x_1M,N) \cong \underline{\mathrm{Hom}}_{R/x_1R}(M/x_1M,N/x_1N)$. This completes the proof for $r = 1$.

Set $\overline{R} = R/x_1R$, $\overline{M} = M/x_1M$ and $\overline{N} = N/x_1N$. Note that $\overline{R}$, $\overline{M}$ and $\overline{N}$ are again Cohen-Macaulay of equal dimension, see Proposition 1.89 and Corollary 3.11. Since $\underline{\mathrm{Ext}}^i_{\overline{R}}(\overline{M},\overline{N}) = \underline{\mathrm{Ext}}^{i+1}_R(\overline{M},N)(-a) = 0$ for $i > 0$ and since $\mathbf{x}' = x_2,\dots,x_r$ is an $\overline{R}$-sequence, we may apply our induction hypothesis and obtain

$$\begin{aligned}
&\underline{\mathrm{Hom}}_R(M,N)/\mathbf{x}\underline{\mathrm{Hom}}_R(M,N) \cong \\
&(\underline{\mathrm{Hom}}_R(M,N)/x_1\underline{\mathrm{Hom}}_R(M,N))/\mathbf{x}'(\underline{\mathrm{Hom}}_R(M,N)/x_1\underline{\mathrm{Hom}}_R(M,N)) \cong \\
&\underline{\mathrm{Hom}}_{\overline{R}}(\overline{M},\overline{N})/\mathbf{x}'\underline{\mathrm{Hom}}_{\overline{R}}(\overline{M},\overline{N}) \cong \underline{\mathrm{Hom}}_{\overline{R}/\mathbf{x}'\overline{R}}(\overline{M}/\mathbf{x}'\overline{M},\overline{N}/\mathbf{x}'\overline{N}) \cong \\
&\underline{\mathrm{Hom}}_{R/\mathbf{x}R}(M/\mathbf{x}M,N/\mathbf{x}N).
\end{aligned}$$

Moreover, $\mathbf{x}'$ is a regular sequence on $\underline{\mathrm{Hom}}_{\overline{R}}(\overline{M},\overline{N})$. Hence $\mathbf{x}$ is a regular sequence on $\underline{\mathrm{Hom}}_R(M,N)$. □

Theorem 3.58

Let R be Cohen–Macaulay of dimension d, and let $M \in \mathscr{M}_R$ be Cohen-Macaulay of dimension t. Then

(a) $\underline{\mathrm{Ext}}^i_R(M,\omega_R) = 0$ for $i \neq d-t$, and $\underline{\mathrm{Ext}}^{d-t}_R(M,\omega_R)$ is Cohen–Macaulay of dimension t.
(b) $\mu(\underline{\mathrm{Ext}}^{d-t}_R(M,\omega_R)) = r(M)$ and $r(\underline{\mathrm{Ext}}^{d-t}_R(M,\omega_R)) = \mu(M)$. In particular, $r(R) = \mu(\omega_R)$ and $r(\omega_R) = \mu(R) = 1$.
(c) $M \cong \underline{\mathrm{Ext}}^{d-t}_R(\underline{\mathrm{Ext}}^{d-t}_R(M,\omega_R),\omega_R)$.

Proof. By Theorem 3.53, we have $\underline{\mathrm{Ext}}^i_R(M,\omega_R) \cong \underline{H}^{d-i}_{\mathfrak{m}}(M)^\vee$ and the vanishing theorem implies that $\underline{H}^{d-i}_{\mathfrak{m}}(M)^\vee = 0$ for $d-i \neq t$. These facts together imply that $\underline{\mathrm{Ext}}^i_R(M,\omega_R) \neq 0$ for $i \neq d-t$.

We have $\mathrm{height}(\mathrm{Ann}(M)) = d-t$ and since R is Cohen-Macaulay it follows that $\mathrm{grade}(\mathrm{Ann}(M)) = \mathrm{height}(\mathrm{Ann}(M)) = d-t$. Thus we may choose a homogeneous R-sequence $\mathbf{x} = x_1,\dots,x_{d-t}$ with $x_i \in \mathrm{Ann}(M)$.

Before proceeding with the proof of (a),(b) and (c), we show by induction on $d-t$, that

$$\underline{\operatorname{Ext}}_R^{d-t}(M,\omega_R)\cong \underline{\operatorname{Hom}}_{R/\mathbf{x}R}(M,\omega_{R/\mathbf{x}R}).$$

If $d-t=0$, then there is nothing to prove. Suppose now that $d-t>0$, and let $a=\deg(x_1)$. We set $\overline{R}=R/x_1R$. Theorem A.44 together with Theorem 3.54(d) implies that

$$\begin{aligned}\underline{\operatorname{Ext}}_R^{d-t}(M,\omega_R)&\cong \underline{\operatorname{Ext}}_{\overline{R}}^{d-t-1}(M,\omega_R/x_1\omega_R)(a)\cong \underline{\operatorname{Ext}}_{\overline{R}}^{d-t-1}(M,\omega_{\overline{R}}(-a))(a)\\ &= \underline{\operatorname{Ext}}_{\overline{R}}^{d-t-1}(M,\omega_{\overline{R}}).\end{aligned}$$

Since $\omega_{\overline{R}}$ is a maximal Cohen-Macaulay $\overline{R}$-module, by Proposition 3.22 $\mathbf{x}'=x_2,\dots,x_{d-t}$ is an $\omega_{\overline{R}}$-sequence. We may apply our induction hypothesis and obtain that

$$\begin{aligned}\underline{\operatorname{Ext}}_R^{d-t}(M,\omega_R)&\cong \underline{\operatorname{Ext}}_{\overline{R}}^{d-t-1}(M,\omega_{\overline{R}})\cong \underline{\operatorname{Hom}}_{\overline{R}/\mathbf{x}'\overline{R}}(M,\omega_{\overline{R}/\mathbf{x}'\overline{R}})\\ &\cong \underline{\operatorname{Hom}}_{R/\mathbf{x}R}(M,\omega_{R/\mathbf{x}R}).\end{aligned}$$

Thus from now on we may assume for the proof of (a), (b) and (c) that $d=t$. Thus, $\underline{\operatorname{Ext}}_R^{d-t}(M,\omega_R)=\underline{\operatorname{Hom}}_R(M,\omega_R)$. Let $\mathbf{y}=y_1,\dots,y_t$ be a homogeneous R-sequence with $\deg(y_i)=b_i$ for $i=1,\dots,t$, and set $\widetilde{R}=R/\mathbf{y}R$. Since M and ω_R are maximal Cohen-Macaulay R-modules, Proposition 3.22 and Lemma 3.57 imply that $\mathbf{y}$ is regular on $\underline{\operatorname{Hom}}_R(M,\omega_R)$. This proves (a). Moreover, together with Theorem 3.54, we obtain that

$$\begin{aligned}\underline{\operatorname{Hom}}_R(M,\omega_R)/\mathbf{y}\underline{\operatorname{Hom}}_R(M,\omega_R)&\cong \underline{\operatorname{Hom}}_{\widetilde{R}}(M/\mathbf{y}M,\omega_R/\mathbf{y}\omega_R)\\ &\cong \underline{\operatorname{Hom}}_{\widetilde{R}}(M/\mathbf{y}M,\widetilde{R}^\vee)(-b),\end{aligned}\tag{3.11}$$

where $b=\sum_{i=1}^d b_i$. Hence (3.11) and Proposition 3.38 imply that

$$\begin{aligned}\mu(\underline{\operatorname{Hom}}_R(M,\omega_R))&=\mu(\underline{\operatorname{Hom}}_R(M,\omega_R)/\mathbf{y}\underline{\operatorname{Hom}}_R(M,\omega_R))\\ &=\mu(\underline{\operatorname{Hom}}_{\widetilde{R}}(M/\mathbf{y}M,\widetilde{R}^\vee))=r(M/\mathbf{y}M)=r(M).\end{aligned}$$

For the last equality see Remark 3.37.

(c) Lemma 3.57, Theorem 3.54 and Proposition 3.38 imply that

$$\begin{aligned}&\underline{\operatorname{Hom}}_R(\underline{\operatorname{Hom}}_R(M,\omega_R),\omega_R)/\mathbf{y}\underline{\operatorname{Hom}}_R(\underline{\operatorname{Hom}}_R(M,\omega_R),\omega_R)\cong\\ &\underline{\operatorname{Hom}}_{\widetilde{R}}(\underline{\operatorname{Hom}}_{\widetilde{R}}(M/\mathbf{y}M,\omega_{\widetilde{R}})(-b),\omega_{\widetilde{R}})(-b)=\\ &\underline{\operatorname{Hom}}_{\widetilde{R}}(\underline{\operatorname{Hom}}_{\widetilde{R}}(M/\mathbf{y}M,\omega_{\widetilde{R}}),\omega_{\widetilde{R}})=\underline{\operatorname{Hom}}_{\widetilde{R}}(\underline{\operatorname{Hom}}_{\widetilde{R}}(M/\mathbf{y}M,\widetilde{R}^\vee),\widetilde{R}^\vee)\cong M/\mathbf{y}M.\end{aligned}$$

This isomorphism

$$M/\mathbf{y}M\to \underline{\operatorname{Hom}}_R(\underline{\operatorname{Hom}}_R(M,\omega_R),\omega_R)/\mathbf{y}\underline{\operatorname{Hom}}_R(\underline{\operatorname{Hom}}(M,\omega_R),\omega_R),$$

which we just proved, is induced by the natural R-module homomorphism $\varphi : M \to \underline{\mathrm{Hom}}_R(\underline{\mathrm{Hom}}_R(M, \omega_R), \omega_R)$. Thus, Lemma 3.56 implies that φ is an isomorphism. □

The properties of ω_R listed in the next result follow immediately from Theorem 3.58.

Corollary 3.59

(a) $\mathrm{End}(\omega_R) = R$.
(b) ω_R is a faithful R-module, that is, $\mathrm{Ann}(\omega_R) = 0$.
(c) Let R be a domain. Then $\mathrm{rank}\,\omega_R = 1$.

Proof. (a) It follows from Theorem 3.58 that

$$R = \underline{\mathrm{Hom}}_R(\underline{\mathrm{Hom}}_R(R, \omega_R), \omega_R) = \underline{\mathrm{Hom}}_R(\omega_R, \omega_R) = \mathrm{End}(\omega_R),$$

(b) Let $x \in \mathrm{Ann}(\omega_R)$. Then $0 = x\mathrm{End}(\omega_R) = xR$. Therefore $x = 0$, and hence $\mathrm{Ann}(\omega_R) = 0$.

(c) Let $S = R \setminus \{0\}$, and let $S^{-1}\omega_R = Q(R)^r$. By (a) and Proposition A.39 we have

$$\begin{aligned} Q(R) = \; & S^{-1}R = S^{-1}\underline{\mathrm{Hom}}_R(\omega_R, \omega_R) \cong \underline{\mathrm{Hom}}_{S^{-1}R}(S^{-1}\omega_R, S^{-1}\omega_R) \\ & \cong \underline{\mathrm{Hom}}_{Q(R)}(Q(R)^r, Q(R)^r) \cong Q(R)^{r^2}. \end{aligned}$$

It follows that $r = 1$, as desired. □

The canonical module ω_R can be characterized as follows.

Theorem 3.60

Let R be Cohen–Macaulay of dimension d, and let $C \in \mathscr{M}_R$. Then $C \cong \omega_R$ if and only if

$$\underline{\mathrm{Ext}}^i_R(R/\mathfrak{m}, C) \cong \begin{cases} R/\mathfrak{m}, & \text{if } i = d, \\ 0, & \text{if } i \neq d. \end{cases} \tag{3.12}$$

Proof. By Theorem 3.53 we have $\underline{\mathrm{Ext}}^i_R(R/\mathfrak{m}, \omega_R) = \underline{H}^{d-i}_{\mathfrak{m}}(R/m)^\vee$. Thus, Theorem 3.52 implies that $\underline{\mathrm{Ext}}^i_R(R/\mathfrak{m}, \omega_R) = 0$, if $i \neq d$, while

$$\underline{\mathrm{Ext}}^d_R(R/\mathfrak{m}, \omega_R) = \underline{H}^0_{\mathfrak{m}}(R/\mathfrak{m})^\vee = \underline{\Gamma}_{\mathfrak{m}}(R/\mathfrak{m})^\vee = (R/\mathfrak{m})^\vee = R/\mathfrak{m}.$$

This shows that if $C \cong \omega_R$, then C satisfies (3.12).

Conversely, suppose that C satisfies (3.12). Then C is Cohen–Macaulay of dimension d, and hence $C \cong \underline{\mathrm{Hom}}_R(\underline{\mathrm{Hom}}_R(C, \omega_R), \omega_R)$, by Theorem 3.58(c). Thus it suffices to show that $\underline{\mathrm{Hom}}_R(C, \omega_R) \cong R$, because then

$$C \cong \underline{\mathrm{Hom}}_R(\underline{\mathrm{Hom}}_R(C, \omega_R), \omega_R) \cong \underline{\mathrm{Hom}}_R(R, \omega_R) = \omega_R.$$

Let $D = \underline{\mathrm{Hom}}_R(C, \omega_R)$. Since $r(C) = 1$, part (a) and part (b) of Theorem 3.58 imply that D is Cohen-Macaulay of dimension d with $\mu(D) = 1$. Therefore, $D \cong (R/I)(a)$, where $I \subsetneq R$ is a graded ideal, R/I is Cohen–Macaulay of dimension d and a is an integer. We want to show that $I = 0$ and $a = 0$.

Theorem 3.58(c) implies that

$$C \cong \underline{\mathrm{Hom}}_R((R/I)(a), \omega_R) = \underline{\mathrm{Hom}}_R(R/I, \omega_R)(-a),$$

so that $C(a) \cong \underline{\mathrm{Hom}}_R(R/I, \omega_R)$. Hence, the exact sequence $0 \to I \to R \to R/I \to 0$ induces the exact sequence

$$0 \to C(a) \to \omega_R \to W \to 0,$$

where $W = \underline{\mathrm{Hom}}_R(I, \omega_R)$. The exactness follows since $\underline{\mathrm{Ext}}^1_R(R/I, \omega_R) = 0$ by Theorem 3.58(a).

This short exact sequence yields the exact sequence

$$\begin{aligned}&\underline{\mathrm{Ext}}^{d-1}_R(R/\mathfrak{m}, W) \to \underline{\mathrm{Ext}}^d_R(R/\mathfrak{m}, C(a)) \to \underline{\mathrm{Ext}}^d_R(R/\mathfrak{m}, \omega_R) \to \\ &\underline{\mathrm{Ext}}^d_R(R/\mathfrak{m}, W) \to \underline{\mathrm{Ext}}^{d+1}_R(R/\mathfrak{m}, C(a)).\end{aligned}$$

Note that $\underline{\mathrm{Ext}}^d_R(R/\mathfrak{m}, C(a)) = (R/\mathfrak{m})(a)$ and $\underline{\mathrm{Ext}}^d_R(R/\mathfrak{m}, \omega_R) = R/\mathfrak{m}$. Moreover, we have $\underline{\mathrm{Ext}}^{d+1}_R(R/\mathfrak{m}, C(a)) = 0$. Therefore we obtain the exact sequence

$$\underline{\mathrm{Ext}}^{d-1}_R(R/\mathfrak{m}, W) \to (R/\mathfrak{m})(a) \to R/\mathfrak{m} \to \underline{\mathrm{Ext}}^d_R(R/\mathfrak{m}, W) \to 0.$$

If $I = 0$, then $\underline{\mathrm{Ext}}^{d-1}_R(R/\mathfrak{m}, W) = 0$, since in this case $W = 0$. If $I \neq 0$, then I is a Cohen–Macaulay module of dimension d, since R and R/I are Cohen–Macaulay of dimension d. Hence, Theorem 3.58(a) implies that W is Cohen–Macaulay of dimension d, as well. It follows that $\underline{\mathrm{Ext}}^{d-1}_R(R/\mathfrak{m}, W) = 0$. Thus, in any case we obtain the exact sequence

$$0 \to (R/\mathfrak{m})(a) \to R/\mathfrak{m} \to \underline{\mathrm{Ext}}^d_R(R/\mathfrak{m}, W) \to 0.$$

Suppose that $a \neq 0$. Then it follows that $R/\mathfrak{m} = 0$, a contradiction. Thus, $a = 0$, and then $\underline{\mathrm{Ext}}^d_R(R/\mathfrak{m}, W) = 0$.

Suppose that $I \neq 0$. Then $W \neq 0$, and W is Cohen–Macaulay of dimension d, which implies that $\underline{\mathrm{Ext}}_R^d(R/\mathfrak{m}, W) \neq 0$, a contradiction. □

Exercise 3.61

Let R be Cohen–Macaulay of dimension d, let $C \in \mathscr{M}_R$ be a maximal Cohen–Macaulay module, and let $\overline{C} = C/\mathbf{x}C$ where $\mathbf{x} = x_1, \dots, x_d$ is a homogeneous C-sequence with $\sum_{i=1}^d \deg x_i = a$. Show that the following statements are equivalent:

(i) $C \cong \omega_R$.
(ii) $\overline{C}(a) \cong \overline{R}^{\vee}$.

The next corollary slightly extends the statement given in Example 3.55, and provides an alternative proof.

Corollary 3.62

Let $S = K[x_1, \dots, x_n]$ be the polynomial ring with $\deg(x_i) = a_i$ for $i = 1, \dots, n$. Then

$$\omega_S \cong S(-a) \quad \text{where} \quad a = \sum_{i=1}^n a_i.$$

Proof. The sequence $\mathbf{x} = x_1, \dots, x_n$ is an S-sequence with $S/\mathbf{x}S = K$. Therefore, $\operatorname{depth}(S) = n$, and hence $\underline{\mathrm{Ext}}_S^i(S/\mathfrak{m}, S(-a)) = 0$ for $i < n$. Furthermore, by Theorem A.44, for $i \geq n$ we have

$$\underline{\mathrm{Ext}}_S^i(S/\mathfrak{m}, S(-a)) \cong \underline{\mathrm{Ext}}_{S/\mathbf{x}S}^{i-n}(K, (S/\mathbf{x}S)(-a))(a) = \underline{\mathrm{Ext}}_K^{i-n}(K, K).$$

Clearly, $\underline{\mathrm{Ext}}_K^{i-n}(K, K) = 0$ for $i > n$, while $\underline{\mathrm{Ext}}_K^{i-n}(K, K) = \underline{\mathrm{Hom}}_K(K, K) = K$ for $i = n$. Thus the desired result follows from Theorem 3.60. □

3.2.4 Gorenstein Graded K-Algebras

Now we come to an important definition. The ring R is called *Gorenstein*, if there exists an integer b such that $R(b) \cong \omega_R$.

It is clear that if R is Gorenstein, then $b = a(R)$. Moreover, by Theorem 3.60, R is Gorenstein if and only if there exists an integer b such that

$$\underline{\mathrm{Ext}}_R^i(R/\mathfrak{m}, R(b)) \cong \begin{cases} R/\mathfrak{m}, & \text{if } i = d, \\ 0, & \text{if } i \neq d. \end{cases}$$

For example, the polynomial ring S with the grading $\deg(x_i) = a_i$ for i as given in Corollary 3.62 is a Gorenstein ring with a-invariant $a(S) = -\sum_{i=1}^{n} a_i$.

Exercise 3.63

(a) Let R be Cohen-Macaulay, and let $\mathbf{x}$ be a homogeneous R-sequence. Show that R is Gorenstein if and only if $R/\mathbf{x}R$ is Gorenstein. How are the a-invariants of R and $R/\mathbf{x}R$ related to each other?

Examples of Gorenstein rings are given by the following

Proposition 3.64

Let R be complete intersection. Then R is Gorenstein.

Proof. Let $R = S/I$, where I is generated by a homogeneous S-sequence. By Corollary 3.62, S is Gorenstein. Thus, Exercise 3.63 implies that R is Gorenstein. □

Our definition of Gorenstein rings implies that they have finite graded injective dimension, namely $\underline{\mathrm{id}}(R) = \dim R$. Actually this property characterizes Gorenstein rings as shown by Bass.

Another characterization of Gorenstein rings is given in

Theorem 3.65

The following conditions are equivalent:

(i) R is Gorenstein.
(ii) R is Cohen–Macaulay and $r(R) = 1$.

Proof. (i)⇒(ii) follows immediately from our definition of Gorenstein rings, because ω_R is Cohen–Macaulay of type 1.

(ii)⇒(i): Let $\mathbf{x}$ be a maximal homogeneous regular sequence in R. By Exercise 3.63, R is Gorenstein if and only if $R/\mathbf{x}R$ is Gorenstein. Furthermore, Remark 3.37 implies that $r(R) = r(R/\mathbf{x}R)$. Thus we may assume that $\dim R = 0$. Since $r(R) = 1$, it follows that $R^\vee$ is generated by a single element, and since $\ell(R) = \ell(R^\vee)$, we deduce that, up to a shift, $R^\vee$ is isomorphic to R. Hence R is Gorenstein. □

By what we have seen before, we obtain the following implications among finitely generated graded K-algebras.

polynomial ring $\Rightarrow$ complete intersection $\Rightarrow$ Gorenstein $\Rightarrow$ Cohen-Macaulay

The next theorem and its corollary allows us in many cases to compute ω_R.

Theorem 3.66
Let R and R' be Cohen–Macaulay graded finitely generated positively graded Cohen-Macaulay K-algebras. Furthermore, let $\varphi\ R \to R'$ be a graded K-algebra homomorphism of graded K-algebras such that R', viewed as an R-module via φ, is finitely generated. Then

$$\omega_{R'} \cong \underline{\mathrm{Ext}}^t_R(R', \omega_R),$$

where $t = \dim R - \dim R'$.

Proof. We first notice that the Krull dimension s of R' as a ring is the same as the Krull dimension of R', viewed as an R-module. Indeed, to see this we may replace R by its image in R'. We denote this image by S. Since R' is a finitely generated S-module, it follows from Lemma 1.38 that $S \subseteq R'$ is an integral extension. Therefore, Corollary 1.95 implies that the Krull dimension of R' coincides with the Krull dimension of S. On the other hand, R' is a finitely generated S-module with $\mathrm{Ann}_S(R') = 0$, since $1 \in R'$. Thus the dimension of R' as an S-module coincides with the Krull dimension of S. This implies our assertion.

Since $R' \in \mathscr{M}_R$ we may apply Theorem 3.53 and find that $\underline{\mathrm{Ext}}^t_R(R', \omega_R) \cong \underline{H}^s_{\mathfrak{m}}(R')^\vee$. Let $\mathfrak{n}$ be the graded maximal ideal of R'. Since R' is a finitely generated R-module, it follows from Nakayama's lemma that $R'/\varphi(\mathfrak{m})R'$ is a finitely generated K-vector space which implies that $\mathfrak{n}^k \subseteq \varphi(\mathfrak{m})R'$ for some k. Therefore, since $\mathfrak{m}$ operates on R' in the same way as $\varphi(\mathfrak{m})$ does, Corollary 3.47 implies that

$$\omega_{R'} = \underline{H}^s_{\mathfrak{n}}(R')^\vee = \underline{H}^s_{\varphi(\mathfrak{m})}(R')^\vee = \underline{H}^s_{\mathfrak{m}}(R')^\vee \cong \underline{\mathrm{Ext}}^t_R(R', \omega_R). \qquad \square$$

Let R be Cohen-Macaulay of dimension d, minimally generated over K by the elements $r_1, \ldots, r_n$ with $\deg(r_i) = a_i$ for $i = 1, \ldots, n$. Then $R \cong S/I$, where $S = K[x_1, \ldots, x_n]$ is the polynomial ring with the grading given by $\deg(x_i) = a_i$ for $i = 1, \ldots, n$, and $I \subsetneq S$ is a graded ideal. Let

$$F_\bullet : 0 \to F_p \to F_{p-1} \to \cdots \to F_1 \to F_0 \to 0,$$

be the minimal graded free S-resolution of R. Then $H_0(F_\bullet) \cong R$ and by Theorem 3.23, $p = n - d$. We denote by $G_\bullet$ the S-dual complex $\underline{\mathrm{Hom}}_S(F_\bullet, S)$. Then

$$G_\bullet : 0 \to G_p \to G_{p-1} \to \cdots \to G_1 \to G_0 \to 0, \tag{3.13}$$

with $F^*_{p-i} := \underline{\mathrm{Hom}}_S(F_{p-i}, S) = G_i$ for $i = 0, \ldots, p$.

Corollary 3.67
With the notation and assumptions introduced, the complex $G_\bullet$ is acyclic with $H_0(G_\bullet) = \omega_R(a)$, where $a = \sum_{i=1}^n a_i$. Hence $\omega_R(a) \cong \mathrm{Coker}(F^*_{p-1} \to F^*_p)$, and $G_\bullet$ is a graded minimal free S-resolution of $\omega_R(a)$. In particular, $r(R) = \beta_p(R)$.

Proof. By Corollary 3.62 we have $\omega_S = S(-a)$. Hence together with Theorem 3.66 we get

$$\omega_R \cong \underline{\mathrm{Ext}}^p_S(R, S(-a)) = \underline{\mathrm{Ext}}^p_S(R, S)(-a) = H_0(G_\bullet)(-a).$$

$G_\bullet$ is acyclic, since $H_i(G_\bullet) = \underline{\mathrm{Ext}}^{p-i}_S(R, S) = \underline{\mathrm{Ext}}^{n-d-i}(R, S) = 0$ for $i \neq 0$, see Theorem 3.58(a). The minimality of $G_\bullet$ follows from the fact that the matrices representing the differentials of $G_\bullet$ are just the transpose of the matrices representing the differentials of $F_\bullet$, see Exercise 2.5. It follows from the free presentation of ω_R that $\mu(\omega_R) = \beta_p(R)$. Thus, the equation $r(R) = \beta_p(R)$ follows from Theorem 3.58(b). $\square$

Exercise 3.68
Let $R = S/I$ be Cohen–Macaulay, where $S = K[x_1, \ldots, x_n]$ is the polynomial ring with $\deg(x_i) = a_i$ for $i = 1, \ldots, n$, and where $I \subsetneq S$ is a graded ideal. We set $a = \sum_{i=1}^n a_i$. Let $F_\bullet : 0 \to F_p \to F_{p-1} \to \cdots \to F_1 \to F_0 \to 0$ be the graded minimal free S-resolution of R with $F_i = \bigoplus_{j=1}^{\beta_i} S(-d_{i,j})$ for $i = 1, \ldots, p$. Then the following conditions are equivalent:

(i) R is Gorenstein.
(ii) $\beta_i = \beta_{p-i}$ for $i = 0, \ldots, p$.
(iii) $\beta_p = 1$

If the equivalent conditions hold, then $a(R) = d_{p,1} - a$, and for $i = 0, \ldots, p$ we have

$$\{d_{i,j}\ j = 1, \ldots, \beta_i\} = \{d_{p,1} - d_{p-i,j}\ j = 1, \ldots, \beta_{p-i}\}.$$

Numerical Semigroups and Their Relations

4

This chapter provides an introduction to the fundamental concepts on numerical semigroups and their semigroup rings. Symmetric numerical semigroups and the gluing construction, which is a method for building new symmetric numerical semigroups from existing ones are considered. The semigroup ring of a numerical semigroup is introduced and its defining ideal is described. The algebraic structure of this ring very well reflects and explains in many cases properties of the underlying numerical semigroup. Special attention is given to relations for families of numerical semigroups, including those arising from gluing, complete intersections, and 3-generated cases. We close the chapter with the study of numerical semigroups of maximal embedding dimension.

4.1 Introduction to Numerical Semigroups

4.1.1 The Coin-Exchange Problem of Frobenius

Suppose we live in some (strange) country where they would have only 5 cent and 7 cent coins. Then obviously not all amounts can be paid with these coins without change. For example it is impossible to pay 16 cents. In this example the biggest amount that cannot be paid with these coins is 23 cents, while any amount ≥ 24 cents can be paid. How can we see this? We find:

$$24 = 2 \cdot 5 + 2 \cdot 7,\ \ 25 = 5 \cdot 5,\ \ 26 = 1 \cdot 5 + 3 \cdot 7,\ \ 27 = 4 \cdot 5 + 1 \cdot 7 \text{ and } 28 = 4 \cdot 7.$$

So for sure these 5 amounts can be paid. But then it is easy to see that any amount ≥ 24 cents can be paid as well. Because, if we let $a \geq 24$ be any integer, then we

J. Herzog et al., *Numerical Semigroups*, Compact Textbooks in Mathematics,
https://doi.org/10.1007/978-3-032-05424-1_4

find an integer $c \geq 0$ such that $a - c \cdot 5 \in \{24, \ldots, 28\}$. Then $a = c \cdot 5 + b$ with $b \in \{24, \ldots, 28\}$.

Now, we consider the more general situation that we have m coins of denomination $a_1, a_2, \cdots, a_m$, where the a_i are pairwise distinct positive integers. The amounts that can be paid with these coins is the set of numbers

$$H = \{c_1 a_1 + c_2 a_2 + \cdots + c_m a_m \;\; c_i \in \mathbb{Z}_{\geq 0} \text{ for } i = 1, \ldots, m\}. \tag{4.1}$$

The question is now the following: when does there exist an integer c such that all integers $a \geq c$ belong to the set H? Obviously, we should have $\gcd(a_1, \ldots, a_m) = 1$. Because if $d > 1$ is the greatest common divisor of these integers, then all elements of H are divisible by d, and so the infinitely many integers which are not divisible by d do not belong to H. Surprisingly, the converse is also true.

Theorem 4.1
Suppose $\gcd(a_1, \ldots, a_m) = 1$. Then there exists an integer c such that

$$\{a \in \mathbb{Z}_{\geq 0} \;\; a \geq c\} \subseteq H.$$

Proof. Since $\gcd(a_1, \ldots, a_n) = 1$, there exist integers $c_1, \ldots, c_m \in \mathbb{Z}$ (of course not all non-negative, unless some $a_i = 1$) such that

$$1 = c_1 a_1 + c_2 a_2 + \cdots + c_m a_m. \tag{4.2}$$

We let $c = \sum_{i=1}^{m} (a_1 - 1)|c_i| a_i$. Clearly, $c \in H$, because c is a linear combination of the a_i with non-negative integer coefficients. As we have seen in the example before, we only need to show that $c + j \in H$ for $j = 0, \ldots, a_1 - 1$. By using (4.2) we see that

$$c + j = \sum_{i=1}^{m} [(a_1 - 1)|c_i| + j c_i] a_i.$$

If $c_i \geq 0$, then $(a_1 - 1)|c_i| + j c_i = (a_1 - 1 + j) c_i \geq 0$, and if $c_i < 0$, then $(a_1 - 1)|c_i| + j c_i = (a_1 - 1 - j)|c_i| \geq 0$. Thus, $c + j$ is a linear combination of the a_i with non-negative integer coefficients, and so $c + j \in H$, as desired. □

We call the set H in (4.1) a *numerical semigroup*, if $\gcd(a_1, \ldots, a_m) = 1$. (The letter H stands for the "Halbgruppe", which is the German word for "Semigroup".) The set H is indeed a semigroup, because it is additively closed, that is, if $s_1, s_2 \in H$, then $s_1 + s_2 \in H$. The elements $a_1, \ldots, a_m$ are called *generators* of H, because any other element in H can be written as a suitable sum of these integers. We write

$$H = \langle a_1, \ldots, a_m \rangle.$$

Theorem 4.1 shows that for a numerical semigroup H, the set $\mathbb{Z}_{\geq 0} \setminus H$ is a finite set. Indeed, this property characterizes numerical semigroups.

Proposition 4.2
A semigroup H of $\mathbb{Z}_{\geq 0}$ is a numerical semigroup if and only if $\mathbb{Z}_{\geq 0} \setminus H$ is a finite set.

Proof. The "only if" statement follows from Theorem 4.1. Now, suppose that $H \neq \mathbb{Z}_{\geq 0}$ is a semigroup of $\mathbb{Z}_{\geq 0}$ such that $\mathbb{Z}_{\geq 0} \setminus H$ is a finite set. Let b be the maximum element of $\mathbb{Z}_{\geq 0} \setminus H$, and let $\{a \in H : a < b\} = \{a_1, \ldots, a_m\}$. We show that the set

$$A = \{a_1, \ldots, a_m\} \cup \{b+1, b+2, \ldots, b+a_1\}$$

generates H. To see this, let $c > b + a_1$ be an integer. Then $c = a_1 + b + d$ for an integer $d > 0$. Write $d = a_1 q + r$ for integers $q \geq 0$ and $0 \leq r \leq a_1 - 1$. We have $c = a_1 + b + d = a_1(q+1) + (b+r)$. If $r > 0$, then from $a_1, b+r \in H$, we obtain that c belongs to the semigroup generated by A. If $r = 0$, then $c = (a_1 + b) + a_1 q$ with $a_1 + b, a_1 \in H$. Hence, in this case c belongs to the semigroup generated by A, as well. This shows that A generates H. Moreover, it is clear that $\gcd(A) = 1$. □

Exercise 4.3
Let H_1 and H_2 be numerical semigroups. Show that $H_1 \cap H_2$ is a numerical semigroup.

Exercise 4.4
Consider the numerical semigroup H generated by $a_i = a_1 + (i-1)$ for $i = 1, \ldots, m$, where $1 \leq m \leq a_1$. Show that H is the union of the integer intervals

$$[ja_1, ja_m] = \{a \in \mathbb{Z} \ \ ja_1 \leq a \leq ja_m\}.$$

A set of generators $\{a_1, \ldots, a_m\}$ of H is called a *minimal set of generators* of H, if none of the generators can be dropped, that is, for each i, the set $\{a_1, \ldots, a_m\} \setminus \{a_i\}$ is not a set of generators of H. So for example $\{4, 5, 6\}$ is a minimal set of generators of $\langle 4, 5, 6 \rangle$, while $\{2, 3, 4\}$ is not a minimal set of generators of $\langle 2, 3, 4 \rangle$.

Exercise 4.5
Show that the minimal set of generators of a numerical semigroup is uniquely determined.

The cardinality of this unique minimal set of generators of H is called the *embedding dimension* of H and is denoted by $\mathrm{emb}(H)$. The smallest element among the generators of H is called the *multiplicity* of H and is denoted by $e(H)$. This terminology is borrowed from commutative algebra where the embedding dimension is the minimal number of generators of the graded maximal ideal of a graded K-algebra. In Sect. 4.2 we will attach to each numerical semigroup H a K-algebra $K[H]$ whose embedding dimension is $\mathrm{emb}(H)$ and whose multiplicity is $e(H)$.

In what follows we always assume that $\gcd(a_1, \ldots, a_m) = 1$.

By Theorem 4.1, the set $\mathcal{G}(H) = \mathbb{Z}_{\geq 0} \setminus H$ is finite for any numerical semigroup H. The elements of the set $\mathcal{G}(H)$ are called the *gaps* of H. These are the amounts which cannot be paid by the coins with denomination $a_1, \ldots, a_m$. In our above example with $H = \langle 5, 7 \rangle$ we have

$$\mathcal{G}(H) = \{1, 2, 3, 4, 6, 8, 9, 11, 13, 16, 18, 23\}. \tag{4.3}$$

The largest gap of a numerical semigroup H is the so-called *Frobenius number* of H, denoted by $F(H)$. It is a challenging problem to compute $F(H)$ in terms of the generators of H. This is known as the Frobenius Coin-Exchange Problem.

Exercise 4.6
(a) Let f be a positive integer. Show that there are only finitely many numerical semigroups with $F(H) = f$.
(b) Determine all numerical semigroups H with $F(H) = 7$.

An excellent tool that helps to compute the relevant data for numerical semigroups is the package *numericalsgps* which is part of the GAP System for Computational Discrete Algebra [21]. This package has been developed by Manuel Delgado, Pedro A. García-Sánchez and José João Morais. For example, the Frobenius number of $\langle 12, 25, 35 \rangle$ can be computed by the following commands in *numericalsgps*.

```
gap>  LoadPackage("numericalsgp");
true
gap> s:=NumericalSemigroup(12,25,35);
<Numerical semigroup with 3 generators>
gap> FrobeniusNumberOfNumericalSemigroup(s);
163
```

Sylvester [73] in 1882 noticed the following formula for the Frobenius number of a 2-generated numerical semigroup.

Theorem 4.7
Let $H = \langle a_1, a_2 \rangle$. Then

$$F(H) = (a_1 - 1)(a_2 - 1) - 1.$$

This formula applied to our semigroup $H = \langle 5, 7 \rangle$ gives us $F(H) = 4 \cdot 6 - 1 = 23$, as we noticed before.

Proof (of Theorem 4.7). We first show that $(a_1 - 1)(a_2 - 1) - 1$ is a gap. Indeed, suppose $(a_1 - 1)(a_2 - 1) - 1 \in S$. Then there exist integers $c_i \geq 0$ such that $c_1a_1 + c_2a_2 = (a_1 - 1)(a_2 - 1) - 1 = a_1a_1 - a_1 - a_2$. This implies that

$$(c_1 + 1)a_1 + (c_2 + 1)a_2 = a_1a_2.$$

Since $\gcd(a_1, a_2) = 1$, this equation implies that $a_1|(c_2 + 1)$ and $a_2|(c_1 + 1)$. Therefore, $c_2 + 1 = a_1b_1$ and $c_1 + 1 = a_2b_2$ with positive integers b_1 and b_2. Hence we get $(b_1 + b_2)a_1a_2 = a_1a_2$, which is a contradiction.

In order to prove that $(a_1 - 1)(a_2 - 1) - 1$ is the largest gap of H we may assume that $a_1 < a_2$, and we have to show that $a_1a_2 - a_1 - a_2 + j \in S$ for $j = 1, \dots, a_1$. Since $\gcd(a_1, a_2) = 1$, there exist integers c_1 and c_2 such that $j = c_2a_2 - c_1a_1$. Then $j = (c_2 - ka_1)a_2 - (c_1 - ka_2)a_1$ for any integer k. For a suitable choice of k, $0 \leq c_1 - ka_2 < a_2$. Hence, replacing c_1 by $c_1 - ka_2$ we may assume from the very beginning that $0 \leq c_1 < a_2$. Then necessarily $c_2 > 0$. Actually we must have $c_1 > 0$, too. Because otherwise a_2 would divide j, which is a contradiction. With these choices of c_1 and c_2 we get

$$a_1a_2 - a_1 - a_2 + j = (a_2 - c_1 - 1)a_1 + (c_2 - 1)a_2,$$

which shows that $a_1a_2 - a_1 - a_2 + j \in S$, since $a_2 - c_1 - 1 \geq 0$ and $c_2 - 1 \geq 0$. □

Exercise 4.8
Determine $F(H)$ for the numerical semigroup defined in Exercise 4.4.

In the following and later sections we will present formulas for $F(H)$ in more difficult situations. Here, for the proof of Sylvester's theorem we only needed elementary number theory. For the cases considered later we will have to use some tools from commutative algebra, demonstrating the power of algebraic methods.

4.1.2 Gaps and Non-Gaps, Symmetric Numerical Semigroups

Let H be a numerical semigroup. It is of interest to know how the gaps of H are distributed among the integers. Of course the negative numbers do not belong to H while all integers bigger than $F(H)$ belong to H. Interesting is what happens between 0 and $F(H)$. In the previous section we called the elements belonging to $\mathcal{G}(H) = \mathbb{Z}_{\geq 0} \setminus H$ the gaps of H. Where there are gaps, there should be also non-gaps. The elements in H which are smaller than $F(H)$ are called the *non-gaps*. We denote the set of non-gaps of H by $\mathcal{N}(H)$. In our running example (4.3) we have

$$\mathcal{N}(H) = \{0, 5, 7, 10, 14, 15, 17, 19, 20, 21, 22\}.$$

We denote by $g(H)$ the number of gaps and by $n(H)$ the number of non-gaps of H. The number $g(H)$ is also called the *genus* of H. Obviously, one has

$$g(H) + n(H) = F(H) + 1.$$

For 2-generated numerical semigroups we notice an interesting pattern which for the case $H = \langle 5, 7\rangle$ looks as follows:

$$\cdots - -0 - - - -5 - 7 - -10 - 12 - 14, 15 - 17 - 19, 20, 21, 22 - 24 \cdots$$

Here the gaps are indicated as bars.

At a first glance the distributions of the gaps looks pretty weird. Looking more closely at it, we notice that at the beginning there are more gaps. This is explained by a surprising property of this gap sequence, see Exercise 4.12. In this example we have

$$a \in \mathbb{Z}_{\geq 0} \text{ is a gap if and only if } F(H) - a \text{ is a non-gap.} \tag{4.4}$$

For example 8 is a gap and $23 - 8 = 15$ is a non-gap, and 14 is a non-gap and $23 - 14 = 9$ is a gap.

A numerical semigroup H satisfying (4.4) is called *symmetric*.

The numerical semigroup $H = \langle 3, 4, 5\rangle$ is not symmetric, while $H = \langle 4, 5, 6\rangle$ is symmetric.

Exercise 4.9
Let a be a positive integer and $H = \langle 2a + 1, 2a + 2, 2a + 3\rangle$. Show that H is not symmetric.

Symmetric numerical semigroups can be characterized as follows.

Proposition 4.10
Let H be a numerical semigroup. Then $g(H) \geq n(H)$. Equality holds if and only if H is symmetric.

Proof. If $a \in \mathcal{N}(H)$, then $F(H) - a \in \mathcal{G}(H)$. Indeed, suppose that $F(H) - a \notin \mathcal{G}(H)$. Then $F(H) - a \in \mathcal{N}(H)$, and hence $F(H) = (F(H) - a) + a \in \mathcal{N}(H)$, a contradiction.

Thus we have a map $\varphi : \mathcal{N}(H) \to \mathcal{G}(H)$ with $\varphi(a) = F(H) - a$. Since φ is injective we get

$$n(H) = |\mathcal{N}(H)| \leq |\mathcal{G}(H)| = g(H).$$

Moreover, $n(H) = g(H)$ if and only if φ is bijective, and this is the case if and only if H is symmetric. □

It is now clear that $1 \leq n(H) \leq (F(H) + 1)/2$ and that $n(H) < (F(H) + 1)/2$, if and only if H is not symmetric.

Exercise 4.11
Let n and f be integers with $1 \leq n \leq (f+1)/2$. Show that there exists a numerical semigroup H with $f = F(H)$ and $n = n(H)$.

Exercise 4.12
Let H be a symmetric numerical semigroup. Show that

(a) $F(H)$ is odd.

(b) The number of gaps $< F(H)/2$ is bigger than or equal to the number of gaps $> F(H)/2$.

Is statement (b) also true when H is not symmetric?

The next result shows that the symmetry in the example $\langle 5, 7 \rangle$ was not accidental.

Theorem 4.13
(Sylvester) Let H be numerical semigroup generated by 2 elements. Then H is symmetric.

Proof. Referring to the proof of Proposition 4.10 we need to show that the map $\varphi : \mathcal{N}(H) \to \mathcal{G}(H)$ is surjective. In other words, we have to show that if a is a gap of H, then $F(H) - a$ is a non-gap. Since $\gcd(a_1, a_2) = 1$, any $a \in \mathbb{Z}$ can be written uniquely as $a = c_1 a_1 + c_2 a_2$ with $c_i \in \mathbb{Z}$ and $0 \leq c_2 < a_1$, see Exercise 4.14.

Now since a is a gap, we must have $c_1 < 0$. Then, by Theorem 4.7,

$$F(H) - a = (a_1 - 1 - c_2)a_2 + (-c_1 - 1)a_1,$$

which shows that $F(H) - a \in S$, because the coefficients of a_1 and a_2 are non-negative. This completes the proof. □

Exercise 4.14
Let a_1, a_2 be positive integers with $\gcd(a_1, a_2) = 1$. Show that any $a \in \mathbb{Z}$ has a unique presentation $a = c_1 a_1 + c_2 a_2$ with integers c_1 and c_2 such that $0 \leq c_2 < a_1$.

4.1.3 The Gluing Construction

Theorem 4.13 is a very special case of a general result. It can be deduced from a construction, called *gluing*, which allows to build a new numerical semigroup from a given one. So let H be a numerical semigroup. We choose $a \in H$ and a positive integer b, and let

$$H' = \langle a, bH \rangle.$$

In other words, if $H = \langle a_1, \ldots, a_m \rangle$, then

$$H' = \langle a, ba_1, \ldots, ba_m \rangle.$$

The choice of a and b should be such that $\gcd(a, ba_1, \ldots, ba_m) = 1$, which is the case if and only if $\gcd(a, b) = 1$. Then H' is a numerical semigroup, and we call H' the *gluing* of H induced by the pair (a, b).

For example the semigroup $\langle 4, 5, 6 \rangle$ is the gluing of $\langle 2, 3 \rangle$ induced by $(5, 2)$, and the 2-generated numerical semigroup $H = \langle a_1, a_2 \rangle$ can be viewed as the gluing of the numerical semigroup $\mathbb{Z}_{\geq 0} = \langle 1 \rangle$ induced by (a_1, a_2).

For the gluing construction we have the following result.

Theorem 4.15
Let H be a numerical semigroup and $H' = \langle a, bH \rangle$ be the gluing of H induced by (a, b). Then

(a) *(Johnson, Brauer and Shockley)* $F(H') = bF(H) + (b - 1)a$.
(b) H' is symmetric if and only if H is symmetric.

Let us show how we recover Theorems 4.7 and 4.13 from this result: as we noticed above, $H = \langle a_1, a_2 \rangle$ is the gluing of the numerical semigroup $\langle 1 \rangle$ induced by (a_1, a_2). Observe that $F(\langle 1 \rangle) = -1$. Actually, this is the only case when the Frobenius number is negative. Anyway, $\langle 1 \rangle$ is symmetric by the trivial reason that $\langle 1 \rangle$ has no gaps. Thus, Theorem 4.15(b) implies that H is symmetric and Theorem 4.15(a) implies that

$$F(H) = a_2 F(\langle 1 \rangle) + (a_2 - 1)a_1 = -a_2 + (a_2 - 1)a_1 = (a_1 - 1)(a_2 - 1) - 1,$$

which is Sylvester's formula.

In Chap. 5 we will give a proof of this theorem by using methods from commutative algebra. This will demonstrate the power of algebraic techniques and will be a good motivation to get deeper in it.

An interesting family of numerical semigroups one obtains by applying this gluing operation recursively. In order to construct a numerical semigroup in this family, we choose for $i = 1, \ldots, r$ a sequence of pairs of positive integers (a_i, b_i) with $\gcd(a_i, b_i) = 1$ and $b_i > 1$ which produce a sequence of numerical semigroups H_i, starting with $H_0 = \langle 1 \rangle$. Suppose H_i with $i < r$ is already constructed. Then we let $H_{i+1} = \langle a_{i+1}, b_{i+1} H_i \rangle$ be the gluing of H_i induced by (a_{i+1}, b_{i+1}). Of course in each step it is required that $a_{i+1} \in H_i$.

The following example illustrates this construction. We let $(a_1, b_1) = (5, 7)$. Then $H_1 = \langle 5, 7 \rangle$. Now we choose $(a_2, b_2) = (12, 5)$. The choice of $a_2 = 12$ is okay, because $12 \in H_1$. Then $H_2 = \langle 12, 25, 35 \rangle$. We may proceed in this way to get semigroups with more and more generators.

Numerical semigroups of this type are precisely the complete intersections in the sense as explained in Sect. 4.3.

By using Theorem 4.15 we can compute the Frobenius number of any complete intersection numerical semigroup. In the above example with $H = H_2 = \langle 12, 25, 35\rangle$ we have $F(H_1) = 23$, $a_2 = 12$ and $b_2 = 5$, so that

$$F(H) = 5 \cdot 23 + 4 \cdot 12 = 163.$$

This is in accordance with our calculation with *numericalsgps*, see Sect. 4.1.1.

Exercise 4.16
Let a be an odd positive integer and $H = \langle 2a, 2a+4, 3a+4\rangle$. Then H is symmetric.

Exercise 4.17
Let $H = \langle a_1, \dots, a_m\rangle$ be a numerical semigroup, and assume that

$$\operatorname{lcm}(\gcd(a_1, \dots, a_i), a_{i+1}) \in \langle a_1, \dots, a_i\rangle$$

for $i = 1, \dots, m-1$. Then H is symmetric, and

$$F(H) = \sum_{i=1}^{m-1} \operatorname{lcm}(\gcd(a_1, \dots, a_i), a_{i+1}) - \sum_{i=1}^{m} a_i.$$

Exercise 4.18
Let $\{b_1, \dots, b_m\}$ be a set of pairwise coprime numbers, and let $H = \langle a_1, \dots, a_m\rangle$ be the numerical semigroup with $a_i = \prod_{j \neq i} b_j$. Then H is a symmetric semigroup with

$$F(H) = (m-1)\prod_{j=1}^{m} b_j - \sum_{i=1}^{m} a_i.$$

Notes

According to Brauer [9], Frobenius mentioned the Coin Exchange Problem occasionally in his lectures. This may be the reason why the largest gap in a numerical semigroup is called the Frobenius number of the numerical semigroup. There have been many efforts to compute this number. In general a simple closed formula cannot be expected. There are various computer programs that do the job for us, see for example [81]. Here we recommend the package "numericalsgps" in GAP [21] developed by Delgado et al., because it also allows us to compute many other relevant data of a numerical semigroup. In some special cases, a formula or a bound for the Frobenius number is known. Notable are the cases when the generators of H are in arithmetic progression, see for example [74], or when H is generated by 3 elements. An explicit general formula for computing the Frobenius number in terms of the least representable multiples of the three generators is given in the papers by Denham [23], Ramírez Alfonsín [61], and Tripathi and Vijay [76].

For a 3-generated symmetric numerical semigroup the first author of this book gives a formula for the Frobenius number in his dissertation [34] and conjectures the Frobenius number in the case that the semigroup is not symmetric. The conjectured formula is proved in Chap. 5 and its proof follows the arguments given in the paper [54] by Nari, Numata and K. Watanabe. An explicit formula in the symmetric case is given in Theorem 5.9.

Let $f(n)$ be the number of numerical semigroups H for which $F(H) = n$. In [80, Problem 2.1(b)] Wilf asks what is the order of magnitude of $f(n)$ for $n \to \infty$? In [67] a table is shown which gives these numbers for $n < 40$. It can be seen that $f(n)$ grows quickly as n gets bigger, although it is not a strictly increasing function. The growth of $f(n)$ is actually exponential. In fact, Backelin [4] showed that $2^{\lfloor (n-1)/2 \rfloor} \leq f(n) \leq 4 \cdot 2^{\lfloor (n-1)/2 \rfloor}$.

Here are a few more papers [9,10,30,55,75] dealing with the Frobenius number, where the interested reader may find further references.

In 1978, Wilf [80] conjectured the following inequality

$$n(H) \geq (F(H) + 1)/\mathrm{emb}(H).$$

This conjecture is still widely open, although it is proved in several special cases. The most exciting result in this direction is that of Eliahou [27] who proved that Wilf's conjecture holds if $F(H) + 1 \leq 3e(H)$. He also observed that, combined with a result of Zhai [82], the conjecture is asymptotically true as $g(H)$ tends to infinity. Bruns et al. [14] developed an algorithm to check Wilf's conjecture for all numerical semigroups with a given multiplicity $e(H)$. They use this algorithm to show with the help of the program Normaliz [16] (developed by Bruns and his team) that Wilf's conjecture holds true whenever $e(H) \leq 18$. The algorithm is based on results of Kunz [48] and uses techniques of polyhedral geometry.

The gluing construction, as introduced here, appeared first in the paper [77] by K. Watanabe, where he also proved part (b) of Theorem 4.15. In [30] you find an elementary and direct proof of Theorem 4.15(a). Part (a) of Theorem 4.15 goes back to Johnson [44] and Brauer and Shockley [10].

The gluing construction was later extended by Delorme [22] and Rosales [63]. The generalization is defined as follows: given numerical semigroups H_1 and H_2 and integers a_1 and a_2, the numerical semigroup $H = \langle a_1 H_1, a_2 H_2 \rangle$ is called a gluing of H_1 and H_2, if $a_1 \in H_2$ and $a_2 \in H_1$ and $\gcd(a_1, a_2) = 1$. If H_1 and H_2 are complete intersections, resp. symmetric, then H is a complete intersection, resp. symmetric, see [65, Proposition 9.9, Proposition 9.11].

4.2 The Semigroup Ring of a Numerical Semigroup and Its Relations

In this section, when studying relations of numerical semigroups, the first time commutative algebra comes into play. Indeed, for a fixed field K we will attach to a numerical semigroup H a K-algebra, denoted by $K[H]$, whose relations reflect the semigroup relations.

4.2.1 Factorizations of Semigroup Elements

Let us first consider semigroup relations itself.

We start with an example, and consider the numerical semigroup $H = \langle 3, 5, 7\rangle$. Then 20 belongs to H and it can be written in different ways as a linear combination of the generators with non-negative coefficients:

$$20 = 4 \cdot 5 = 5 \cdot 3 + 1 \cdot 5 = 2 \cdot 3 + 2 \cdot 7 = 1 \cdot 3 + 2 \cdot 5 + 1 \cdot 7.$$

Now let H be any numerical semigroup minimally generated by $a_1, \ldots, a_m$, and let $a \in \mathbb{Z}_{\geq 0}$. Then any presentation $a = \sum_{i=1}^{m} b_i a_i$ with $b_i \in \mathbb{Z}_{\geq 0}$ is called a *factorization* of a with respect to H. In our example we have 4 factorizations of 20 with respect to $\langle 3, 5, 7\rangle$. We denote by $\rho_H(a)$ the number of factorizations of a with respect to H.

Exercise 4.19
Let $H = \langle a_1, a_2\rangle$ and $a \in \mathbb{Z}_{\geq 0}$. Determine $\rho_H(a)$.

Is there a way to determine the number of factorizations of a in general? More precisely, what can we say about the number $\rho_H(a)$, which for $a \in \mathbb{Z}$ counts the number of factorizations of a with respect to H? Of course, $\rho_H(a) \neq 0$, if and only if $a \in H$. In particular, $F(H) = \max\{a \;\; \rho_H(a) = 0\}$.

In our example, when $H = \langle 3, 5, 7\rangle$, the values of $\rho_H(a)$ for $0 \leq a \leq 15$ are:

a	0	1	2	3	4	5	6	7	8	9	10	11	12	13	14	15
$\rho_H(a)$	1	0	0	1	0	1	1	1	1	1	2	1	2	2	2	3

The *enumerating function* for the sequence $\rho_H(a)$, $a = 0, 1, 2, \ldots$ is the formal power series

$$\rho_H(\lambda) = \sum_{a=0}^{\infty} \rho_H(a)\lambda^a.$$

We have the following

Lemma 4.20
Let $H = \langle a_1, \ldots, a_m\rangle$. Then

$$\rho_H(\lambda) = \frac{1}{\prod_{i=1}^{m}(1 - \lambda^{a_i})}.$$

Proof. For the proof we simply observe that $\frac{1}{1-\lambda^{a_i}} = \sum_{k=0}^{\infty} \lambda^{ka_i}$, so that

$$\frac{1}{\prod_{i=1}^{m}(1-\lambda^{a_i})} = \prod_{i=1}^{m}(\sum_{k=1}^{\infty} \lambda^{ka_i}) = \sum_{a=0}^{\infty} \rho_H(a)\lambda^a = \rho_H(\lambda).$$ □

The set of all factorizations of an element in a numerical semigroup can be determined by linear programming. Indeed, the factorizations of $a \in S$ can be identified with the coordinates $(c_1, \ldots, c_n)$ of integer points of the convex polytope determined by the equation $\sum_{i=1}^{n} c_i a_i = a$ and the inequalities $0 \leq c_i$ for all i.

4.2.2 Semigroup Relations

Here we are interested in a different aspect: let $H = \langle a_1, \ldots, a_m \rangle$ be a numerical semigroup. We consider the vector $\mathbf{a} = (a_1, \ldots, a_m) \in \mathbb{Z}_{\geq 0}^m$. A *relation* of H is a pair $(\mathbf{v}, \mathbf{w})$ with $\mathbf{v}, \mathbf{w} \in \mathbb{Z}_{\geq 0}^m$ and such that

$$\mathbf{v} \cdot \mathbf{a} = \mathbf{w} \cdot \mathbf{a}.$$

Here, this product denotes the standard scalar product of two vectors in $\mathbb{R}^m$. The common value of the above scalar products on the left and right side of the equation is called the *degree of the relation* $(\mathbf{v}, \mathbf{w})$.

So whenever we have two presentations of an element in H, then we have a relation. For example, for $20 \in \langle 3, 5, 7 \rangle$ we had the presentations: $20 = 5 \cdot 3 + 1 \cdot 5$ and $20 = 1 \cdot 3 + 2 \cdot 5 + 1 \cdot 7$. This gives rise to the relation $((5, 1, 0), (1, 2, 1))$ of degree 20.

Obviously, the set $\mathscr{R}$ of relations of H is a set of equivalence relations on $\mathbb{Z}_{\geq 0}^m$. In other words, we have:

(1) $(\mathbf{v}, \mathbf{v}) \in \mathscr{R}$, for all $\mathbf{v} \in \mathbb{Z}_{\geq 0}^m$,

(2) if $(\mathbf{v}, \mathbf{w}) \in \mathscr{R}$, then $(\mathbf{w}, \mathbf{v}) \in \mathscr{R}$,

(3) if $(\mathbf{v}, \mathbf{w}), (\mathbf{w}, \mathbf{u}) \in \mathscr{R}$, then $(\mathbf{v}, \mathbf{u}) \in \mathscr{R}$.

In addition, the relations of H satisfy property (4) which says that if $(\mathbf{v}, \mathbf{w}) \in \mathscr{R}$ and $\mathbf{u} \in \mathbb{Z}_{\geq 0}^m$, then $(\mathbf{v} + \mathbf{u}, \mathbf{w} + \mathbf{u}) \in \mathscr{R}$.

For any set $\mathscr{S} \subset \mathbb{Z}_{\geq 0}^m \times \mathbb{Z}_{\geq 0}^m$ there exists a smallest set $\widetilde{\mathscr{S}} \subset \mathbb{Z}_{\geq 0}^m \times \mathbb{Z}_{\geq 0}^m$ containing $\mathscr{S}$, which satisfies the conditions (1), (2), (3) and (4). A subset $\mathscr{S} \subset \mathscr{R}$ with $\widetilde{\mathscr{S}} = \mathscr{R}$ is called a *set of generators of the relations* of H. Rédei [62] proved that for a finitely generated commutative semigroup, finitely many relations generate its set of relations. In particular, Redei's theorem applies to numerical semigroups.

With *numericalsgps* we can compute the finitely many generating relations of a numerical semigroup, as the following example demonstrates:

```
gap> LoadPackage("numericalsgp");
gap> s:=NumericalSemigroup(20,27,38,69);
<Numerical semigroup with 4 generators>
```

```
gap> MinimalPresentationOfNumericalSemigroup(s);
[ [ [ 0, 7, 0, 0 ], [ 6, 0, 0, 1 ] ],
[ [ 1, 0, 2, 0 ],[ 0, 1, 0, 1 ] ],
[ [ 3, 2, 0, 0 ], [ 0, 0, 3, 0 ] ],
[ [ 4, 1, 0, 0 ], [ 0, 0, 1, 1 ] ],
[ [ 5, 0, 1, 0 ], [ 0, 0, 0, 2 ] ],
[ [ 10, 0, 0, 0 ],[ 0, 6, 1, 0 ] ] ]
```

4.2.3 Numerical Semigroup Rings and Their Defining Ideal

The theorem of Rédei is most comfortably proved by using commutative algebra. For this purpose we fix a field K. In most cases it is not relevant which field we choose, although occasionally the field's characteristic or its finiteness is pertinent. Nonetheless, in most cases, we may assume K to be the field of rational numbers, $\mathbb{Q}$.

Now, let $H = \langle a_1, \ldots, a_m \rangle$ be a numerical semigroup, and let $K[H]$ be the K-subalgebra of the polynomial ring $K[t]$ which is generated over K by the monomials t^{a_i} for $i = 1, \ldots, m$. Thus,

$$K[H] = K[t^{a_1}, \ldots, t^{a_m}].$$

The K-algebra $K[H]$ is called the *semigroup ring* of H (over K). It is a graded K-algebra with grading

$$K[H]_a = \begin{cases} 0, & \text{if } a \notin H, \\ Kt^a, & \text{if } a \in H. \end{cases}$$

The ideal $\mathfrak{m} = (t^{a_1}, \ldots, t^{a_m})$ is the graded maximal ideal of $K[H]$. The minimal number of generators $\mu(\mathfrak{m})$ of $\mathfrak{m}$ is called the *embedding dimension* of $K[H]$. It is also the embedding dimension of H, as defined in Sect. 4.1.1.

The Krull dimension of $K[H]$ is given in

Lemma 4.21
Let H be a numerical semigroup. Then $\dim K[H] = 1$ and $K[H]$ is a Cohen-Macaulay ring.

Proof. Since $K[H]$ is a subring of $K[t]$, it is a domain. Moreover, $K[H]$ has the same quotient field as $K[t]$, namely $K(t)$. Therefore, by Corollary 1.92, $\dim K[H] = \operatorname{trdeg}(K(t)/K) = 1$.

Since $K[H]$ is a domain, we have $0 < \operatorname{depth} K[H] \leq \dim K[H] = 1$. This implies that $K[H]$ is Cohen-Macaulay. □

The ideal $C = (t^a \ a \geq F(H)+1)$ is called the *conductor* of $K[H]$. With respect to inclusion it is the largest ideal in $K[H]$ with the property that $CK[t] = C$.

We have

$$n(H) = \dim_K K[H]/C \quad \text{and} \quad g(H) = \dim_K K[t]/K[H],$$

so that

$$\dim_K K[H]/C \leq \dim_K K[t]/K[H],$$

with equality if and only if H is symmetric, see Proposition 4.10.

The semigroup ring $R = K[H]$, like any other K-algebra, can be written as a factor ring of a polynomial ring over K. Let H be minimally generated by $a_1, \dots, a_m$. We let $S = K[x_1, \dots, x_m]$ be the polynomial ring over K in m variables, and define the K-algebra homomorphism

$$\varphi \ S \to R \quad \text{with} \quad x_i \mapsto t^{a_i} \quad \text{for} \quad i = 1, \dots, m.$$

Since φ is surjective, it follows that $R \cong S/I$, where $I = \operatorname{Ker}\varphi$. We call I the *defining ideal* of $K[H]$, and we denote it further on by I_H. Since R is a domain, I_H is a prime ideal. If we assign to each variable x_i the degree a_i, then S becomes a (non-standard) graded K-algebra, φ a graded K-algebra homomorphism and I_H a graded ideal of S. In the given grading, a monomial $u = x_1^{c_1} \cdots x_m^{c_m}$ is of degree d, if $\sum_{i=1}^m c_i a_i = d$, and a polynomial $f \in S$ is homogeneous of degree d if all monomials appearing in f are of degree d.

Exercise 4.22

Let $H = \langle a_1, \dots, a_m \rangle$ be a numerical semigroup, and $S = K[x_1, \dots, x_m]$ be the polynomial ring with $\deg x_i = a_i$ for $i = 1, \dots, m$. Show that for the Hilbert function of S we have $H(S, a) = \rho_H(a)$ for all a.

In order to describe I_H, we consider the hyperplane $\mathbf{a}^{\top}$ consisting of all vectors $\mathbf{v} \in \mathbb{Q}^m$ with $\mathbf{a} \cdot \mathbf{v} = 0$. A vector $\mathbf{v} = (c_1, \dots, c_m) \in \mathbb{Q}^m$, can be written as $\mathbf{v}^+ - \mathbf{v}^-$, where $\mathbf{v}^+ = (d_1, \dots, d_m)$ with $d_i = c_i$ for $c_i \geq 0$ and $d_i = 0$ if $c_i < 0$, and where $\mathbf{v}^- = \mathbf{v}^+ - \mathbf{v}$. Then both vectors $\mathbf{v}^+$ and $\mathbf{v}^-$ are vectors with non-negative entries. For example, if $\mathbf{v} = (-1, 0, 2, 3, -4)$, then $\mathbf{v}^+ = (0, 0, 2, 3, 0)$ and $\mathbf{v}^- = (1, 0, 0, 0, 4)$.

To an integer vector $\mathbf{v} = (c_1, \dots, c_m) \in \mathbb{Z}^m$, we assign the binomial

$$f_{\mathbf{v}} = \prod_{\substack{i \\ c_i > 0}} x_i^{c_i} - \prod_{\substack{i \\ c_i < 0}} x_i^{-c_i}.$$

By using the notation $\mathbf{x}^{\mathbf{c}} = \prod_{i=1}^m x_i^{c_i}$ for $\mathbf{c} = (c_1, \dots, c_m) \in \mathbb{Z}^m_{\geq 0}$, the binomial $f_{\mathbf{v}}$ can be rewritten as $f_{\mathbf{v}} = \mathbf{x}^{\mathbf{v}^+} - \mathbf{x}^{\mathbf{v}^-}$.

With the grading of S given by $\deg(x_i) = a_i$ for $i = 1, \dots, m$, the binomial $f_{\mathbf{v}}$ is homogeneous of degree d if and only if $d = \mathbf{v}^+ \cdot \mathbf{a} = \mathbf{v}^- \cdot \mathbf{a}$, where $\mathbf{a} = (a_1, \dots, a_m)$.

Note that any binomial $f \in S$ can be written as $f = u f_{\mathbf{v}}$ for a unique vector $\mathbf{v} \in \mathbb{Z}^m$ and a unique monomial $u \in S$.

Theorem 4.23
Let $H = \langle a_1, \dots, a_m \rangle$ be a numerical semigroup, and let $\mathbf{a} = (a_1, \dots, a_m)$. Then the following statements hold:

(a) I_H is generated by finitely many binomials $f_\mathbf{v}$ with $\mathbf{v} \in \mathbf{a}^\top$;
(b) the binomials $f_{\mathbf{v}_1}, \dots, f_{\mathbf{v}_r}$ form a set of generators of I_H if and only if the pairs

$$(\mathbf{v}_1^+, \mathbf{v}_1^-), \dots, (\mathbf{v}_r^+, \mathbf{v}_r^-)$$

form a set of generators for the set of relations $\mathscr{R}$ of H.

Proof. (a) First we show that the binomials $f_\mathbf{v}$ with $\mathbf{v} \in \mathbf{a}^\top$ generate I_H. Since I_H is a graded ideal, it is enough to show that if we take a homogeneous polynomial $f \in I_H$, then f is a linear combination of the binomials $f_\mathbf{v}$. So let $f = \sum_{i=1}^r \lambda_i u_i$, where each $\lambda_i \in K$ and each of the u_i is a monomial in S of degree a (in the given grading). We may assume that $\lambda_i \neq 0$ for all i. Since $0 = \varphi(f)$ and $\varphi(u_i) = t^a$ for all i, we see that $\sum_{i=1}^r \lambda_i = 0$. Thus we may rewrite f as follows:

$$f = \sum_{i=1}^r \lambda_i (u_i - u_1).$$

Let $w_i = \gcd(u_1, u_i)$. Then $u_i - u_1 = w_i f_{\mathbf{v}_i}$ with $\mathbf{v}_i \in \mathbf{a}^\top$, as desired. But why does $\mathbf{v}_i$ belong to $\mathbf{a}^\top$? To see this, note that

$$0 = t^a - t^a = \varphi(u_i) - \varphi(u_1) = \varphi(w_i)\varphi(f_{\mathbf{v}_i}),$$

which implies that $t^{\mathbf{v}_i^+ \cdot \mathbf{a}} - t^{\mathbf{v}_i^- \cdot \mathbf{a}} = \varphi(f_{\mathbf{v}_i}) = 0$, since $\varphi(w_i) \neq 0$. Therefore, $\mathbf{v}_i^+ \cdot \mathbf{a} = \mathbf{v}_i^- \cdot \mathbf{a}$, and hence $\mathbf{v} \in \mathbf{a}^\top$.

Now, we show that I_H is generated by finitely many $f_\mathbf{v}$ with $\mathbf{v} \in \mathbf{a}^\top$. Let $\mathfrak{m}$ be the graded maximal ideal of S. Since the binomials $f_\mathbf{v}$ with $\mathbf{v} \in \mathbf{a}^\top$ generate I_H, their residue classes in the K-vector space $I_H/\mathfrak{m}I_H$ generate this K-vector space. Since S is Noetherian, the ideal I_H is finitely generated, and so $I_H/\mathfrak{m}I_H$ is a finitely generated K-vector space. From linear algebra we know that a basis of a finitely generated K-vector space can be chosen among the elements of any set of generators. From this and the Lemma of Nakayama (Corollary 1.24) we conclude that there exist $\mathbf{v}_1, \dots, \mathbf{v}_r \in \mathbf{a}^\top$ such that $I_H = (f_{\mathbf{v}_1}, \dots, f_{\mathbf{v}_r})$.

Solving Exercise 1.13 gives another proof of the fact that I_H is generated by finitely many $f_\mathbf{v}$ with $\mathbf{v} \in \mathbf{a}^\top$.

(b) Note that $(\mathbf{u}, \mathbf{v}) \in \mathscr{R}$ if and only if the binomial $\mathbf{x}^\mathbf{u} - \mathbf{x}^\mathbf{v} \in I_H$. By [36, Lemma 2.8], $\mathbf{x}^\mathbf{u} - \mathbf{x}^\mathbf{v} \in I_H$ if and only if there exists a subset $\{i_1, \dots, i_s\} \subset [r]$ and monomials $\mathbf{x}^{\mathbf{b}_k}$ for $k = 1, \dots, s$ such that

$$\mathbf{x}^{\mathbf{u}} - \mathbf{x}^{\mathbf{v}} = \sum_{k=1}^{s} \epsilon_k \mathbf{x}^{\mathbf{b}_k} f_{\mathbf{v}_{i_k}}.$$

with $\epsilon_k \in \{1, -1\}$. This yields the desired conclusion. □

The proof of Theorem 4.23 shows that Redei's theorem applied to the relations of a numerical semigroup follows from Nakayama's Lemma and the fact that a polynomial ring is a Noetherian ring.

Corollary 4.24
The ideal I_H is generated by homogeneous binomials.

For the numerical semigroup $H = \langle 20, 27, 38, 69\rangle$, whose semigroup relations we determined with *numericalsgps*, our computation together with Theorem 4.23 tells us that the defining ideal of $R = K[t^{20}, t^{27}, t^{38}, t^{69}]$ is minimally generated by

$$x_2^7 - x_1^6x_4, \quad x_1x_3^2 - x_2x_4, \quad x_1^4x_2 - x_3x_4, \quad x_1^5x_3 - x_4^2, \quad x_1^{10} - x_2^6x_3.$$

When looking at these relations one may ask how is it possible to find them? In general this is indeed a difficult problem. It is one thing to prove, as we did, that there exists a minimal system of generators of binomials of the form $f_{\mathbf{v}}$ for the defining ideal of a numerical semigroup ring and that such a system is finite, but another thing is to actually find such a system of generators. In the following example we consider a very simple case.

Example 4.25
Let $H = \langle a_1, a_2\rangle$ and $R = K[x_1, x_2]/I_H$. Then $I_H = (x_1^{a_2} - x_2^{a_1})$. Indeed, $\varphi(x_1^{a_2} - x_2^{a_1}) = (t^{a_1})^{a_2} - (t^{a_2})^{a_1} = 0$, so that $x_1^{a_2} - x_2^{a_1} \in I_H$. Since $\gcd(a_1, a_2) = 1$, the polynomial $f = x_1^{a_2} - x_2^{a_1}$ is irreducible. Therefore, f generates a prime ideal $P \subseteq I_H$ which by Krull's principal ideal theorem is of height 1. Since $\dim R = 1$, (see Lemma 4.21), and since $\dim K[x_1, x_2] = 2$, we see that height$I_H = 1$, as well. Therefore, $P = I_H$, and this is what we wanted to show.

Exercise 4.26
(a) Show that $x_1^{a_2} - x_2^{a_1}$ is indeed an irreducible polynomial in $K[x_1, x_2]$ when $\gcd(a_1, a_2) = 1$, as we claimed in Example 4.25.

(b) Try to prove directly that $K[t^{a_1}, t^{a_2}] \cong K[x_1, x_2]/(x_1^{a_2} - x_2^{a_1})$ without using Krull's principal ideal theorem.

Exercise 4.27
Let H be a numerical semigroup generated by m elements. Show that all minimal sets of generators of I_H have the same cardinality, say ℓ, and that $\ell \geq m - 1$.

Hint: Use Corollary 1.24 and Krull's generalized principal ideal theorem (Theorem 1.69). A direct proof in terms of semigroup relations would be hard.

Exercise 4.28
Let $H = \langle 6, 10, 15\rangle$. Show that there is not a unique minimal set of binomials generators of I_H.

Exercise 4.29
Let H be a numerical semigroup. Show that I_H does not contain any monomial.

Theorem 4.23 describes the generators of $I_H \subset S = K[x_1, \dots, x_n]$ for the numerical semigroup $H = \langle a_1, \dots, a_n\rangle$. But we can also describe a K-basis of each graded component

$$I_a = \{f \in I_H \quad f \text{ is homogeneous with } \deg(f) = a\}.$$

Here, f is homogeneous with respect to the grading given by $\deg x_i = a_i$ for $i = 1, \dots, n$.

Proposition 4.30
With the notation introduced, let $\{u_1, \dots, u_r\}$ be the set of all monomials of degree a in S. Then $r = \rho_H(a)$ and $u_1 - u_2, \dots, u_1 - u_r$ is a K-basis of I_a.

Proof. We have the following exact sequence of finite dimensional K-vector spaces

$$0 \to I_a \to S_a \to K[H]_a \to 0.$$

By Exercise 4.22 we know that $r = \rho_H(a)$. If $a \notin H$, then $\dim_K S_a = \rho_H(a) = 0$, and hence $I_a = 0$.

Now let $a \in H$. Then $\dim_K K[H]_a = 1$, and the above exact sequence shows that $\dim_K I_a = \rho_H(a) - 1 = r - 1$. Since the $r - 1$ binomials $u_1 - u_2, \dots, u_1 - u_r$ are K-linearly independent, the desired conclusion follows. □

Notes

The formula for $\rho_H(a)$ given in Lemma 4.20, which counts the number of factorizations of a in the numerical semigroup, tells us something about the nature of the function $\rho_H \; \mathbb{Z}_{\geq 0} \to \mathbb{Z}_{\geq 0}$. Lemma 4.20 together with Corollary 1.5 in Stanley's book [72] implies that $\rho_H(a)$ is a quasipolynomial of quasiperiod N, where $N = \operatorname{lcm}(a_1, \dots, a_m)$.

Recall that a function $f \; \mathbb{Z}_{\geq 0} \to \mathbb{C}$ is called a *quasipolynomial of quasiperiod* N, if there exist polynomials $f_0, \dots, f_{N-1} \in \mathbb{C}[x]$ such that

$$f(a) = f_i(a) \quad \text{if} \quad a \equiv i \mod N.$$

The polynomials f_i are called the *constituents* of f. It would be interesting to determine explicitly the constituents describing the quasipolynomial $\rho_H(a)$ for certain classes of numerical semigroups.

Non-unique factorization properties have been studied by many authors. For the interested reader we refer to [18, 19, 31, 66].

4.3 The Relations for Some Families of Numerical Semigroups

As we mentioned already, it is a difficult task to determine the relations of a numerical semigroup. Certainly with *numericalsgps* we can compute very efficiently the relations in each concrete case. But if we are dealing with infinite families of numerical semigroups, we should find an abstract description of their relations. In this section we consider few such cases.

4.3.1 Relations Arising from the Gluing Construction

We start with a numerical semigroup $H = \langle a_1, \ldots, a_m \rangle$ and a pair (a, b) of integers with $a \in H, b > 1$ and $\gcd(a, b) = 1$, and consider the gluing $H' = \langle a, bH \rangle$ of H with respect to (a, b). Since $a \in H$ we can write $a = \sum_{i=1}^{m} c_i a_i$ with $c_i \in \mathbb{Z}_{\geq 0}$ for all i. Then $ba = \sum_{i=1}^{m} c_i(ba_i)$.

The ideal $I_{H'}$ is the kernel of the K-algebra homomorphism

$$\varphi'\ S' = K[x_1, \ldots, x_{m+1}] \to R' = K[H']$$

with $\varphi'(x_i) = t^{ba_i}$ for $i = 1, \ldots, m$ and $\varphi'(x_{m+1}) = t^a$.

Theorem 4.31
With the notation introduced, we have

$$I_{H'} = (I_H, f)S', \quad \text{where} \quad f = x_{m+1}^b - \prod_{i=1}^{m} x_i^{c_i}.$$

Proof. Since any $g \in I_H$ belongs to $I_{H'}$ and since $f \in I_{H'}$, we conclude that $(I_H, f)S' \subseteq I_{H'}$.

Conversely, let $g \in I_{H'}$. We know that $I_{H'}$ is generated by binomials. Thus we may assume that g is a binomial, and that there is no common factor among the monomial terms in g, see Theorem 4.23.

If $g \in I_H$, we are done. Otherwise, $g = x_{m+1}^c u - v$ with $c > 0$ and with u, v monomials in $S = K[x_1, \ldots, x_m]$. Let $u = \prod_{i=1}^m x_i^{b_i}$ and $v = \prod_{i=1}^m x_i^{d_i}$. Then

$$ac + \sum_{i=1}^{m} b_i(ba_i) = \sum_{i=1}^{m} d_i(ba_i).$$

Since $\gcd(a, b) = 1$, it follows that $c = be$ for some positive integer e. Note that f divides the binomial

$$f^{[e]} := (x_{m+1}^b)^e - (\prod_{i=1}^{m} x_i^{c_i})^e$$

and that $g - f^{[e]}u \in I_H$, because both g and $f^{[e]}u$ are homogeneous of same degree. Therefore, $g \in (I_H, f)S'$, as desired. □

This theorem has the following nice consequence.

Corollary 4.32
Let H be the numerical semigroup which is obtained from $\mathbb{Z}_{\geq 0}$ by iterating the gluing construction with respect to the pairs of positive integers (a_i, b_i) with $\gcd(a_i, b_i) = 1$ and $b_i > 1$ for $i = 1, \ldots, r$. Then I_H is generated by binomials of the form

$$x_2^{b_1} - u_1,\ x_3^{b_2} - u_2,\ \ldots,\ x_{r+1}^{b_r} - u_r,$$

where u_i is a monomial in $K[x_1, \ldots, x_i]$ for $i = 1, \ldots, r$.

For a numerical semigroup H the minimal number of generators of I_H determines whether $K[H]$ is a complete intersection.

Proposition 4.33
Let $H = \langle a_1, \ldots, a_{r+1}\rangle$ be a numerical semigroup. Then $K[H]$ is a complete intersection if and only if I_H is generated by r elements.

Proof. We have $K[H] \cong S/I_H$, where $S = K[x_1, \ldots, x_{r+1}]$. Since $\dim K[H] = 1$ (Lemma 4.21), and $\dim S = r + 1$ (Corollary 1.73), it follows from Theorem 3.20 that I_H is of height r. Thus I_H is generated by r elements if and only if I_H is of principal class. Using Lemma 4.21 and Proposition 3.21 it follows that $K[H]$ is a complete intersection if and only if I_H is of principal class, and the latter is equivalent to I_H being generated by r elements. □

The numerical semigroup H is called a *complete intersection*, if $K[H]$ is a complete intersection.

The following theorem gives a characterization of complete intersection numerical semigroups.

Theorem 4.34
A numerical semigroup H is a complete intersection if and only if H is obtained from $\mathbb{Z}_{\geq 0}$ by a sequence of iterating gluings.

The 'if' part of the proof follows from Proposition 4.33 together with Corollary 4.32. The reader, who is curious to see the complete proof, may consult the original paper [22] of Delorme, where this result is proved.

Exercise 4.35
Let $H = \langle 16, 17, 18, 30 \rangle$. Show that H is a complete intersection and determine the generators of I_H.

Theorem 4.34 together with Theorem 4.15, whose proof will be given in Chap. 5, shows that if a numerical semigroup is a complete intersection, then it is symmetric. The converse is not true, as the example in Exercise 4.36 shows.

Exercise 4.36
Let $H = \langle 5, 6, 7, 8 \rangle$. Show that H is symmetric but not a complete intersection.

One special case of Theorem 4.34 is notable.

Corollary 4.37
Let H be a numerical semigroup minimally generated by a_1, a_2, a_3. Then H is a complete intersection if and only if there exists $1 \leq i \leq 3$ such that $a_i \in \langle a_j/b, a_k/b \rangle$, where $\{j, k\} = [3] \setminus \{i\}$ and $b = \gcd(a_j, a_k)$.

Proof. By Theorem 4.34, H is a complete intersection if and only if H is obtained by gluing from a numerical semigroup H_1 as $H = \langle a_i, bH_1 \rangle$ for some $i \in [3]$ and some integer b with $\gcd(a_i, b) = 1$. This is equivalent to say that b divides a_j and a_k, $H_1 = \langle a_j/b, a_k/b \rangle$ and $a_i \in H_1$. To have H_1 as a numerical semigroup is equivalent to $\gcd(a_j/b, a_k/b) = 1$ which means $\gcd(a_j, a_k) = b$. □

4.3.2 The Generating Relations of 3-Generated Numerical Semigroup Rings

By Theorem 4.34, a numerical semigroup, minimally generated by 3 elements, is a complete intersection if and only if it is of the form as described in Corollary 4.37. But what can we say in general about the number of generators of I_H, when H is a numerical semigroup generated by 3 elements?

In order to answer this question, we let H be a numerical semigroup, minimally generated by a_1, a_2, a_3. For $i = 1, 2, 3$, let c_i be the smallest positive integer for which there exist non-negative integers r_{ij} such that

$$c_i a_i = r_{il} a_l + r_{ik} a_k \quad \text{with} \quad k, l \neq i. \tag{4.5}$$

Theorem 4.38
With the notation introduced one has:

(a) If $r_{ij} = 0$ for some i and j, then H is a complete intersection.
(b) If $r_{ij} > 0$ for all i and j, then the following holds:

 (i) Let $\mathbf{v}_1 = (-c_1, r_{12}, r_{13})$, $\mathbf{v}_2 = (r_{21}, -c_2, r_{23})$ and $\mathbf{v}_3 = (r_{31}, r_{32}, -c_3)$. Then $\mathbf{v}_1 + \mathbf{v}_2 + \mathbf{v}_3 = 0$.
 (ii) The integers r_{ij} are uniquely determined.
 (iii) The binomials $f_{\mathbf{v}_1}$, $f_{\mathbf{v}_2}$ and $f_{\mathbf{v}_3}$ form a minimal set of generators of I_H.

Before proving Theorem 4.38, we give an example to demonstrate it. Let $H = \langle 3, 5, 7\rangle$. Then the smallest multiple of 3 which belongs to $\langle 5, 7\rangle$ is $12 = 4\cdot 3$. Indeed, $4\cdot 3 = 1\cdot 5 + 1\cdot 7$ and no smaller multiple of 3 does have such a presentation. Similarly, we find $2\cdot 5 = 1\cdot 3 + 1\cdot 7$ and $2\cdot 7 = 3\cdot 3 + 1\cdot 5$. Hence the corresponding semigroup relations are

$$\mathbf{v}_1 = (-4, 1, 1), \quad \mathbf{v}_2 = (1, -2, 1), \quad \mathbf{v}_3 = (3, 1, -2),$$

and you can see that $\mathbf{v}_1 + \mathbf{v}_2 + \mathbf{v}_3 = 0$. Theorem 4.38 tells us that

$$I_H = (x_1^4 - x_2x_3, x_2^2 - x_1x_3, x_3^2 - x_1^3x_2).$$

Proof (of Theorem 4.38). (a) We claim that $(0, -c_2, c_3)$ or $(c_1, 0, -c_3)$ or $(-c_1, c_2, 0)$ belong to $\{\mathbf{v}_1, \mathbf{v}_2, \mathbf{v}_3\}$. Without loss of generality we may assume that $r_{21} = 0$. Therefore $\mathbf{v}_2 = (0, -c_2, r_{23})$. If $r_{23} = c_3$, then our assertion follows. Thus we may assume that $r_{23} > c_3$, since c_3 is minimal. Consider $\mathbf{v}_2 + \mathbf{v}_3 = (r_{31}, r_{32} - c_2, r_{23} - c_3)$.

Since $r_{31} \geq 0$ and $r_{23}-c_3 > 0$, we get $r_{32}-c_2 < 0$. This implies that $r_{32}-c_2 \leq -c_2$, and hence $r_{32} \leq 0$. However, $r_{32} \geq 0$, and therefore $r_{32} = 0$ and $r_{31} \geq c_1$. If $r_{31} = c_1$, then $(c_1, 0, -c_3) = \mathbf{v}_3$ and our assertion follows again. Thus, let us assume that $r_{31} > c_1$. Now consider $\mathbf{v}_1 + \mathbf{v}_3 = (r_{31} - c_1, r_{12}, r_{13} - c_3)$. Since $r_{31} - c_1 > 0$ and $r_{12} \geq 0$, it follows that $r_{13} - c_3 < 0$. By the minimality of c_3 this means that $r_{13} - c_3 \leq -c_3$, and hence $r_{13} = 0$. Thus, $\mathbf{v}_1 = (-c_1, r_{12}, 0)$ and $r_{12} \geq c_2$. This, however, is a contradiction, since $\mathbf{v}_1 + \mathbf{v}_2 + \mathbf{v}_3 = (r_{31} - c_1, r_{12} - c_2, r_{23} - c_3)$ is a relation with $r_{31} - c_1 > 0$, $r_{12} - c_2 \geq 0$ and $r_{23} - c_3 > 0$. Therefore, $r_{31} = c_1$ and $(c_1, 0, -c_3) = \mathbf{v}_3$, as desired.

For our further discussion we may assume that $\mathbf{v}_1 = (-c_1, r_{12}, r_{13})$ and $\mathbf{v}_2 = (0, -c_2, c_3)$. Then

$$a_1c_1 = r_{12}a_2 + r_{13}a_3 \quad \text{and} \quad c_2a_2 = c_3a_3.$$

By the minimality of the c_i's we have $\gcd(c_2, c_3) = 1$. This implies that $a_2 = c_3b$ and $a_3 = c_2b$ for a positive integer b, and hence $a_1c_1 = (r_{12}b)c_3 + (r_{13}b)c_2$. In particular, b divides a_1c_1. Since $\gcd(a_1, a_2, a_3) = 1$, we see that $\gcd(a_1, b) = 1$ and b divides c_1. We claim that $b = c_1$. The claim implies that $a_1 \in H'$, where $H' = \langle c_2, c_3 \rangle$. Since $H = \langle a_1, bH' \rangle$, $b > 1$ and $\gcd(a_1, b) = 1$, it follows that H is obtained by a gluing. By Example 4.25, H' is a complete intersection and hence by Corollary 4.37, H is a complete intersection, as desired.

Suppose the claim does not hold. Then $b < c_1$, and it follows that $a_1 \notin \langle c_2, c_3 \rangle$. Indeed, if $a_1 \in \langle c_2, c_3 \rangle$, then $a_1 = r'c_2 + s'c_3$ with $r', s' \in \mathbb{Z}_{\geq 0}$. Multiplying this equation with b, we get $a_1b = r'a_3 + s'a_2$. This contradicts the minimality of c_1.

Now, since $\gcd(c_2, c_3) = 1$, Theorems 4.7 and 4.13 then imply that $c_2c_3 - c_2 - c_3 - a_1 = rc_2 + sc_3$ with $r, s \in \mathbb{Z}_{\geq 0}$. The integers r and s may be chosen such that $0 \leq r < c_3$. Then we get $c'c_2 = a_1 + (s+1)c_3$ with $0 \leq c' = c_3 - 1 - r < c_3$. Multiplying the last equation on both sides with b, we obtain $c'a_3 = ba_1 + (s+1)a_2$. This contradicts the minimality of c_3.

(b)(i) Suppose that $\mathbf{v} = \mathbf{v}_1 + \mathbf{v}_2 + \mathbf{v}_3 \neq 0$. Let $\mathbf{v} = (s_1, s_2, s_3)$. Then one of the components of $\mathbf{v}$, say the first component, has a different sign than the others. If $s_1 < 0$, then $-c_1 \leq -c_1 + r_{21} + r_{31} = s_1 < 0$, which by the choice of c_1 implies that $s_1 = -c_1$. This is only possible, if $r_{21} = r_{31} = 0$, a contradiction. If $s_1 > 0$, then the choice of c_1 implies that $s_1 \geq c_1$, from which it follows that $r_{21} + r_{31} \geq 2c_1$. Therefore, $r_{21} \geq c_1$ or $r_{31} \geq c_1$. We may assume that $r_{21} \geq c_1$. Then the first component of the vector $\mathbf{v}_1 + \mathbf{v}_2 = (r_{21} - c_1, -c_2 + r_{12}, r_{13} + r_{23})$ is non-negative and the last component is positive. This implies that $-c_2 + r_{12} < 0$. Since $-c_2 < -c_2 + r_{12}$, this contradicts the choice of c_2.

(b)(ii) Let $\mathbf{v}_1' = (-c_1, r_{12}', r_{13}')$. By (a) we have $\mathbf{v}_1' + \mathbf{v}_2 + \mathbf{v}_3 = 0 = \mathbf{v}_1 + \mathbf{v}_2 + \mathbf{v}_3$, and hence $\mathbf{v}_1' = \mathbf{v}_1$. Therefore, $r_{12}' = r_{12}$ and $r_{13}' = r_{13}$. Similarly, one shows that the other r_{ij} are uniquely determined.

(b)(iii) Since I_H is generated by homogeneous binomials, it suffices to show that if $f_{\mathbf{v}} \in I_H$ is homogeneous, then $f_{\mathbf{v}} \in (f_{\mathbf{v}_1}, f_{\mathbf{v}_2}, f_{\mathbf{v}_3})$. We proceed by induction on $\deg f_{\mathbf{v}}$. Let $\mathbf{v} = (s_1, s_2, s_3)$. We may assume that $s_1 > 0$ and $s_2, s_3 \leq 0$. Then

$f_{\mathbf{v}} = x_1^{s_1} - x_2^{-s_2}x_3^{-s_3}$ with $s_1 \geq c_1$. If $s_1 = c_1$, then (b) implies that $\mathbf{v} = -\mathbf{v}_1$, and we are done. If $s_1 > c_1$, then

$$f_{\mathbf{v}} + x_1^{s_1-c_1} f_{\mathbf{v}_1} = x_1^{s_1-c_1} x_2^{r_{12}} x_3^{r_{13}} - x_2^{-s_2} x_3^{-s_3} = u f_{\mathbf{w}},$$

where $u = \gcd(x_2^{r_{12}} x_3^{r_{13}}, x_2^{-s_2} x_3^{-s_3})$. Note that $u \neq 1$, since $r_{12}, r_{13} > 0$. It follows that $\deg(f_{\mathbf{w}}) < \deg f_{\mathbf{v}}$, so that $f_{\mathbf{w}} \in (f_{\mathbf{v}_1}, f_{\mathbf{v}_2}, f_{\mathbf{v}_3})$, by our induction hypothesis. Since $f_{\mathbf{v}} = u f_{\mathbf{w}} - x_1^{s_1-c_1} f_{\mathbf{v}_1}$, the desired result follows. □

Exercise 4.39

Let $0 < b < a$ be integers, and suppose that $a > 2$ and $\gcd(a, b) = 1$.

(a) Show that $\gcd(a, a+b, a+2b) = 1$ and $H = \langle a, a+b, a+2b \rangle$ is minimally generated by a, $a+b$ and $a+2b$.
(b) For which values of a and b is H a complete intersection?
(c) In all the cases for a and b, determine the ideal I_H.

Notes

The surprising and beautiful result characterizing complete intersection numerical semigroups (Theorem 4.34) was proved by Delorme [22]. Theorem 4.38 is due to the first author of this book and appeared in his paper [35] from 1970. In view of this result one may expect that the number of generators of the defining ideal of a numerical semigroup ring is bounded above by its embedding dimension or at least by a function only depending on the embedding dimension. But this is not the case. Indeed in 1975, Bresinsky [11] showed that for each integer $m \geq 4$ and each integer $r > 0$ there exists a numerical semigroup H with $\mathrm{emb}(H) = m$ and $\mu(I_H) \geq r$. In contrast to this result, Bresinsky [12] proved in the same year that $\mu(I_H) = 3$ or $\mu(I_H) = 5$, if $\mathrm{emb}(H) = 4$ and H is symmetric. This result and also Theorem 4.38 are related to the fact that for Cohen–Macaulay ideals of height 2 and for Gorenstein ideals of height 3 there exist structure theorems. In the first case this is the Hilbert–Burch theorem, see Theorem 3.30, in the second case this is the Buchsbaum–Eisenbud structure theorem [15, 3.4.1]. The proof presented here for Theorem 4.38 and also Bresinsky's proofs however do not depend on these structure theorems. A short proof of Bresinky's theorem by using the Buchsbaum–Eisenbud structure theorem is given by K. Watanabe [78].

Under additional assumptions on the structure of the semigroup H more is known about the generators of I_H. For example, if H is almost symmetric as defined in Sect. 6.2 and $\mathrm{emb}(H) = 4$, the generators of I_H can be described, see [29,42]. Also in the case that the generators of H form an arithmetic sequence, Patil [59] determined a minimal set of generators for I_H. In [32, Theorem 1.1], Gimenez, Sengupta and Srinivasan identified Patil's ideal as a sum of two determinantal ideals.

4.4 Apéry Sets and Semigroup Ideals

4.4.1 Apéry Sets and Selmer's Genus Formula

Let H be a numerical semigroup and let $a \in H$. The set

$$\mathrm{Ap}_H(a) = H \setminus (a + H)$$

is called the *Apéry set* of H with respect to a. Note that

$$\mathrm{Ap}_H(a) = \{b \in H \;\; b - a \notin H\}.$$

We will see in a moment the usefulness of this concept. But let us first prove the following nice property of Apéry sets.

Proposition 4.40
(a) For any $a \in H$ we have $|\mathrm{Ap}_H(a)| = a$. More precisely,

$$\mathrm{Ap}_H(a) = \{b_0, b_1, \ldots, b_{a-1}\},$$

where for each i, b_i is the smallest element in H with $b_i \equiv i \mod a$.
(b) $\max\{b \;\; b \in \mathrm{Ap}_H(a)\} = F(H) + a$.

Proof. (a) Choose $0 \leq i \leq a - 1$. Since $F(H) + k \in H$ for all integers $k > 0$, we may choose $b > F(H)$ with $b \equiv i \mod a$. Now let $c > 0$ be the largest integer such that $b - ca \in H$ and set $b_i = b - ca$. Then $b_i \equiv i \mod a$ and $b_i \notin a + H$ by the choice of c. Our choice of b_i shows that b_i is the smallest element in H with $b_i \equiv i \mod a$.

(b) It is obvious that $F(H)+a \in \mathrm{Ap}_H(a)$. If $b > F(H)+a$, then $b-a > F(H)$, and hence $b - a \in H$. Therefore, $b \in a + H$ and consequently $b \notin \mathrm{Ap}_H(a)$. □

In terms of the set $\mathrm{Ap}_H(a)$ the genus of H can be computed.

Theorem 4.41
(Selmer [70]) Let H be a numerical semigroup and $a \in H$. Then

$$g(H) = \frac{1}{a}\Big(\sum_{b \in \mathrm{Ap}_H(a)} b\Big) - \frac{a-1}{2}.$$

Proof. For an integer $0 \le i \le a - 1$, let $b \in \mathbb{Z}_{\ge 0}$ with $b \equiv i \mod a$. We claim that $b \in H$ if and only if $b \ge b_i$, where b_i is defined as in Proposition 4.40. In fact, the definition of b_i implies that $b \ge b_i$ if $b \in H$.

Conversely, suppose that $b \ge b_i$. Since $b - b_i \equiv 0 \mod a$, there exists an integer $k \ge 0$, such that $b - b_i = ka$. Then $b = b_i + ka$ which implies that $b \in H$.

Let $b_i = k_i a + i$ for $i = 0, \dots, a - 1$. Then the above considerations show that the set of gaps of H is the disjoint union $\bigcup_{i=0}^{a-1}\{q_i a + i \;\; q_i < k_i\}$. Therefore,

$$g(H) = \sum_{i=1}^{a-1} k_i = \frac{1}{a}\sum_{i=1}^{a-1}(b_i - i) = \frac{1}{a}(\sum_{b \in \mathrm{Ap}_H(a)} b) - \frac{a-1}{2}. \qquad \square$$

For any numerical semigroup H and $a \in H$, the package *numericalsgps* returns immediately the Apéry set of H with respect to a.

```
gap> LoadPackage("numericalsgp");
gap> s:=NumericalSemigroup(9,13,16,20);
<Numerical semigroup with 4 generators>
gap> AperyListOfNumericalSemigroupWRTElement(s,9);
[ 0, 46, 20, 39, 13, 32, 33, 16, 26 ]
```

4.4.2 Numerical Semigroups of Maximal Embedding Dimension

Let H be a numerical semigroup minimally generated by $a_1, a_2, \dots, a_m$. Then m is the embedding dimension of H and

$$e(H) = \min\{a_1, a_2, \dots, a_m\}$$

is the multiplicity of H. We will explain below the reason for this naming.

We may assume that $a_1 = e(H)$. It is then obvious that $\{0, a_2, \dots, a_m\} \subseteq \mathrm{Ap}_H(e(H))$. Hence, by Proposition 4.40 we obtain the inequality

$$\mathrm{emb}(H) \le e(H). \tag{4.6}$$

We say that H has *maximal embedding dimension* if $\mathrm{emb}(H) = e(H)$, which is equivalent to saying that H has *minimal multiplicity*. An example of a numerical semigroup with maximal embedding dimension is the semigroup $H = a + \mathbb{Z}_{\ge 0}$ for any $a \ge 2$. Another example of a numerical semigroup with maximal embedding dimension, which looks more irregular, is the semigroup

$$H = \langle 5, 9, 12, 13, 16 \rangle.$$

Exercise 4.42

Let $H = \langle a_1, \dots, a_m \rangle$ be a numerical semigroup with maximal embedding dimension and let $a_1 < a_1 < \cdots < a_m$. Show that

$$g(H) = (1/a_1)(a_2 + \cdots + a_m) - (a_1 - 1)/2 \quad \text{and} \quad F(H) = a_m - a_1.$$

Common to all numerical semigroups of maximal embedding dimension is the following result.

Theorem 4.43
Let H be a numerical semigroup of maximal embedding dimension. Then the defining ideal I_H of $K[H]$ is minimally generated by $\binom{\mathrm{emb}(H)}{2}$ binomials.

Before coming to the proof of this theorem, we generalize inequality (4.6) in the next proposition. First we give the following definitions:

A non-empty subset $L \subseteq H$ is called a *semigroup ideal* of H, if $L + H \subseteq L$. Note that if L_1 and L_2 are semigroup ideals of H, then the set

$$L_1 + L_2 := \{b_1 + b_2 \; b_1 \in L_1, \; b_2 \in L_2\}$$

is also a semigroup ideal of H. It is called the *sum* of L_1 and L_2. A subset $B \subset L$ is called a *set of generators* of the semigroup ideal of L, if $B + H = L$.

The semigroup ideal $M = H \setminus \{0\}$ is called the *maximal ideal* of H. If $H = \langle a_1, \ldots, a_m \rangle$, then M is generated by the set $\{a_1, \ldots, a_m\}$.

Exercise 4.44
Let H be a numerical semigroup, and let $L \subset H$ be a semigroup ideal. Show that L has a unique minimal set of generators.

In addition to the statement of this exercise, we have

Proposition 4.45
Let H be a numerical semigroup, and let $L \subseteq H$ be a semigroup ideal. Then L is minimally generated by at most $e(H)$ elements.

Proof. Let $B = \{b_1, \ldots, b_r\}$ be the minimal set of generators of L with $b_1 < b_2 < \cdots < b_r$. Suppose $b_i \equiv b_j \mod e(H)$ for some $i < j$. Then $b_j = b_i + ke(H)$ for some positive integer k. Since $e(H) \in H$, this shows that b_j is not a minimal generator of L, a contradiction. Hence we see that $b_i \not\equiv b_j \mod e(H)$ for $i \neq j$. Since there exists exactly $e(H)$ different residue classes modulo $e(H)$, the desired conclusion follows. □

The semigroup ideals $L \subseteq H$ are in bijection to the monomial ideals in $K[H]$. In fact, $L \subseteq H$ is a semigroup ideal, if and only if $L' = \bigoplus_{a \in L} Kt^a$ is a monomial ideal in $K[H]$. In particular, the maximal ideal M of the semigroup H corresponds to the graded maximal ideal $\mathfrak{m}$ of $K[H]$.

One more result is needed before we come to the proof of Theorem 4.43.

Theorem 4.46
Let H be a numerical semigroup, and let $\mathfrak{m}$ be the graded maximal ideal of $K[H]$. Then $\mathfrak{m}^2 = t^{e(H)}\mathfrak{m}$ if and only if H has maximal embedding dimension.

Proof. Let us first assume that H has maximal embedding dimension and that H is minimally generated by $a_1, \ldots, a_m$ with $a_1 = e(H)$. The maximal ideal $M = H \setminus \{0\}$ of H is minimally generated by $a_1, \ldots, a_m$, and by assumption $m = a_1$. The corresponding monomial ideal in $K[H]$ is the graded maximal ideal $\mathfrak{m}$ of $K[H]$, and $M + M$ corresponds to $\mathfrak{m}^2$.

We claim that $a_1 + a_1, a_1 + a_2, \ldots, a_1 + a_m$ is the set of generators of $M + M$. This claim implies that $\mathfrak{m}^2 = t^{e(H)}\mathfrak{m}$. In order to prove the claim, let $a_i + a_j$ be any of the generators of $M + M$, and assume that $a_i + a_j \equiv k \mod a_1$. Since H has maximal embedding dimension, we have $a_\ell \equiv k \mod a_1$ for some a_ℓ. Thus, $a_\ell = sa_1 + a_i + a_j$ for some $s \in \mathbb{Z}$. Since $a_1, a_2, \ldots a_m$ is a minimal system of generators of M, it follows that $s < 0$. Then $(a_1 + a_\ell) + (-s - 1)a_1 = a_i + a_j$. Since $-s - 1 \geq 0$, this shows that $a_1 + a_1, a_1 + a_2, \ldots a_1 + a_m$ is indeed the set of generators of $M + M$.

For the converse direction we use some commutative algebra. We want to show $\mu(\mathfrak{m}) = a_1$. By Nakayama's Lemma and our assumption we have $\mu(\mathfrak{m}) = \dim_K(\mathfrak{m}/\mathfrak{m}^2) = \dim_K(\mathfrak{m}/t^{a_1}\mathfrak{m})$. Thus we must show that $\dim_K(\mathfrak{m}/t^{a_1}\mathfrak{m}) = a_1$.

We have $t^{a_1}\mathfrak{m} \subset \mathfrak{m} \subset K[H]$ and $\dim_K K[H]/\mathfrak{m} = 1$. Therefore,

$$\dim_K K[H]/t^{a_1}\mathfrak{m} = \dim_K \mathfrak{m}/t^{a_1}\mathfrak{m} + 1. \tag{4.7}$$

On the other hand we have the chain of inclusions $t^{a_1}\mathfrak{m} \subset (t^{a_1}) \subset K[H]$, so that

$$\dim_K K[H]/t^{a_1}\mathfrak{m} = \dim_K K[H]/(t^{a_1}) + \dim_K(t^{a_1})/t^{a_1}\mathfrak{m}. \tag{4.8}$$

Since (t^{a_1}) is a principal ideal, Nakayama's Lemma implies that $\dim_K(t^{a_1})/t^{a_1}\mathfrak{m} = 1$. Thus, comparing (4.7) with (4.8) we see that

$$\dim_K \mathfrak{m}/t^{a_1}\mathfrak{m} = \dim_K K[H]/(t^{a_1}).$$

Notice that a K-basis of the K-vector space $K[H]/(t^{a_1})$ is given by the elements $t^b + (t^{a_1})$ with $b \in \mathrm{Ap}_H(a_1)$. By Proposition 4.40, $|\mathrm{Ap}_H(a_1)| = a_1$. Thus, the desired conclusion follows. □

Exercise 4.47
Let H be a numerical semigroup, and let M be the maximal ideal of H. Recursively, we define $jM = (j - 1)M + M$ for all $j \geq 2$. Then for all $j \gg 0$, there exists a minimal set of generators $b_0, \ldots, b_{e(H)-1}$ of jM with $b_i \equiv i \mod e(H)$. In particular, $\mu(jM) = e(H)$ for $j \gg 0$.

Now we are ready to prove Theorem 4.43. We certainly could have phrased Theorem 4.46 in semigroup terms. But for the proof of Theorem 4.43 we use commutative algebra, and for this purpose the statement of Theorem 4.46 in algebraic terms is more convenient.

Proof (of Theorem 4.43). Let H be minimally generated by $a_1, \ldots, a_m$ with $e(H) = a_1$. Then we have the exact sequence

$$0 \to I \to S \to R \to 0$$

with $R = K[H]$, $S = K[x_1, \ldots, x_m]$, $x_i \mapsto t^{a_i}$ for $i = 1, \ldots, m$, and $I = I_H$ the defining ideal of R. Since x_1 operates on R as t^{a_1}, it follows that x_1 is a non-zerodivisor on R. Thus the above exact sequence induces the exact sequence

$$0 \to I/x_1 I \to S/x_1 S \to R/t^{a_1} R \to 0,$$

see Exercise 2.14.

Note that $S/x_1 S \cong K[x_2, \ldots, x_m]$. We denote this polynomial ring by $\overline{S}$. Identifying $S/x_1 S$ with $\overline{S}$, the image of $I/x_1 I \subset S/x_1 S$ identifies with the ideal $\overline{I} \subset \overline{S}$ which is generated by the elements $\overline{f}_1, \ldots, \overline{f}_r$, where $f_1, \ldots, f_r$ is a system of generators of I and where $\overline{f} \in \overline{S}$ is obtained from $f \in S$ by substituting x_1 with 0. So for example, if $f = x_1^2 - x_2 x_3$ then $\overline{f} = -x_2 x_3$.

Since x_1 belongs to the graded maximal ideal of S we may apply the graded version of Nakayama's Lemma (Corollary 1.24), and conclude that $\mu(I) = \mu(I/x_1 I) = \mu(\overline{I})$.

So far we did not use that H has maximal embedding dimension. Thus, what we said so far is valid for any numerical semigroup, and it is a good strategy in many cases to get hold on the relations of H by looking at the generators of $\overline{I}$.

Since H has maximal embedding dimension, we know from Theorem 4.46 that $\mathfrak{m}^2 = t^{a_1}\mathfrak{m}$. Thus, if $\overline{\mathfrak{m}}$ denotes the graded maximal ideal of $R/t^{a_1}R$, then $\overline{\mathfrak{m}}^2 = 0$. Since $\overline{S}/\overline{I} \cong R/t^{a_1}R$ and since $\overline{I} \subseteq (x_2, \ldots, x_m)^2$, this implies that $\overline{I} = (x_2, \ldots, x_m)^2$ and shows that $\mu(\overline{I}) = \binom{m}{2}$, as desired. □

When we look more closely at the proof of Theorem 4.43, we see that for a numerical semigroup H with maximal embedding dimension which is minimally generated by $a_1, \ldots, a_m$, the $\binom{m}{2}$ generators of the defining ideal I_H of $K[H]$ are of the form $x_i x_j - u_{ij}$ with $2 \leq i \leq j \leq m$ and u_{ij} monomials in $S = K[x_1, \ldots, x_m]$ which are all divisible by x_1.

Let us compute the generators of the defining ideal I_H for $H = \langle 4, 5, 6, 7 \rangle$ under the K-algebra homomorphism

$$\varphi\ K[x_1, x_2, x_3, x_4] \to R \quad \text{with} \quad x_1 \mapsto t^4, \quad x_2 \mapsto t^5, \quad x_3 \mapsto t^6, \quad x_4 \mapsto t^7.$$

Here we have $\binom{4}{2} = 6$ generators. One of these generators is $x_2 x_4 - u_{24}$. But what is u_{24}? The degree of this relation is $5 + 7 = 12$, and in order to find u_{24} we must rewrite 12 as a sum of the generators involving 4. For example, $12 = 3 \cdot 4$. Thus we may choose $u_{24} = x_1^3$, and $x_2x_4 - x_1^3$ is then one of our generators. In the same way we find all the other generators and obtain

$$I = (x_2^2 - x_1x_3,\ x_2x_3 - x_1x_4,\ x_2x_4 - x_1^3,\ x_3^2 - x_1^3,\ x_3x_4 - x_1^2x_2,\ x_4^2 - x_1x_2^2).$$

Let us double check this calculation with *numericalsgps*:

```
gap> LoadPackage("numericalsgp");
gap> s:=NumericalSemigroup(4,5,6,7);
<Numerical semigroup with 4 generators>
gap> MinimalPresentationOfNumericalSemigroup(s);
[ [ [ 0, 1, 1, 0 ], [ 1, 0, 0, 1 ] ],
[ [ 0, 2, 0, 0 ], [ 1, 0, 1, 0 ] ],
[ [ 1, 2, 0, 0 ], [ 0, 0, 0, 2 ] ],
[ [ 2, 1, 0, 0 ], [ 0, 0, 1, 1 ] ],
[ [ 3, 0, 0, 0 ], [ 0, 0, 2, 0 ] ],
[ [ 3, 0, 0, 0 ], [ 0, 1, 0, 1 ] ] ]
```

We notice that, up to signs, the relations coincide. On the other hand, we should also observe that in general, even up to signs, a minimal system of binomial generators of the defining ideal of a numerical semigroup is not uniquely determined. In our example we could, if we wished, replace $x_2x_4 - x_1^3$ by $x_2x_4 - x_3^2$.

4.4.3 The Reduction Number of a Numerical Semigroup

In Theorem 4.46 we have seen that $\mathfrak{m}^2 = t^{e(H)}\mathfrak{m}$, when H has maximal embedding dimension. In fact, for any numerical semigroup H we have $\mathfrak{m}^{j+1} = t^{e(H)}\mathfrak{m}^j$ for some j. The smallest number j for which this equality holds is called the *reduction number* of H. Why does there exist such a number j? To see this, we should consider the Hilbert-Samuel function of R, which by definition is the function

$$HS(R, j) = \mu(\mathfrak{m}^j).$$

see Sect. 1.3. By Theorem 1.108, $HS(R, j)$ is a polynomial function of degree $d - 1$ for $j \gg 0$, where $d = \dim R$. Moreover, the leading coefficient of this polynomial is $e(R)/(d-1)!$, where $e(R)$ is the multiplicity of R.

In our situation, since $\dim R = 1$, we conclude that $\mu(\mathfrak{m}^j) = e(R)$ for $j \gg 0$. By Exercise 4.47, the semigroup ideals jM are minimally generated by $e(H)$ elements for $j \gg 0$. Thus we see that

$$e(H) = e(R),$$

and this justifies that we call $e(H)$ the multiplicity of H.

Proposition 4.48
Let H be a numerical semigroup. Then $\mathfrak{m}^{j+1} = t^{e(H)}\mathfrak{m}^j$ for $j \gg 0$.

Proof. For $j \gg 0$, let $b_0, \dots, b_{e(H)-1}$ be the minimal set of generators of jM with $b_i \equiv i \mod e(H)$ for all i, and similarly let $b'_0, \dots, b'_{e(H)-1}$ be the minimal set of generators of $(j+1)M$ with $b'_i \equiv i \mod e(H)$ for all i. We claim that

$$b'_i = e(H) + b_i \quad \text{for all } i \tag{4.9}$$

Indeed, since $b'_i - b_i \equiv 0 \mod e(H)$ it follows that $b'_i = ke(H) + b_i$. If $k \leq 0$, then $b_i = b'_i - ke(H) \in (j+1)M$, a contradiction. On the other hand, if $k > 1$, then

$$b'_i = (k-1)e(H) + (e(H) + b_i).$$

Since $e(H)+b_i \in (j+1)M$, it follows that b'_i is not a minimal generator of $(j+1)M$, which is again a contradiction. Thus, $b'_i = e(H) + b_i$.

The desired identity $t^{e(H)}\mathfrak{m}^j = \mathfrak{m}^{j+1}$ follows now from the Eq. (4.9). □

Exercise 4.49
Determine the reduction number of $H = \langle a_1, a_2 \rangle$.

Notes

Apéry sets are named after Roger Apéry who in his paper [1], among other facts, noticed that the value semigroup of a plane algebroid curve is symmetric. This generalizes Sylvester's theorem (Theorem 4.13). Later, in a famous paper, Kunz [48] showed that the one-dimensional analytically local rings with symmetric value semigroup are precisely the Gorenstein rings.

Apéry sets have turned out to be a powerful tool in analysing the structure of numerical semigroups. A beautiful example for this is Selmer's Theorem 4.41 from [70]. Another nice example for the use of Apéry sets is the theorem of Barucci and Fröberg [6] which, applied to numerical semigroup rings, characterizes those of them whose associated graded ring is Cohen–Macaulay.

We have seen that the multiplication map $t^{e(H)}\ \mathfrak{m}^j/\mathfrak{m}^{j+1} \to \mathfrak{m}^{j+1}/\mathfrak{m}^{j+2}$ is an isomorphism of K-vector spaces for $j \gg 0$. One may ask whether this map is injective for all $j \geq 0$, which in particular would imply that $\mu(\mathfrak{m}^j) \leq \mu(\mathfrak{m}^{j+1})$ for all $j \geq 0$. But in general this is not the case. The first example, showing that this inequality is not always satisfied was presented in [42]. The example given there is the following: Let

$$R = K[t^{30}, t^{35}, t^{42}, t^{47}, t^{148}, t^{153}, t^{157}, t^{169}, t^{181}, t^{193}].$$

Then

$$\mathfrak{m}^2 = (t^{60}, t^{65}, t^{70}, t^{72}, t^{77}, t^{82}, t^{84}, t^{89}, t^{94}),$$

and so $\mu(\mathfrak{m}) = 10$, while $\mu(\mathfrak{m}^2) = 9$. Later some other and simpler examples were given, see for example the semigroup ring

$$K[t^{15}, t^{21}, t^{23}, t^{47}, t^{48}, t^{49}, t^{50}, t^{52}, t^{54}, t^{55}, t^{56}, t^{58}]$$

by Eakin and Sathaye [25]. In 1980, Orecchia [58] showed that for each integer $m \geq 5$ there exists a reduced one-dimensional local ring of embedding dimension m and decreasing Hilbert function. On the other hand, J. Sally [69, p. 40] conjectured that the Hilbert function of a one-dimensional Cohen-Macaulay local rings of embedding dimension three is not decreasing. This was then indeed proved by J. Elias [28]. Later M.E. Rossi in [63, Problem 4.9] asked whether the Hilbert function of one-dimensional Gorenstein local rings have non-decreasing Hilbert function. In [57] Oneto, Strazzanti and Tamone give infinitely many counterexamples to this question. These counterexamples are numerical semigroups with very large embedding dimension. One may ask which is the smallest possible embedding dimension for such a counterexample. In [57] you find other interesting references dealing with this sort of problems.

Pseudo-Frobenius Numbers via Minimal Graded Free Resolutions

5

We study pseudo-Frobenius numbers, which are maximal gaps in a numerical semigroup H in the sense that adding any nonzero element of H with them produce non-gaps. We see that these numbers can be read from the minimal graded free resolution of the numerical semigroup ring $K[H]$. The set of pseudo-Frobenius numbers of a numerical semigroup H determine the canonical ideal and the type of H, key invariants that measure how far the semigroup H is from being symmetric. A proof of the Johnson-Brauer-Shockley theorem, on the Frobenius number of numerical semigroups constructed by gluing, is presented. This is followed by the study of pseudo-Frobenius numbers for 3-generated numerical semigroups. The chapter concludes by examining the gaps in non-symmetric 3-generated numerical semigroups.

5.1 The Frobenius Number and the Type of a Numerical Semigroup Ring

5.1.1 The Frobenius Number as Maximal Shift in the Minimal Graded Free Resolution

The free resolution of the defining ideal of a graded K-algebra A encodes all the important numerical invariants of A. In this section, we will demonstrate how the Frobenius number of a numerical semigroup H can be determined in terms of the minimal graded free resolution of I_H.

Let K be a field, and let $H = \langle a_1, \ldots, a_m \rangle$ be a numerical semigroup. The semigroup ring $K[H]$ is a graded K-algebra, as explained in Sect. 4.2. We let $S = K[x_1, \ldots, x_m]$ be the polynomial ring with $\deg x_i = a_i$. Then $K[H]$ becomes a graded S-module via the K-algebra homomorphism $\epsilon\colon S \to K[H]$ with $\epsilon(x_i) = t^{a_i}$.

J. Herzog et al., *Numerical Semigroups*, Compact Textbooks in Mathematics,
https://doi.org/10.1007/978-3-032-05424-1_5

Thus we may consider the minimal graded free S-resolution

$$F_{.}: 0 \to \bigoplus_j S(-j)^{b_{pj}} \xrightarrow{\varphi_p} \cdots \xrightarrow{\varphi_2} \bigoplus_j S(-j)^{b_{1j}} \xrightarrow{\varphi_1} S \to 0 \tag{5.1}$$

of $K[H]$. The numbers b_{ij}, also denoted by $\beta_{i,j}(K[H])$ are the *graded Betti numbers* of $K[H]$, while the numbers $\beta_i = \sum_j \beta_{i,j}$ are the (total) Betti numbers of $K[H]$. Furthermore, p is the projective dimension of $K[H]$. For details we refer to Chap. 2.

Since $K[H]$ is a one-dimensional integral domain, we have $\operatorname{depth}(K[H]) = 1$. Thus the Auslander-Buchsbaum formula implies that $p = m - 1$, see Theorem 3.23.

Notice that for the numerical semigroup $H = \langle a_1, \ldots, a_m \rangle$, the K-algebra $K[H]/(t^{a_1})$ is zero-dimensional.

Theorem 5.1
Let K be a field, and let $H = \langle a_1, \ldots, a_m \rangle$ be a numerical semigroup. Then

$$\beta_{m-1,j}(K[H]) = \dim_K \operatorname{Soc}(K[H]/(t^{a_1}))_{j-a_2-\cdots-a_m}.$$

Proof. Let $\overline{S} = S/(x_1) = K[x_2, \ldots, x_m]$. The K-algebra $K[H]/(t^{a_1})$ is a $\bar{S}$-module which is obtained from $K[H]$ by reduction modulo x_1 (because x_1 acts on $K[H]$ like t^{a_1}). Because x_1 is a non-zerodivisor of $K[H]$, it follows that $\operatorname{Tor}_i^S(S/(x_1), K[H]) = 0$ for $i > 0$ (see Exercise 2.25). Thus, if $F_.$ is the minimal graded free S-resolution of $K[H]$, then $H_i(F_./x_1F_.) = 0$ for $i \neq 0$ and $H_0(F_./x_1F_.) \cong K[H]/(t^{a_1})$. Hence $F_./x_1F_.$ is the minimal graded free $\bar{S}$-resolution of $K[H]/(t^{a_1})$. It follows that

$$\beta_{i,j}(K[H]) = \beta_{i,j}(K[H]/(t^{a_1})) \quad \text{for all } i \text{ and } j, \tag{5.2}$$

Therefore, Corollary 2.27 yields the desired conclusion. □

The $\overline{S}$-module $K[H]/(t^{a_1})$ has a K-basis consisting of the elements $t^a + (t^{a_1})$ with $a \in \operatorname{Ap}_H(a_1)$. It follows that $\operatorname{Soc}(K[H]/(t^{a_1}))$ is the K-vector space spanned by the elements $t^a + (t^{a_1})$ with $\{a \in \operatorname{Ap}_H(a_1) \colon\ a + M \subseteq a_1 + H\}$, where $M = H \setminus \{0\}$.

The elements of the set

$$\operatorname{PF}(H) = \{a \in \mathcal{G}(H) \colon\ a + M \subseteq H\}$$

are called the *pseudo-Frobenius numbers* of H. Obviously, the Frobenius number $F(H)$ is the maximal element in this set.

From the above considerations we deduce that

$$\{t^a + (t^{a_1}) \colon\ a \in a_1 + \operatorname{PF}(H)\}$$

is the K-basis of $\operatorname{Soc}(K[H]/(t^{a_1}))$. Combining this with Theorem 5.1, we obtain

Corollary 5.2
Let $H = \langle a_1, \ldots, a_m \rangle$ be a numerical semigroup. Then

$$\beta_{m-1,j}(K[H]) = \begin{cases} 1, & \text{if } j - a_1 - a_2 - \cdots - a_m \in \mathrm{PF}(H), \\ 0, & \text{otherwise.} \end{cases}$$

The GAP package *numericalsgps* provides a routine to compute the pseudo-Frobenius numbers of a numerical semigroup. Here is an example.

```
gap> LoadPackage("numericalsgp");
gap> s:=NumericalSemigroup(20,27,38,69);
gap> PseudoFrobeniusOfNumericalSemigroup(s);
[ 49, 142, 191 ]
```

From Corollary 5.2 we obtain the following formula for the Frobenius number of a numerical semigroup in terms of the graded Betti numbers of $K[H]$.

Corollary 5.3
Let $H = \langle a_1, \ldots, a_m \rangle$ be a numerical semigroup. Then

$$F(H) = \max\{j\colon\ \beta_{m-1,j}(K[H]) \neq 0\} - \sum_{i=1}^{m} a_i.$$

Let us demonstrate this formula with a simple example.

Let $H = \langle a_1, a_2 \rangle$ be a numerical semigroup. As shown in Sect. 4.2,

$$K[H] \cong K[x_1, x_2]/(x_1^{a_2} - x_2^{a_1}).$$

Since $\deg(x_1^{a_2} - x_2^{a_1}) = a_1 a_2$ and since $(x_1^{a_2} - x_2^{a_1})$ is a free S-submodule for $S = K[x_1, x_2]$, the graded minimal free S-resolution of $K[H]$ is given by

$$0 \longrightarrow S(-a_1 a_2) \longrightarrow S \to K[H] \longrightarrow 0.$$

Thus, Theorem 5.1 implies that $F(H) = a_1 a_2 - a_1 - a_2$, in accordance with the theorem of Sylvester (Theorem 4.7).

5.1.2 Pseudo-Frobenius Numbers and the Type of a Numerical Semigroup

The cardinality of the set $\mathrm{PF}(H)$ is called the *type* of H, and is denoted by $t(H)$.

Corollary 5.2 implies that

$$t(H) = \beta_{m-1}(K[H]). \tag{5.3}$$

In Sect. 3.2 we defined the Cohen-Macaulay type of a Cohen–Macaulay module M, denoted $r(M)$. We have

Proposition 5.4
Let H be a numerical semigroup. Then $t(H) = r(K[H])$.

Proof. The result follows from (5.3) and Corollary 3.67. □

Symmetric numerical semigroups can be characterized in terms of their type as follows.

Theorem 5.5
(Kunz [47]) Let H be a numerical semigroup. Then the following conditions are equivalent:

(i) H is symmetric.
(ii) $t(H) = 1$.
(iii) $K[H]$ is Gorenstein.

Proof. (i) $\Rightarrow$ (ii): Suppose $t(H) > 1$. Then there exists $a \in \mathrm{PF}(H)$ with $a \neq F(H)$. By definition, $a \in \mathscr{G}(H)$. Since H is symmetric and $a \neq F(H)$, it follows that $F(H) - a \in M$. But then $F(H) = a + (F(H) - a) \in H$, a contradiction.

(ii) $\Rightarrow$ (i): Suppose H is not symmetric. Then there exists $a \in \mathscr{G}(H)$ such that $F(H) - a \in \mathscr{G}(H)$. Let $a \in \mathscr{G}(H)$ be the largest element with this property. We claim that $a \in \mathrm{PF}(H)$. Indeed, suppose that $a + b \notin H$ for some $b \in M$. Then by the choice of a,

$$(F(H) - a) - b = F(H) - (a + b) \in H.$$

Since $b \in H$, we obtain $F(H) - a \in H$ which is a contradiction. So we have $a \in \mathrm{PF}(H)$. Also $a \neq F(H)$, since by assumption $F(H) - a \in \mathscr{G}(H)$. This shows that $t(H) > 1$, a contradiction. Therefore, H must be symmetric.

(ii) $\iff$ (iii): By Proposition 5.4 we have $t(H) = r(K[H])$. Therefore, the desired result follows from Theorem 3.65 and Lemma 4.21. □

5.1.3 Proof of the Theorem of Johnson, Brauer and Shockley

As an application of Theorem 5.1 we give the proof of Theorem 4.15(a) which says that if H is a numerical semigroup and $H' = \langle a, bH\rangle$ is the gluing of H induced by (a, b), then $F(H') = bF(H) + (b-1)a$.

Proof. Theorem 4.15(a) Let $I = I_H \subsetneq S = K[x_1, \ldots, x_m]$ and $I' = I_{H'} \subsetneq S' = K[x_1, \ldots, x_{m+1}]$. By Theorem 4.31 we know that $I' = (I, f)S'$, where $f = x_{m+1}^b - \prod_{i=1}^m x_i^{c_i}$ and where the exponents c_i satisfy the equation $a = \sum_{i=1}^m c_i a_i$.

We first prove that f is a regular element modulo IS'. Indeed, suppose that $gf \in IS'$ for some $g \in S'$. We must show that $g \in IS'$. Note that $S'/IS' \cong K[H][x_{m+1}]$. We denote by $\bar{g}$ and $\bar{f}$ the image of g and f in S'/IS'. Then $\bar{f} = x_{m+1}^b - u'$ for some monomial $u' \in K[H]$. This shows that $\bar{f} \neq 0$. Since $fg \in IS'$ we have $\bar{g}\bar{f} = 0$, and since $K[H][x_{m+1}]$ is a domain and $\bar{f} \neq 0$, we conclude that $\bar{g} = 0$. Therefore, $g \in IS'$, as desired.

Next observe that in S we have $\deg x_i = a_i$ for $i = 1, \ldots, m$, while in S' we have $\deg x_i = a_i b$ for $i = 1, \ldots, m$ and $\deg x_{m+1} = a$. In particular, f is homogeneous of degree ab.

Let $F_{.}$ be the minimal graded free S-resolution of S/I with $F_i = \bigoplus_j S(-j)^{\beta_{i,j}}$ for all i. Then $F'_{.} = F_{.} \otimes_S S'$ is the minimal graded free S'-resolution of S'/IS'. Since the degree of the x_i in S' is b-times the degree of the x_i in S, we see that $F'_i = \bigoplus_j S'(-jb)^{\beta_{i,j}}$ for all i. Let $G'_{.}$ be the minimal graded free S'-resolution of S'/fS'. Then we have

$$G'_{.}\colon 0 \longrightarrow S'(-ab) \xrightarrow{f} S' \longrightarrow 0.$$

Since, as noticed before, f is regular modulo IS', the resolution of $S'/(I, f)S'$ is the tensor product $F'_{.} \otimes_{S'} G'_{.}$ of the resolutions $F'_{.}$ and $G'_{.}$, see Corollary 2.28. From this it follows that

$$\begin{aligned}(F'_{.} \otimes_{S'} G'_{.})_m &= \bigoplus_j S'(-jb)^{\beta_{m-1,j}} \otimes_{S'} S'(-ab) \\ &= \bigoplus_j S'(-jb-ab)^{\beta_{m-1,j}}.\end{aligned} \tag{5.4}$$

Now, we apply Corollary 5.3 and obtain

$$\begin{aligned}F(H') &= \max\{jb + ab\colon\ \beta_{m-1,j} \neq 0\} - \sum_{i=1}^m a_i b - a \\ &= b(\max\{j\colon\ \beta_{m-1,j} \neq 0\} - \sum_{i=1}^m a_i) - a + ab = bF(H) + (b-1)a,\end{aligned}$$

as desired. □

Exercise 5.6
Let H be a numerical semigroup, and $a \in \mathscr{G}(H)$. Then there exists $b \in H$ such that $a + b \in \mathrm{PF}(H)$.

Exercise 5.7
Let H be a numerical semigroup, and assume that $\mathscr{G}(H) = \mathrm{PF}(H)$. Show that $a \in H$ if and only if $a > F(H)$.

Combining the identity (5.4) with Corollary 3.67 and Theorem 5.5 we obtain the following result, which in particular provides the proof of Theorem 4.15(b).

Corollary 5.8
Let H be a numerical semigroup and $H' = \langle a, bH \rangle$ be the gluing of H induced by (a, b). Then

$$t(H) = t(H').$$

In particular, H is symmetric if and only if H' is symmetric.

5.2 3-Generated Numerical Semigroups

5.2.1 The Pseudo-Frobenius Numbers for 3-Generated Numerical Semigroups

Next we describe the set $\mathrm{PF}(H)$, when H is a 3-generated numerical semigroup. Let $H = \langle a_1, a_2, a_3 \rangle$. There are two cases to be considered.

Case 1: H is symmetric.
Case 2: H is not symmetric.

First we discuss case 1. When H is symmetric we know from Theorem 5.5 that $t(H) = 1$ and hence the only pseudo-Frobenius number of H is $F(H)$.

In the following theorem we obtain $F(H)$ when H is symmetric.

Theorem 5.9
Let $H = \langle a_1, a_2, a_3 \rangle$ be a symmetric numerical semigroup. Then the following hold:

(a) Two of the numbers a_1, a_2, a_3 have a greatest common divisor > 1, say $b = \gcd(a_1, a_2) > 1$, and $a_3 \in H_0 = \langle a_1/b, a_2/b \rangle$. In other words, $H = \langle a_3, bH_0 \rangle$.
(b) $F(H) = ((a_1a_2)/b) - a_1 - a_2 + (b - 1)a_3$.

Proof. (a) Corollary 5.2 and Theorem 5.5 imply that

$$\beta_2(K[H]) = t(H) = 1.$$

This shows that the minimal graded free S-resolution of $K[H]$ is of the form

$$0 \to S(-d) \longrightarrow S(-d_1) \oplus S(-d_2) \longrightarrow S \longrightarrow 0 \tag{5.5}$$

for suitable numbers d_1, d_2 and d. Here we used that the projective dimension of $K[H]$ is 2 (see Theorem 3.23 and Lemma 4.21) and that for any numerical semigroup H, which is minimally generated by m elements, one has

$$\sum_{i=0}^{m-1}(-1)^i \beta_i(K[H]) = 0. \tag{5.6}$$

This formula is a special case of Corollary 2.32. The exact sequence (5.5) shows that $\mu(I_H) = 2$. Since $height(I_H) = 2$, it follows that I_H is of principal class. Thus, by Proposition 4.33, H is a complete intersection. Then by Corollary 4.37 we conclude (a).

(b) We use (a) and Theorem 4.15(a) and obtain that $F(H) = bF(H_0) + (b-1)a_3$. The desired formula follows then by Sylvester formula in Theorem 4.7, $F(H_0) = ((a_1/b) - 1)((a_2/b) - 1) - 1$. □

For example, let $H = \langle 9, 15, 25\rangle$. Then with the notation of Theorem 5.9, $a_1 = 15$, $a_2 = 25$, $b = 5$ and $a_3 = 9$. Thus $F(H) = (15 \cdot 25)/5 - 15 - 25 + 4 \cdot 9 = 71$.

Exercise 5.10
With the assumptions and notation of Theorem 5.9 determine the numbers d_1, d_2 and d in (5.5).

Now we study case 2, when $H = \langle a_1, a_2, a_3\rangle$ is not symmetric. By Theorem 4.38, I_H is minimally generated by $f_{\mathbf{v}_1}, -f_{\mathbf{v}_2}, f_{\mathbf{v}_3}$, where

$$\mathbf{v}_1 = (-c_1, r_{12}, r_{13}),\ \mathbf{v}_2 = (r_{21}, -c_2, r_{23}),\ \mathbf{v}_3 = (r_{31}, r_{32}, -c_3),$$

and these vectors satisfy $\mathbf{v}_1 + \mathbf{v}_2 + \mathbf{v}_3 = 0$. In particular,

$$c_1 = r_{21} + r_{31},\quad c_2 = r_{12} + r_{32},\quad c_3 = r_{13} + r_{23}. \tag{5.7}$$

By (5.6), the minimal graded free S-resolution of $K[H]$ is of the form

$$0 \to S(-\mu_1) \oplus S(-\mu_2) \to S(-d_1) \oplus S(-d_2) \oplus S(-d_3) \to S \to 0.$$

Here, $d_i = \deg f_{\mathbf{v}_i}$ for $i = 1, 2, 3$ and by Corollary 5.2 we know that

$$\mathrm{PF}(H) = \{\mu_1 - a_1 - a_2 - a_3, \mu_2 - a_1 - a_2 - a_3\},$$

and

$$F(H) = \max\{\mu_1 - a_1 - a_2 - a_3, \mu_2 - a_1 - a_2 - a_3\}.$$

In order to get more explicit formulas for the pseudo-Frobenius numbers we should compute μ_1 and μ_2. The graded minimal free S-resolution of $K[H]$ tells us that I_H has two relations. If we write the relations of I_H as column vectors, then we obtain a 3×2 relation matrix A describing the map

$$S(-\mu_1) \oplus S(-\mu_2) \to S(-d_1) \oplus S(-d_2) \oplus S(-d_3).$$

Two relations among the generators $f_{\mathbf{v}_1}, -f_{\mathbf{v}_2}, f_{\mathbf{v}_3}$ can be seen from (5.7). Indeed,

$$\rho_1 = (x_3^{r_{23}}, x_1^{r_{31}}, x_2^{r_{12}})^{\mathsf{T}} \text{ and } \rho_2 = (x_2^{r_{32}}, x_3^{r_{13}}, x_1^{r_{21}})^{\mathsf{T}},$$

are relations, since

$$\begin{aligned} x_3^{r_{23}} f_{\mathbf{v}_1} + x_1^{r_{31}}(-f_{\mathbf{v}_2}) + x_2^{r_{12}} f_{\mathbf{v}_3} &= 0 \text{ and} \\ x_2^{r_{32}} f_{\mathbf{v}_1} + x_3^{r_{13}}(-f_{\mathbf{v}_2}) + x_1^{r_{21}} f_{\mathbf{v}_3} &= 0. \end{aligned}$$

Assuming these are the generating relations, the transpose of the relation matrix A is

$$A^{\mathsf{T}} = \begin{pmatrix} x_3^{r_{23}} & x_1^{r_{31}} & x_2^{r_{12}} \\ x_2^{r_{32}} & x_3^{r_{13}} & x_1^{r_{21}} \end{pmatrix}. \tag{5.8}$$

But are the relations ρ_1 and ρ_2 the generating relations? Here Theorem 3.30 helps us. Let $\Delta_i(A)$ be the minor of A^{T} which is obtained from A^{T} by skipping the ith column of A^{T}. By using (5.7) one sees that $(-1)^{i+1} f_{\mathbf{v}_i} = \Delta_i(A)$ for $i = 1, 2, 3$. For example,

$$\Delta_1(A) = x_1^{r_{21}} x_1^{r_{31}} - x_3^{r_{13}} x_2^{r_{12}} = x_1^{c_1} - x_3^{r_{13}} x_2^{r_{12}} = f_{\mathbf{v}_1}.$$

Now, Theorem 3.30 applied to our case says that A is the relation matrix of the generators $f_{\mathbf{v}_1}, -f_{\mathbf{v}_2}, f_{\mathbf{v}_3}$ of I_H.

Our discussion so far shows that

$$\mu_1 = \deg(\rho_1) \text{ and } \mu_2 = \deg(\rho_2).$$

Since the relations ρ_1 and ρ_2 are homogenous (with respect to the grading $\deg(x_i) = a_i$ for $i = 1, 2, 3$), it follows that $\deg(\rho_1)$ is equal to the degree of each of the summands of $x_3^{r_{23}} f_{\mathbf{v}_1} + x_1^{r_{31}}(-f_{\mathbf{v}_2}) + x_2^{r_{12}} f_{\mathbf{v}_3}$, and $\deg(\rho_2)$ is equal to the degree of each of the summands of $x_2^{r_{32}} f_{\mathbf{v}_1} + x_3^{r_{13}}(-f_{\mathbf{v}_2}) + x_1^{r_{21}} f_{\mathbf{v}_3}$.

Thus, we obtain

Theorem 5.11

Let $H = \langle a_1, a_2, a_3 \rangle$ be a numerical semigroup, and assume that H is not symmetric. Then $\mathrm{PF}(H) = \{f, f'\}$, where

$$f = a_1c_1 - a_1 - a_2 - a_3 + a_3r_{23} \quad \text{and} \quad f' = a_1c_1 - a_1 - a_2 - a_3 + a_2r_{32}.$$

Proof. We have seen that $\mathrm{PF}(H) = \{\mu_1 - a_1 - a_2 - a_3, \mu_2 - a_1 - a_2 - a_3\}$. Furthermore, $\mu_1 = \deg(x_3^{r_{23}} f_{\mathbf{v}_1}) = r_{23}a_3 + a_1c_1$. This gives us the formula for f. In the same way we get the formula for f'. □

Exercise 5.12
Compute $\mathrm{PF}(H)$ for $H = \langle 9, 13, 17\rangle$.

5.2.2 The Number of Gaps of a Non-symmetric 3-Generated Numerical Semigroup

The relations ρ_1 and ρ_2 and Theorem 5.11 allow us to compute the number of gaps of any 3-generated non-symmetric numerical semigroup. The result is the following

Theorem 5.13
Let $H = \langle a_1, a_2, a_3\rangle$ be a numerical semigroup, and assume that H is not symmetric. Then

$$2g(H) = F(H) + 1 + r_{13}r_{32}r_{21}, \text{ if } a_3r_{23} > a_2r_{32},$$

and

$$2g(H) = F(H) + 1 + r_{12}r_{23}r_{31}, \text{ if } a_2r_{32} > a_3r_{23}.$$

Proof. Let us consider the case when $a_3r_{23} > a_2r_{32}$. (The case $a_2r_{32} > a_3r_{23}$ is treated similarly.) Theorem 5.11 implies that $f > f'$, and in particular it follows that $f = F(H)$. Let

$$A = \{f - a\colon\ a < f, a \in H\} \text{ and } A' = \{f' - a\colon\ a < f', a \in H\}.$$

Then $\mathcal{G}(H) = A \cup A'$, and hence $g(H) = |A| + |A' \setminus A|$. Furthermore, $|A| = n(H) = F(H) + 1 - g(H)$. It follows that $2g(H) = F(H) + 1 + |A' \setminus A|$. Since $f' - a \notin A$ if and only $f - (f' - a) \notin H$, we see that

$$A' \setminus A = \{f' - a\colon\ a \in H, (f - f') + a \notin H\}.$$

Hence, $|A' \setminus A|$ is equal to the cardinality of the set $B = \{a \in H\colon\ (f - f') + a \notin H\}$. Thus we must show that the set B has cardinality $r_{13}r_{32}r_{21}$.

We actually prove the following: let $p, q, r \in \mathbb{Z}_{\geq 0}$. Then

$$(f - f') + pa_1 + qa_2 + ra_3 \notin H \iff p < r_{21}, q < r_{32}, r < r_{13}. \quad (5.9)$$

Let $p, p', q, q', r, r' \in \mathbb{Z}_{\geq 0}$ with $p, p' < r_{21}$, $q, q' < r_{32}$, $r, r' < r_{13}$ and $pa_1 + qa_2 + ra_3 = p'a_1 + q'a_2 + r'a_3$. Then $f_{\mathbf{v}} \in I_H$ for $\mathbf{v} = (p - p', q - q', r - r')$.

Suppose $\mathbf{v} \neq 0$. Then Theorem 4.38(b)(iii) implies that $|p-p'| \geq c_1$ or $|q-q'| \geq c_2$ or $|r-r'| \geq c_3$. This is a contradiction, since $|p-p'| < r_{21} < c_1, |q-q'| < r_{32} < c_2$ and $|r-r'| < r_{13} < c_3$, as follows from Theorem 4.38(b)(i). Thus we see that

$$|\{pa_1 + qa_2 + ra_3\colon\ 0 \leq p < r_{21}, 0 \leq q < r_{32}, 0 \leq r < r_{13}\}| = r_{13}r_{32}r_{21}.$$

Hence, once we have shown the statement in (5.9), we have $|B| = r_{13}r_{32}r_{21}$ and the formula for $2g(H)$ follows.

Proof of (5.9): We first observe that $f - f' = \deg(\rho_1) - \deg(\rho_2)$ which is equal to the difference of the degrees of the corresponding summands ρ_1 and ρ_2. Thus we find that

$$f - f' = r_{31}a_1 - r_{13}a_3 = r_{23}a_3 - r_{32}a_2 = r_{12}a_2 - r_{21}a_1,$$

which yields that

$$f - f' + r_{13}a_3 = r_{31}a_1, \quad f - f' + r_{32}a_2 = r_{23}a_3 \quad \text{and} \quad f - f' + r_{21}a_1 = r_{12}a_2.$$

It follows that $(f - f') + pa_1 + qa_2 + ra_3 \in H$, if $p \geq r_{21}$ or $q \geq r_{32}$ or $r \geq r_{13}$.

Conversely, suppose that $p < r_{21}, q < r_{32}$ and $r < r_{13}$, and suppose further that

$$(f - f') + pa_1 + qa_2 + ra_3 = ua_1 + va_2 + wa_3 \tag{5.10}$$

with $u, v, w \in \mathbb{Z}_{\geq 0}$. We show that this is not possible.

Since $f - f' = r_{12}a_2 - r_{21}a_1$, the above equation yields the relation

$$(p - r_{21} - u)a_1 + (q + r_{12} - v)a_2 + (r - w)a_3 = 0.$$

Suppose that $q + r_{12} - v \leq 0$. Then $(r - w)a_3 = (r_{21} + u - p)a_1 + (v - q - r_{12})a_2$. Since $r_{21} + u - p > 0$, this implies that $r - w \geq r_{13} + r_{23}$ (see Theorem 4.38(b)(iii)). This is a contradiction, since $r - w < r_{13}$. Therefore we have $q + r_{12} - v > 0$. If $w \geq r$, then $(q + r_{12} - v)a_2 = (r_{21} + u - p)a_1 + (w - r)a_3$ together with Theorem 4.38(b)(iii) implies that $q + r_{12} - v \geq r_{12} + r_{32}$. This contradicts to $q < r_{32}$. Hence, we obtain $r - w > 0$. The equation

$$(r_{21} + u - p)a_1 = (q + r_{12} - v)a_2 + (r - w)a_3$$

and Theorem 4.38(b)(iii) imply that $r_{21} + u - p \geq r_{21} + r_{31}$. So we must have $u - p \geq r_{31} > 0$. Replacing $f - f'$ by $r_{23}a_3 - r_{32}a_2$ (5.10) we obtain

$$(r - w + r_{23})a_3 = (u - p)a_1 + (r_{32} - q + v)a_2.$$

Since $u - p > 0$, $r_{32} - q + v > 0$ and $r - w + r_{23} > 0$, Theorem 4.38(b)(iii) implies that $r - w + r_{23} \geq r_{13} + r_{23}$. This contradicts to $r - w < r_{13}$. $\square$

Notes

A Gorenstein ring is characterized by the property that it is Cohen–Macaulay and of Cohen–Macaulay type 1, see Theorem 3.65. Therefore, Theorem 5.5 is a special case of the more general theorem of Kunz [47]. Theorem 5.5 and other characterizations for symmetric numerical semigroups first appeared in [41]. In [35] it was shown that for a non-symmetric numerical semigroup $H = \langle a_1, a_2, a_3 \rangle$, the integers f and f' defined in Theorem 5.11 belong to $\mathrm{PF}(H)$. That these are no other pseudo-Frobenius numbers of H has been shown by Rosales and García-Sánchez in [64, Corollary 12]. Theorem 5.13 is due to Nari, Numata and Watanabe [54]. In this presentation we follow closely their line of arguments.

Almost Symmetric and Nearly Symmetric Numerical Semigroups

6

This chapter focuses on the canonical ideal of a numerical semigroup and symmetry properties. We begin by presenting the canonical ideal and key aspects of relative ideals. The notions of pseudo-symmetric and almost symmetric numerical semigroups are then introduced, with a particular focus on their characterization and examples involving 3 generators. Finally, we discuss nearly symmetric numerical semigroups and their connection to almost symmetric numerical semigroups, concluding with an analysis of the canonical trace for 3-generated numerical semigroups.

6.1 The Canonical Ideal of a Numerical Semigroup

6.1.1 The Canonical Module of a Numerical Semigroup Ring

Let K be a field, and let H be a numerical semigroup. By Corollary 3.59 the canonical module ω_R of the numerical semigroup ring $R = K[H]$ is a graded module of rank 1, and hence can be identified with a graded fractionary ideal, as explained below. The canonical ideal ω_R is uniquely determined by the property that

$$\operatorname{Ext}^i_R(R/\mathfrak{m}, \omega_R) \cong \begin{cases} R/\mathfrak{m}, & \text{if } i = 1, \\ 0, & \text{if } i \neq 1, \end{cases} \tag{6.1}$$

where $\mathfrak{m}$ is the graded maximal ideal of R, see Theorem 3.60.

This definition is quite abstract and does not help us so easily to identify the fractionary ideal ω_R. First of all let us explain what is a fractionary ideal.

Let R be any integral domain and let $Q(R)$ be the quotient field of R. A finitely generated R-submodule $I \subset Q(R)$ is called a *fractionary ideal*. Let $f_1, \ldots, f_r \in Q(R)$ be the generators of I. These elements are fractions. If we multiply I with

J. Herzog et al., *Numerical Semigroups*, Compact Textbooks in Mathematics,
https://doi.org/10.1007/978-3-032-05424-1_6

the product g of the denominators of these fractions, then gI is a submodule of R, and hence an ideal. Thus we see that any fractionary ideal of R is isomorphic to an ideal.

Now, we come back to $R = K[H]$. Since $t^z \in Q(R)$ for all $z \in \mathbb{Z}$, any R-submodule of $Q(R)$ with generators $t^{z_1}, \ldots, t^{z_r}$ is a fractionary ideal, and all graded fractionary ideals are of this form. We also call a graded fractionary ideal a *monomial fractionary ideal.* The fractionary ideal generated by $t^{z_1}, \ldots, t^{z_r}$ is denoted by $(t^{z_1}, \ldots, t^{z_r})$. For example, if $H = \langle 3, 4, 5\rangle$, then $I = (t^{-2}, t^{-1})$ is a monomial fractionary ideal of $K[H]$ and the K-basis of I is given by: t^{-2}, $t^{-1}, t, t^2, \ldots$.

To the monomial fractionary ideals of $K[H]$ correspond the relative ideals of H. A non-empty proper subset I of $\mathbb{Z}$ is called a *relative ideal* of H, if $I + H = I$. Here, for any two subsets $A, B \subset \mathbb{Z}$ we set $A + B = \{a + b\colon\ a \in A, b \in B\}$. If I is a relative ideal of H, then we set

$$I' = (t^z\colon\ z \in I).$$

The ideal I' is a fractionary monomial ideal of $K[H]$. In particular, I is finitely generated since I' is finitely generated. In other words, there exist elements $z_1, \ldots, z_r \in I$ such that $I = \{z_1, \ldots, z_r\} + H$.

The explicit determination of ω_R relies on Theorem 3.66. Let $a \in H$. Then R is a finitely generated free $K[t^a]$-module. Indeed, it follows from the definition of $\mathrm{Ap}_H(a)$ that

$$R \cong \bigoplus_{b \in \mathrm{Ap}_H(a)} K[t^a]t^b. \tag{6.2}$$

Now we may apply Theorem 3.66 and since $\dim R = \dim K[t^a] = 1$, we obtain the isomorphism

$$\omega_R \cong \underline{\mathrm{Hom}}_{K[t^a]}(R, \omega_{K[t^a]}) \tag{6.3}$$

of graded R-modules.

Theorem 6.1
Let K be a field, let H be a numerical semigroup, and let $R = K[H]$. Then there is an isomorphism

$$\omega_R \cong (t^{-c}\colon\ c \in \mathcal{G}(H))$$

of graded R-modules.

Proof. Let $a \in H \setminus \{0\}$. It follows from Corollary 3.62 that $\omega_{K[t^a]} \cong K[t^a](-a)$. Therefore, together with (6.2) and (6.3), we obtain

$$\begin{aligned}\omega_R \cong & \underline{\mathrm{Hom}}_{K[t^a]}(R, K[t^a](-a)) = \underline{\mathrm{Hom}}_{K[t^a]}(R, K[t^a])(-a) \\ = & \underline{\mathrm{Hom}}_{K[t^a]}\Big(\bigoplus_{b \in \mathrm{Ap}_H(a)} K[t^a]t^b, K[t^a]\Big)(-a) \cong \Big(\bigoplus_{b \in \mathrm{Ap}_H(a)} K[t^a]t^{-b}\Big)(-a) \\ = & \bigoplus_{b \in \mathrm{Ap}_H(a)} K[t^a]t^{-b+a} \cong (t^{-b+a} \colon\ b \in \mathrm{Ap}_H(a)).\end{aligned}$$

For the first equality see Exercise 2.3. Next, we show that

$$(t^{-b+a} \colon\ b \in \mathrm{Ap}_H(a)) = (t^{-c} \colon\ c \in \mathcal{G}(H)).$$

The inclusion $(t^{-b+a} \colon\ b \in \mathrm{Ap}_H(a)) \subseteq (t^{-c} \colon\ c \in \mathcal{G}(H))$ follows from the fact that for any $b \in \mathrm{Ap}_H(a)$, $b - a \in \mathcal{G}(H)$. Now, consider $c \in \mathcal{G}(H)$. Let $k > 0$ be the smallest integer such that $c + ka \in H$. We set $b = c + ka$. Then $b - a = c + (k-1)a \in \mathcal{G}(H)$. This means that $b \in \mathrm{Ap}_H(a)$. Moreover,

$$t^{-c} = t^{-b+a}t^{(k-1)a} \in (t^{-b+a} \colon\ b \in \mathrm{Ap}_H(a)),$$

as desired. □

6.1.2 Relative Ideals and the Canonical Ideal of a Numerical Semigroup

We identify ω_R with $(t^{-c} \colon\ c \in \mathcal{G}(H))$ and call it the *canonical ideal* of $K[H]$. The corresponding relative ideal

$$\Omega_H = (-c \colon\ c \in \mathcal{G}(H)) \tag{6.4}$$

is called the *canonical ideal* of H.

Which are the generators of Ω_H? Here is the answer.

Proposition 6.2
Ω_H is minimally generated by the set $\{-c \colon\ c \in \mathrm{PF}(H)\}$.

Proof. For an element $c \in \mathcal{G}(H) \setminus \mathrm{PF}(H)$ there exists $a \in H \setminus \{0\}$ such that $c + a \in \mathcal{G}(H)$. Assume that a is the maximum integer in H with this property. Then $c + a \in \mathrm{PF}(H)$. Thus $c = b - a$ for some $b \in \mathrm{PF}(H)$ and hence $-c = -b + a \in (-n \colon\ n \in \mathrm{PF}(H))$. This shows that $\Omega_H = (-c \colon\ c \in \mathrm{PF}(H))$. Notice that for any distinct elements $c, c' \in \mathrm{PF}(H)$, we have $c - c' \notin H$, otherwise $c' + a = c \in \mathcal{G}(H)$

for some $a \in H \setminus \{0\}$, which is a contradiction. Hence, $\{-c\colon\ c \in \mathrm{PF}(H)\}$ is a minimal generating set of Ω_H. □

Observe that $-F(H)$ is the smallest element in Ω_H. Thus, Proposition 6.2 together with Theorem 5.5 implies

Corollary 6.3
The minimal number of generators of Ω_H is equal to the type of H, i.e., $\mu(\Omega_H) = t(H)$. In particular, $\mu(\Omega_H) = 1$ if and only if H is symmetric, and this is the case, if and only if $\Omega_H = (-F(H))$.

We call the relative ideal $C_H = F(H) + \Omega_H$ the *normalized canonical ideal* of H. Its minimal element is 0, and $C_H = H$ if and only if H is symmetric, as follows from Corollary 6.3.

C_H can be computed by *numericalsgps* as follows:

```
gap> LoadPackage("numericalsgp");
gap> s:=NumericalSemigroup(7,9,11,13);
gap> c:=CanonicalIdealOfNumericalSemigroup(s);
<Ideal of numerical semigroup>
gap> GeneratorsOfIdealOfNumericalSemigroup(c);
[ 0, 2, 4 ]
```

Corollary 6.3 is a special case of a more general result. To explain this we should consider colon ideals, which we already introduced in Chap. 1.

Let R be any integral domain, and let I and J be two fractionary ideals of R. Then

$$I : J = \{f \in Q(R)\colon\ fJ \subseteq I\}$$

is again a fractionary ideal. It is called the *colon ideal* of I with respect to J.

Exercise 6.4
Let R be an integral domain, and let I, J and L be fractionary ideals of R.
(a) By definition, $I^{-1} = R : I$. Show that $I \subseteq (I^{-1})^{-1}$.
(b) Show that $I : JL = (I : J) : L$.

The canonical ideal establishes a nice duality among fractionary ideals. Before explaining this, let us first observe some more basic facts about colon ideals and relative ideals.

Lemma 6.5
Let R be an integral domain, and let I and J be fractionary ideals of R with $J \neq (0)$. Then

$$I : J \cong \mathrm{Hom}_R(J, I).$$

Proof. Let $f \in I : J$ and denote by $\mu_f : J \to I$ the map with $\mu_f(g) = fg$ for all $g \in J$. Then $\mu_f \in \mathrm{Hom}_R(J, I)$. Thus we obtain a map $\mu\colon I : J \to \mathrm{Hom}_R(J, I)$ with $f \mapsto \mu_f$. Note that μ is an R-module homomorphism.

Since R is a domain it follows that $\mu_f = \mu_{f'}$ if and only if $f = f'$. This shows that μ is injective.

Now let $\varphi \in \mathrm{Hom}_R(J, I)$. We choose $g \in J, g \neq 0$. We let $f = \varphi(g)/g$. Note that $f \in I : J$. Indeed, for any $g' \in J$, we may write $g'f = g'\varphi(g)/g = \varphi(g'g)/g = g\varphi(g')/g = \varphi(g') \in I$. Hence, $fJ \subseteq I$. We claim that $\varphi = \mu_f$. This then implies that $\mu\colon I : J \to \mathrm{Hom}_R(J, I)$ is also surjective, and hence an isomorphism.

In order to prove the claim let $h \in J$. By using that φ is R-linear we obtain

$$g\varphi(h) = \varphi(hg) = h\varphi(g),$$

so that $\varphi(h) = (\varphi(g)/g)h = fh = \mu_f(h)$, as desired. □

As a consequence of Lemma 6.5 and Theorem 2.2 we obtain

Corollary 6.6
Let $R = K[H]$, and let I and J be monomial fractionary ideals of R. Then $I : J$ is again a monomial fractionary ideal. In particular,

$$I : J = (t^a\colon\ a \in \mathbb{Z},\ t^a J \subseteq I\}.$$

Let I and J be relative ideals of H. Then we set

$$I - J = \{z \in \mathbb{Z}\colon\ z + J \subseteq I\}.$$

Then $I - J$ is again a relative ideal of H, and

$$I' : J' = (t^z\colon\ z \in I - J). \tag{6.5}$$

For any two relative ideals I and J, the relative ideal $I - J$ can be computed with *numericalsgps*. The following example demonstrates this.

```
gap> LoadPackage("numericalsgp");
gap> s:=NumericalSemigroup(7,9,11,13);
```

```
gap> I:=[20,36,49]+s;
gap> J:=[-5,3]+s;
gap> L:=I-J;
gap> GeneratorsOfIdealOfNumericalSemigroup(L);
[ 38, 39, 41, 43 ]
```

Exercise 6.7
Let H be a numerical semigroup, and let I be a relative ideal of H.

(a) Show that $H^* := I - I$ is a numerical semigroup with $H \subseteq H^*$.

(b) Suppose that $H \neq \mathbb{Z}_{\geq 0}$. Show that $H - M = M - M$, and conclude that $M - M = H \cup \mathrm{PF}(H)$.

(c) Show that $I \subseteq H - (H - I)$, and give an example for which this containment is strict.

Let I be a relative ideal of H and $c \in \mathbb{Z}$. Then $L = c + I$ is again a relative ideal of H, and we have $L' = t^c I'$. This implies that I' and L' are isomorphic $K[H]$-modules.

Let I and J be relative ideals of H. We write $I \sim J$, if there exists $c \in \mathbb{Z}$ such that $J = c + I$. For example, we have $\Omega_H \sim C_H$ The relation $\sim$ is obviously an equivalence relation. The equivalence classes of $\sim$ are called the *isomorphism classes of relative ideals* of H.

Exercise 6.8
Let H be a numerical semigroup.

(a) Show there exists only a finite number of isomorphism classes of relative ideals of H.

(b) For $H = \langle 3, 5, 7\rangle$ determine all the equivalence classes of relative ideals of H.

Exercise 6.9
Let H be a numerical semigroup, and let I, J, E and L be relative ideals of H with $I \sim E$ and $J \sim L$. Show that

$$I + J \sim E + L \quad \text{and} \quad I - J \sim E - L.$$

In the following result we present an important property of the canonical ideal Ω_H of H.

Theorem 6.10
Let H be a numerical semigroup, and let $I \subseteq J$ be two relative ideals of H. Then

(a) $|J \setminus I| < \infty$.
(b) $\Omega_H - J \subseteq \Omega_H - I$ and $|(\Omega_H - I) \setminus (\Omega_H - J)| = |J \setminus I|$.

Proof. (a) Let $z \in I$, Then $z + F(H) + \mathbb{Z}_{>0} \in I$. Therefore, $|H \setminus I| < \infty$. Since $J \setminus I \subseteq H \setminus I$, it also follows that $|J \setminus I| < \infty$.

(b) Let $z \in \Omega_H - J$. Then $z + I \subseteq z + J \subseteq \Omega_H$. Therefore, $z \in \Omega_H - I$. The second part of (b) we prove by induction on $|J \setminus I|$. If $|J \setminus I| = 0$, then the assertion is trivial.

Now let $|J \setminus I| > 0$, and let z be the largest element in $J \setminus I$, and set $L = I \cup \{z\}$. Since $z + H \subseteq L$, it follows that L is a relative ideal of H. Thus,

$$\Omega_H - J \subseteq \Omega_H - L \subseteq \Omega_H - I.$$

By induction hypothesis,

$$|(\Omega_H - L) \setminus (\Omega_H - J)| = |J \setminus L| = |J \setminus I| - 1.$$

Therefore, it suffices to show that

$$|(\Omega_H - I) \setminus (\Omega_H - L)| = 1. \tag{6.6}$$

Let I' and L' be the fractionary monomial ideals of $R = K[H]$ corresponding to I and L, respectively. Then $\dim_K(L'/I') = 1$, and because of (6.5), equation (6.6) follows once we have shown that

$$\dim_K(\omega_R : I')/(\omega_R : L') = 1. \tag{6.7}$$

The assumption that $\dim_K(L'/I') = 1$ implies that $L'/I' \cong R/\mathfrak{m}$. Hence, we have the short exact sequence

$$0 \to I' \to L' \to R/\mathfrak{m} \to 0,$$

which induces the long exact sequence

$$\begin{aligned} 0 &\to \operatorname{Hom}_R(R/\mathfrak{m}, \omega_R) \to \operatorname{Hom}_R(L', \omega_R) \to \operatorname{Hom}_R(I', \omega_R) \\ &\to \operatorname{Ext}^1_R(R/\mathfrak{m}, \omega_R) \to \operatorname{Ext}^1_R(L', \omega_R). \end{aligned} \tag{6.8}$$

By (6.1) we have $\operatorname{Hom}_R(R/\mathfrak{m}, \omega_R) = 0$ and $\operatorname{Ext}^1_R(R/\mathfrak{m}, \omega_R) = R/\mathfrak{m}$. Moreover, $\operatorname{Hom}_R(L', \omega_R) = \omega_R : L'$ and $\operatorname{Hom}_R(I', \omega_R) = \omega_R : I'$, as we have seen in Lemma 6.5. Thus, (6.8) gives us the exact sequence

$$0 \to (\omega_R : I')/(\omega_R : L') \to R/\mathfrak{m} \to \operatorname{Ext}^1_R(L', \omega_R).$$

By Theorem 3.53, $\operatorname{Ext}^1_R(L', \omega_R) \cong H^0_{\mathfrak{m}}(L')^\vee = 0$, since depth $L' \neq 0$. It follows that $(\omega_R : I')/(\omega_R : L') \cong R/\mathfrak{m}$. This completes the proof. □

Theorem 6.10 has several nice consequences.

Corollary 6.11
Let H be a numerical semigroup, and let I be a relative ideal of H. Then

$$I = \Omega_H - (\Omega_H - I).$$

In particular, if H is symmetric, then $I = H - (H - I)$ for all relative ideals I of H.

Proof. If $z \in I$, then $z + (\Omega_H - I) \subseteq \Omega_H$. Hence $z \in \Omega_H - (\Omega_H - I)$, and so $I \subseteq \Omega_H - (\Omega_H - I)$. Suppose this inclusion is strict. Then Theorem 6.10 implies that

$$\Omega_H - (\Omega_H - (\Omega_H - I)) \subsetneq \Omega_H - I.$$

On the other hand, let $J = \Omega_H - I$. Then

$$\Omega_H - I = J \subseteq \Omega_H - (\Omega_H - J) = \Omega_H - (\Omega_H - (\Omega_H - I)),$$

a contradiction.

If H is symmetric, then $\Omega_H = (-F(H))$, see Corollary 6.3. Hence, by the first statement we have

$$I = (-F(H)) - (-F(H)) - I) = H - (H - I). \qquad \square$$

Since $\Omega_H - H = \Omega_H$, Corollary 6.11 implies

Corollary 6.12
$\Omega_H - \Omega_H = H$.

Let I be a relative ideal of the numerical semigroup H. Then I has a unique minimal set of generators which is actually given by the set $I \setminus (I + M)$. Why?

We let

$$\mu(I) = |I \setminus (I + M)|.$$

Thus, $\mu(I)$ is the minimal number of generators of I.

We also define the number

$$t(I) = |(I - M) \setminus I|,$$

and call this number the *type* of I. For $I = H$, this gives us the type of H, as defined in Sect. 5.

The following result generalizes Corollary 6.3.

Corollary 6.13
Let H be a numerical semigroup, and let I be a relative ideal of H. Then

$$t(I) = \mu(\Omega_H - I) \quad \text{and} \quad \mu(I) = t(\Omega_H - I).$$

Proof. By using Theorem 6.10 and Exercise 6.4 we obtain

$$\begin{aligned}\mu(I) &= |I \setminus (I + M)| = |(\Omega_H - (I + M)) \setminus (\Omega_H - I)| \\ &= |((\Omega_H - I) - M) \setminus (\Omega_H - I)| = t(\Omega_H - I).\end{aligned}$$

Similarly, one shows that $t(I) = \mu(\Omega_H - I)$ □

Notes

Relative ideals of numerical semigroups and their isomorphism classes have been studied systematically by Barucci and Khouja [7]. In that paper an upper bound for the number of isomorphism classes of relative ideals is given. Moreover, the reduction number of relative ideals is studied. In the Springer Lecture Notes [40, 2. Vortrag] by Kunz and the first author of this book, the canonical ideal of a one-dimensional local Cohen-Macaulay ring has been introduced. There one finds most of the results presented in this section, but more generally for one-dimensional local Cohen-Macaulay rings and formulated in the language of commutative algebra. See also our Sect. 3.2. As a consequence of Corollary 6.11 we noticed that if H is symmetric, then $I = H - (H - I)$ for all nonzero relative ideals. This result corresponds to the fact that any nonzero ideal in a one-dimensional Gorenstein domain is a reflexive module. This fact was first shown by Berger in [8].

6.2 Almost Symmetric Numerical Semigroups

In this section we consider numerical semigroup rings which in a certain sense, as specified later, are close to being symmetric.

6.2.1 Pseudo-symmetric and Almost Symmetric Numerical Semigroups

One distinguishes two types of gaps of H. The elements of the set

$$\mathscr{G}_1(H) = \{a \in \mathscr{G}(H) \colon \ F(H) - a \in H\}$$

are called *gaps of type* 1, and the elements of the set

$$\mathscr{G}_2(H) = \{a \in \mathscr{G}(H) \colon \ F(H) - a \notin H\}$$

are called *gaps of type* 2.

Note that $a \in \mathscr{G}_2(H)$ if and only if $F(H) - a \in \mathscr{G}_2(H)$.

We set $g_1(H) = |\mathscr{G}_1(H)|$ and $g_2(H) = |\mathscr{G}_2(H)|$. Then $g(H) = g_1(H) + g_2(H)$ and $g_1(H) = n(H)$.

Since $F(H) + 1 = n(H) + g(H)$, it follows that

$$2g(H) = F(H) + 1 + (g(H) - n(H)) = F(H) + g_2(H) + 1.$$

Let $a \in \mathrm{PF}(H)$ with $a \neq F(H)$. Suppose that $F(H) - a \in H$. Then $F(H) - a \in M$, and hence $F(H) = a + (F(H) - a) \in H$, a contradiction. Thus we have shown that

$$\mathrm{PF}(H) \setminus \{F(H)\} \subseteq \mathscr{G}_2(H). \tag{6.9}$$

As a result of these discussions we obtain

Proposition 6.14
Let H be a numerical semigroup. Then

(a) $2g(H) \geq F(H) + t(H)$.
(b) The following conditions are equivalent:

  (i) $2g(H) = F(H) + t(H)$.
  (ii) $\mathscr{G}_2(H) = \mathrm{PF}(H) \setminus \{F(H)\}$.

Proof. (a) It follows from (6.9) that $t(H) \leq g_2(H) + 1$. Hence, $2g(H) = F(H) + g_2(H) + 1 \geq F(H) + t(H)$.

(b) follows from (a) and it is the case, when the inequalities in the proof of (a) become equalities. □

A numerical semigroup satisfying the equivalent conditions of Proposition 6.14(b) is called *almost symmetric*.

For example, the numerical semigroup $H = \langle 5, 6, 7, 9\rangle$ is almost symmetric, because $g(H) = 5$, $F(H) = 8$ and $\mathrm{PF}(H) = \{4, 8\}$, so that $2g(H) = 10 = 8+2 = F(H) + t(H)$. Note that in this example, $\mathrm{PF}(H) = \{F(H), F(H)/2\}$. This is a special class of almost symmetric numerical semigroups, as we will see below.

A numerical semigroup H is called *pseudo-symmetric*, if $F(H)$ is even and for all $a \in \mathbb{Z}$ with $a \neq F(H)/2$, either $a \in H$ or $F(H) - a \in H$. The semigroup $H = \langle 5, 6, 7, 9\rangle$ is pseudo-symmetric.

Let $a \in H$. If H pseudo-symmetric, then not only $F(H) + a \in \mathrm{Ap}_H(a)$, but also $F(H)/2 + a \in \mathrm{Ap}_H(a)$.

One has the following characterizations of pseudo-symmetric numerical semigroups.

Theorem 6.15

Let H be a numerical semigroup with even Frobenius number. Then the following conditions are equivalent:

(i) H is pseudo-symmetric.
(ii) For all $a \in M$,

$$\mathrm{Ap}_H(a) = \{0 = b_1 < b_2 < \cdots < b_{a-1} = F(H) + a\} \cup \{F(H)/2 + a\}$$

and $b_i + b_{a-1-i} = b_{a-1}$ for $1 \leq i \leq a-1$.
(iii) $\mathrm{PF}(H) = \{F(H)/2, F(H)\}$.
(iv) $2g(H) = F(H) + 2$.

Proof. (i) $\Rightarrow$ (iv): The definition of pseudo-symmetric implies that $\mathscr{G}_2(H) = \{F(H)/2\}$. By (6.9), $\mathrm{PF}(H) \setminus \{F(H)\} \subseteq \mathscr{G}_2(H)$, and hence we have $\mathrm{PF}(H) \subseteq \{F(H), F(H)/2\}$. Since $F(H)$ is even, H is not symmetric, see Exercise 4.12(a). Thus by Theorem 5.5 we have $1 < t(H) = |PF(H)| \leq 2$. Thus $t(H) = 2$ and hence $\mathrm{PF}(H) = \{F(H), F(H)/2\}$. This implies that $\mathscr{G}_2(H) = \mathrm{PF}(H) \setminus \{F(H)\}$. By Proposition 6.14(b), it follows that $2g(H) = F(H) + 2$.

(iv) $\Rightarrow$ (iii): By Proposition 6.14(a), $F(H) + 2 = 2g(H) \geq F(H) + t(H)$. Therefore, $t(H) \leq 2$. Since $F(H)$ is even, it follows that $F(H) - F(H)/2 = F(H)/2 \in \mathscr{G}(H)$, which implies that H is not symmetric. Therefore, $t(H) > 1$, and hence $t(H) = 2$. Thus Proposition 6.14(b) implies that $\mathscr{G}_2(H) = \mathrm{PF}(H) \setminus \{F(H)\}$. Since $F(H)/2 \in \mathscr{G}_2(H)$ and since $|\mathrm{PF}(H)| = 2$, we obtain $\mathrm{PF}(H) = \{F(H)/2, F(H)\}$.

(iii) $\Rightarrow$ (ii): We first show that $\mathscr{G}_2(H) = \{F(H)/2\}$. Suppose this is not the case. Then there exists $b \in \mathscr{G}_2(H)$, $b \neq F(H)/2$. We may assume that $b > F(H)/2$. Otherwise, we replace b by $F(H) - b$ which also belongs to $\mathscr{G}_2(H)$. By assumption, $\mathrm{PF}(H) = \{F(H)/2, F(H)\}$. Therefore, $b \notin \mathrm{PF}(H)$, and hence there exists $c \in M$ such that $b + c \in \mathrm{PF}(H)$, see Exercise 5.6. Since $b > F(H)/2$, it follows that

$b + c = F(H)$. This implies that $F(H) - b = c \in M$, a contradiction. Hence, $\mathscr{G}_2(H) = \{F(H)/2\}$.

Consider $a \in M$. We note that $F(H)/2 + a \in H$, because $F(H)/2 \in \mathrm{PF}(H)$. Since $F(H)/2 \notin H$, it follows that $F(H)/2 + a \in \mathrm{Ap}_H(a)$. Now, let $b \in \mathrm{Ap}_H(a)$ such that $b \neq F(H)/2 + a$. Then $b - a \neq F(H)/2$ and $b - a \in \mathscr{G}(H)$. Since $\mathscr{G}_2(H) = \{F(H)/2\}$, it follows that $(F(H) - b) + a = F(H) - (b - a) \in H$. On the other hand, since $b \in H$, it follows that $F(H) - b \in \mathscr{G}(H)$. This shows that $(F(H) - b) + a \in \mathrm{Ap}_H(a)$. Hence, the map $\alpha\colon \mathrm{Ap}_H(a) \setminus \{F(H)/2 + a\} \to \mathrm{Ap}_H(a) \setminus \{F(H)/2 + a\}$ with $\alpha(b) = (F(H) - b) + a$ is an order reversing bijection with $b + \alpha(b) = F(H) + a$. This proves (ii).

(ii) $\Rightarrow$ (i): Let $a \in \mathscr{G}(H)$, $a \neq F(H)/2$. Since $a + F(H) + 1 \in H$, it follows that $a + F(H) + 1 \in \mathrm{Ap}_H(F(H) + 1)$. Now condition (ii) implies that $F(H) - a = 2F(H) + 1 - (a + F(H) + 1) \in \mathrm{Ap}_H(F(H) + 1)$. In particular, $F(H) - a \in H$. This proves that H is pseudo-symmetric. □

Pseudo-symmetric numerical semigroups are special cases of almost symmetric numerical semigroups. Indeed, Proposition 6.14 together with Theorem 6.15 implies

Corollary 6.16
Let H be a numerical semigroup. The following conditions are equivalent:

(i) H is pseudo-symmetric.
(ii) H is almost symmetric and $t(H) = 2$.

By Theorem 5.11, a 3-generated numerical semigroup is of type ≤ 2. Therefore, we obtain

Corollary 6.17
Let H be a numerical semigroup with 3 generators, which is not symmetric. Then H is pseudo-symmetric if and only if H is almost symmetric.

Exercise 6.18
(a) Let H be a numerical semigroup with $t(H) = 2$. Then H is not necessarily pseudo-symmetric. Find such an example.

(b) For any positive integer c find an almost symmetric numerical semigroup H with $t(H) = c$.

6.2.2 Characterization of Almost Symmetric Numerical Semigroups

Let H be a numerical semigroup. In Sect. 6.1 we introduced the normalized canonical relative ideal of H, namely

$$C_H = \Omega_H + F(H) = (F(H) - a\colon\ a \in \mathcal{G}(H)).$$

Proposition 6.19
Let H be a numerical semigroup. The following conditions are equivalent:

(i) H is almost symmetric.
(ii) $C_H \setminus H = \mathrm{PF}(H) \setminus \{F(H)\}$.
(iii) $C_H \subseteq M - M$.
(iv) If $a \notin H$, then either $F(H) - a \in H$ or $a \in \mathrm{PF}(H)$.

Proof. (i) $\Rightarrow$ (ii): It follows from the definition of C_H that $C_H \setminus H = \mathcal{G}_2(H)$. Since H is almost symmetric, $\mathcal{G}_2(H) = \mathrm{PF}(H) \setminus \{F(H)\}$ (see Proposition 6.14).

(ii) $\Rightarrow$ (iii): It follows from (ii) that $C_H \subseteq H \cup \mathrm{PF}(H)$. By Exercise 6.7 we know that $M - M = H \cup \mathrm{PF}(H)$. Thus, $C_H \subseteq M - M$.

(iii) $\Rightarrow$ (iv): Let $a \notin H$, and suppose that $F(H) - a \notin H$. Then

$$a = F(H) - (F(H) - a) \in C_H \setminus H \subseteq (M - M) \setminus H = \mathrm{PF}(H).$$

(iv) $\Rightarrow$ (i): The condition implies that $\mathcal{G}_2(H) \subseteq \mathrm{PF}(H) \setminus \{F(H)\}$. Since the opposite inclusion always holds (see (6.9)), it follows that H is almost symmetric. $\square$

Almost symmetric numerical semigroups are characterized by a symmetry property of the set $\mathrm{PF}(H)$.

Theorem 6.20
Let H be a numerical semigroup with

$$\mathrm{PF}(H) = \{f_1 < f_2 < \cdots < f_{t(H)} = F(H)\}.$$

Then the following conditions are equivalent:

(i) H is almost symmetric.
(ii) $f_i + f_{t(H)-i} = F(H)$ for $i = 1, \ldots, t(H) - 1$

Proof. We first notice that (ii) is equivalent to the following condition:

(ii') if $f \in \mathrm{PF}(H) \setminus \{F(H)\}$, then $F(H) - f \in \mathrm{PF}(H) \setminus \{F(H)\}$.

Indeed, if (ii') holds, then the map

$$\alpha\colon \mathrm{PF}(H) \setminus \{F(H)\} \to \mathrm{PF}(H) \setminus \{F(H)\}$$

with $\alpha(f) = F(H) - f$ is an order reversing bijection with $f + \alpha(f) = F(H)$. This shows that (ii') is equivalent to (ii).

(i) $\Rightarrow$ (ii'): Let $f \in \mathrm{PF}(H) \setminus \{F(H)\}$ and suppose that $F(H) - f \notin \mathrm{PF}(H) \setminus \{F(H)\}$. Then $f \in \mathcal{G}_2(H)$ (see (6.9)). Since H is almost symmetric $\mathcal{G}_2(H) = \mathrm{PF}(H) \setminus \{F(H)\}$, and so $F(H) - f \in \mathcal{G}_2(H) = \mathrm{PF}(H) \setminus \{F(H)\}$, a contradiction.

(ii') $\Rightarrow$ (i): Suppose H is not almost symmetric. Then Proposition 6.19 implies that there exists $a \notin H$ for which $F(H) - a \notin H$ and $a \notin \mathrm{PF}(H)$. This implies that $a \in \mathcal{G}_2(H) \setminus (\mathrm{PF}(H) \setminus \{F(H)\})$, which implies that H is not almost symmetric. □

6.2.3 Almost Symmetric 3-Generated Numerical Semigroups

Let $H = \langle a_1, a_2, a_3 \rangle$ be a numerical semigroup which is not symmetric. By Theorem 4.38, I_H is minimally generated by $f_{\mathbf{v}_1}, -f_{\mathbf{v}_2}, f_{\mathbf{v}_3}$, where

$$\mathbf{v}_1 = (-c_1, r_{12}, r_{13}),\ \ \mathbf{v}_2 = (r_{21}, -c_2, r_{23}),\ \ \mathbf{v}_3 = (r_{31}, r_{32}, -c_3),$$

and the transpose of the relation matrix A of I_H is

$$A^{\mathsf{T}} = \begin{pmatrix} x_3^{r_{23}} & x_1^{r_{31}} & x_2^{r_{12}} \\ x_2^{r_{32}} & x_3^{r_{13}} & x_1^{r_{21}} \end{pmatrix}.$$

Moreover, $\mathrm{PF}(H) = \{f, f'\}$, where by Theorem 5.11, $f = a_1c_1 - a_1 - a_2 - a_3 + a_3r_{23}$ and $f' = a_1c_1 - a_1 - a_2 - a_3 + a_2r_{32}$.

Theorem 6.21
Let $H = \langle a_1, a_2, a_3 \rangle$ be a 3-generated numerical semigroup, which is not symmetric. Then, with the notation introduced, the following conditions are equivalent:

(i) H is almost symmetric.
(ii) (1) If $f' > f$, then $r_{23}r_{31}r_{12} = 1$, and
     (2) if $f > f'$, then $r_{32}r_{13}r_{21} = 1$.

Proof. First note that since H is not symmetric, it follows from Theorem 5.5 that $f \neq f'$. Let us assume that $f' > f$. Then Theorem 5.13 tells us that

$$2g(H) = F(H) + 1 + r_{23}r_{31}r_{12}.$$

Since $t(H) = 2$, it follows $2g(H) = F(H) + t(H)$ if and only if $r_{23}r_{31}r_{12} = 1$. This proves the assertion. The case $f > f'$ is shown similarly. □

Corollary 6.22
With the assumptions and notation of Theorem 6.21, the following conditions are equivalent:

(a) H is almost symmetric.
(b)

$$A^{\mathsf{T}} = \begin{pmatrix} x_3 & x_1 & x_2 \\ x_2^{r_{32}} & x_3^{r_{13}} & x_1^{r_{21}} \end{pmatrix}, \quad \text{or} \quad A^{\mathsf{T}} = \begin{pmatrix} x_3^{r_{23}} & x_1^{r_{31}} & x_2^{r_{12}} \\ x_2 & x_3 & x_1 \end{pmatrix}.$$

Notes

Gaps of type 1 and type 2 were first considered in [40]. In 1997 Barucci, Dobbs and Fontana introduced pseudo-symmetric numerical semigroups. Theorem 6.15, which characterizes pseudo-symmetric numerical semigroups, is borrowed from their book [5], where this concept was introduced. In the same year, 1997, as a generalization of pseudo-symmetric numerical semigroups, Barucci and Fröberg introduced in [6] one-dimensional almost Gorenstein rings and almost symmetric numerical semigroups. Indeed, as shown by Barucci and Fröberg, a numerical semigroup is pseudo-symmetric if and only if it is almost symmetric of type 2, see Corollary 6.16. In the same article one also finds Proposition 6.19 which characterizes almost symmetric numerical semigroups. Theorem 6.20, which is due to Nari [53], gives a very nice characterization of almost symmetric numerical semigroups in terms of properties of the set $\mathrm{PF}(H)$. Later on, Goto, Takahashi and Taniguchi [33] developed the theory of almost Gorenstein rings much further, and extended this concept to rings of any Krull dimension.

Almost symmetric 3-generated numerical semigroups can be characterized by the relation matrix of the defining ideal of its semigroup ring, as shown by Nari, Numata and K. Watanabe [54]. Their result is the content of Theorem 6.21. Things become more complicated when the numerical semigroup has more than 3 generators. Komeda [46] gave a complete description of I_H, when H is a 4-generated pseudo-symmetric numerical semigroup, and he showed that $\mu(I_H) = 5$ for such semigroups. A simpler proof of Komeda's theorem was given in [42] by using the so-called RF-matrices (row factorization matrices), which were introduced by Moscariello in his paper [52]. The authors also give some structure theorems for 4-generated almost symmetric semigroups in terms of the RF-matrices. Eto [29] classified the almost symmetric numerical semigroups of embedding dimension four, and gave a minimal system of generators and a minimal free resolution for the defining ideals of almost Gorenstein monomial curves in affine four space.

6.3 Nearly Symmetric Numerical Semigroups

6.3.1 Almost Symmetric Numerical Semigroups Are Nearly Symmetric

Let K be a field, let H be a numerical semigroup, and let $R = K[H] \subset K[t]$ be the corresponding semigroup ring. By Theorem 6.1, $t^{-F(H)} \in \omega_R$. Thus we obtain an exact sequence

$$0 \to (t^{-F(H)}) \to \omega_R \to D \to 0$$

of graded R-modules. By Corollary 6.3, H is symmetric if and only if $D = 0$. Proposition 6.19 implies that H is almost symmetric if and only if $\mathfrak{m}D = 0$. In other words, D is "almost" 0, if H is almost symmetric.

In this section we introduce another invariant which measures how far a numerical semigroup H is from being symmetric. The ideal $\omega_R\omega_R^{-1}$ is called the *trace* of the canonical ideal, and is denoted by $\operatorname{tr}(\omega_R)$. The semigroup ideal corresponding to $\operatorname{tr}(\omega_R)$ is the ideal $\Omega_H + (H - \Omega_H)$, which we denote by $\operatorname{tr}(\Omega_H)$.

Exercise 6.23
Show that $\operatorname{tr}(\Omega_H) = C_H + (H - C_H)$.

The crucial observation is the following.

Proposition 6.24
Let H be a numerical semigroup. Then
(a) $\operatorname{tr}(\Omega_H) \subseteq H$.
(b) $\operatorname{tr}(\Omega_H) = H$ if and only if H is symmetric.

Proof. (a) Let $a \in \operatorname{tr}(\Omega_H)$. Then there exists $b \in \Omega_H$ and $c \in H \setminus \Omega_H$ such that $a = b + c$. It follows that $a \in H$.

(b) If H is symmetric, then $\Omega_H \cong H$, see Corollary 6.3. Thus $\operatorname{tr}(\Omega_H) = H$. Conversely, suppose that $\operatorname{tr}(\Omega_H) = H$. Then $0 \in \Omega_H + (H - \Omega_H)$. Hence there exists $a \in \Omega_H$ and $b \in H - \Omega_H$ such that $0 = a + b$. It follows from (6.4) that $a + F(H) \geq 0$. Moreover, since $b \in (H - \Omega_H)$, we have $b + z \in H$ for all $z \in \Omega_H$. In particular, $b - F(H) \in H$, because $-F(H) \in \Omega_H$. This implies that $b - F(H) \geq 0$. Hence the equation $0 = (a + F(H)) + (b - F(H)$ implies that $b - F(H) = 0$. It follows that $F(H) \in H - \Omega_H$, and hence (6.4) yields that $F(H) - c \in H$ for all $c \in \mathcal{G}(H)$. In other words, H is symmetric. □

The number $\operatorname{res}(H) = |H \setminus \operatorname{tr}(\Omega_H)|$ is called the *residuum of* H. By Proposition 6.24, H is symmetric if and only if $\operatorname{res}(H) = 0$. We call H *nearly symmetric*, if $\operatorname{res}(H) \leq 1$, which is equivalent to saying that $M \subseteq \operatorname{tr}(\Omega_H)$.

The class of nearly symmetric numerical semigroups includes the class of almost symmetric numerical semigroups. Indeed, we have

Proposition 6.25
If H is almost symmetric, then H is nearly symmetric.

Proof. Since $\operatorname{tr}(\Omega_H) = C_H + (H - C_H)$, we have to show that $M \subseteq C_H + (H - C_H)$. Indeed, by Proposition 6.19 we have $C_H \subseteq M - M = H - M$. Since $0 \in C_H$, we obtain

$$C_H + (H - C_H) \supseteq H - C_H \supseteq H - (H - M) = M,$$

as desired. □

6.3.2 The Canonical Trace of 3-Generated Numerical Semigroups

For a 3-generated numerical semigroups $H = \langle a_1, a_2, a_3 \rangle$, $\operatorname{tr}(\Omega_H)$ can be computed in terms of the relation matrix A of $R = K[H]$. To show this, we need to consider a free presentation of ω_R.

Let $S = K[x_1, x_2, x_3]$ be the polynomial ring, and let $\epsilon : S \to R$ be the surjective K-algebra homomorphism with $\epsilon(x_i) = t^{a_i}$ for $i = 1, 2, 3$. Let $A^{\top}$ be as given in (5.8). The minimal graded free S-resolution of R is given by

$$0 \to Sf_1 \oplus Sf_2 \xrightarrow{A^{\top}} Se_1 \oplus Se_2 \oplus Se_3 \xrightarrow{\epsilon} R \to 0.$$

Since R is a Cohen-Macaulay ring (see Lemma 4.21), by Corollary 3.67,

$$0 \to Se_1^* \oplus Se_2^* \oplus Se_3^* \xrightarrow{A} Sf_1^* \oplus Sf_2^* \longrightarrow \omega_R \to 0.$$

is the minimal graded free S-resolution of ω_R. Here, the e_i^* and f_j^* denote the dual basis elements of the e_i and f_j, respectively.

Tensorizing this resolution with R we obtain for ω_R the R-free presentation

$$Re_1^* \oplus Re_2^* \oplus Re_3^* \xrightarrow{B} Rf_1^* \oplus Rf_2^* \xrightarrow{\eta} \omega_R \to 0,$$

where B is obtained from A by taking residues modulo $I_H = \operatorname{Ker} \epsilon$. It follows from (5.8) that

$$B = \begin{pmatrix} b_1 & b_2 & b_3 \\ b_1' & b_2' & b_3' \end{pmatrix} = \begin{pmatrix} t^{r_{23}a_3} & t^{r_{31}a_1} & t^{r_{12}a_2} \\ t^{r_{32}a_2} & t^{r_{13}a_3} & t^{r_{21}a_1} \end{pmatrix}.$$

Let $\omega_i = \eta(f_i^*)$ for $i = 1, 2$. Then the elements ω_1, ω_2 generate ω_R, and the relations among the generators are

$$b_1\omega_1 + b_1'\omega_2 = 0, \quad b_2\omega_1 + b_2'\omega_2 = 0, \quad \text{and} \quad b_3\omega_1 + b_3'\omega_2 = 0. \tag{6.10}$$

Let $I_1(B)$ be the ideal generated by the entries of B. Then we have

Theorem 6.26
$\operatorname{tr}(\omega_R) = I_1(B)$.

Proof. Note that $\operatorname{tr}(\omega_R)$ is generated by the elements $x\omega_1, x\omega_2$ with those $x \in Q(R)$ for which $x\omega_1, x\omega_2 \in R$. Fix $i \in \{1, 2, 3\}$, and let $x = b_i/\omega_2$. Then (6.10) implies that $x = -b_i'/\omega_1$. Therefore, $x\omega_1 = -b_i'$ and $x\omega_2 = b_i$. Since $b_i, b_i' \in R$, it follows that $b_i, b_i' \in \operatorname{tr}(\omega_R)$. This shows that $I_1(B) \subseteq \operatorname{tr}(\omega_R)$.

Conversely, let $f \in \operatorname{tr}(\omega_R)$. We want to show that $f \in I_1(B)$. We may assume that f is a generator of $\operatorname{tr}(\omega_R)$, as described above. Say, $f = x\omega_2$ with $x \in Q(R)$. Let $g = x\omega_1 \in R$. Then $f\omega_1 - g\omega_2 = 0$, and therefore $ff_1^* - gf_2^* \in \operatorname{Ker} \eta$. Since the elements $b_i f_1^* + b_i' f_2^*$ generate $\operatorname{Ker} \eta$, it follow that $f \in (b_1, b_2, b_3) \subseteq I_1(B)$. □

Let $r_1 = \min\{r_{21}, r_{31}\}$, $r_2 = \min\{r_{12}, r_{32}\}$ and $r_3 = \min\{r_{23}, r_{13}\}$. Inspecting the matrix B, we see that

$$I_1(B) = (t^{a_1 r_1}, t^{a_2 r_2}, t^{a_3 r_3}).$$

Hence, by Theorem 6.26 we obtain the following

Corollary 6.27
Let $H = \langle a_1, a_2, a_3 \rangle$ be a 3-generated numerical semigroup which is not symmetric. Then

(a) $\operatorname{tr}(\Omega_H) = (a_1 r_1, a_2 r_2, a_3 r_3)$.
(b) H is nearly symmetric if and only if $r_1 = r_2 = r_3 = 1$.

As a consequence of the above considerations we also obtain

Proposition 6.28
Let H be a numerical semigroup as given in Corollary 6.27. Then

$$\operatorname{res}(H) = r_1 r_2 r_3.$$

Proof. The residuum of H is given by

$$\operatorname{res}(H) = |H \setminus \operatorname{tr}(\Omega_H)| = \ell(R/\operatorname{tr}(\omega_R)) = \ell(R/I_1(B)) = \ell(R/(t^{a_1 r_1}, t^{a_2 r_2}, t^{a_3 r_3})).$$

Observe that

$$R/(t^{a_1 r_1}, t^{a_2 r_2}, t^{a_3 r_3}) \cong K[x_1, x_2, x_3]/(I_H, x_1^{r_1}, x_2^{r_2}, x_3^{r_3}).$$

By Theorem 4.38, $I_H = (f_{\mathbf{v}_1}, f_{\mathbf{v}_2}, f_{\mathbf{v}_3})$, where

$$\mathbf{v}_1 = (-c_1, r_{12}, r_{13}), \ \mathbf{v}_2 = (r_{21}, -c_2, r_{23}), \ \mathbf{v}_3 = (r_{31}, r_{32}, -c_3),$$

and $\mathbf{v}_1 + \mathbf{v}_2 + \mathbf{v}_3 = 0$. It follows that $(I_H, x_1^{r_1}, x_2^{r_2}, x_3^{r_3}) = (x_1^{r_1}, x_2^{r_2}, x_3^{r_3})$, and the desired result follows from Exercise 6.3.2. □

Exercise 6.29
$S = K[x_1, \ldots, x_n]$ be the polynomial ring and let $c_1, \ldots, c_n$ be non-negative integers. Then $\ell(S/(x_1^{c_1}, \ldots, x_n^{c_n})) = c_1 c_2 \cdots c_n$.

For any numerical semigroup we have $g(H) \geq n(H)$ with equality if and only if H is symmetric. Thus the difference $g(H) - n(H)$ is also a measure of how far H is from being symmetric. We have the following inequality.

Corollary 6.30
Let H be a 3-generated numerical semigroup. Then

$$\operatorname{res}(H) \leq g(H) - n(H).$$

Proof. When H is symmetric, both sides of the inequality are zero. If H is not symmetric, then Theorem 5.13 implies that $r_1 r_2 r_3 \leq 2g(H) - (F(H) + 1)$. Since $2g(H) - (F(H) + 1) = g(H) - n(H)$, and since by Proposition 6.28 we have $r_1 r_2 r_3 = \operatorname{res}(H)$, the desired inequality follows.

□

For numerical semigroups generated by more than 3 elements the inequality $\operatorname{res}(H) \leq g(H) - n(H)$ does not hold in general. The following example is due to Shinya Kumashiro. Let

$$H = \langle 13, 14, 15, 16, 17, 18, 21, 23 \rangle.$$

Then one can check that $\operatorname{res}(H) = 9$, $g(H) = 17$ and $n(H) = 9$. Therefore, $\operatorname{res}(H) > g(H) - n(H)$.

Notes

The concept of nearly Gorenstein rings was introduced in [37], namely as those Cohen-Macaulay rings for which $\mathfrak{m} \subseteq \operatorname{tr}(\omega_R)$. Already in 1993, Ding [24] studied Cohen-Macaulay local rings with the property that $\operatorname{tr}(\omega_R) = \mathfrak{m}$ and related this property to the index of the ring, and in 2003, Huneke and Vraciu [43] showed that a quotient ring R of a Gorenstein Artinian local rings by its socle is nearly Gorenstein, and they classified the $\mathfrak{m}$-primary monomial ideals I in a polynomial ring S for which S/I is of small type and nearly Gorenstein. In a more recent paper [20] by Dao, Kobayashi and Takahashi, among other results, the behavior of nearly Gorensteinness under reductions by regular sequences is studied.

The results presented here regarding nearly symmetric numerical semigroups are all taken from [38], where $\operatorname{res}(H)$ is introduced. In [39] the invariant $\operatorname{res}(H)$ has been further studied. It is shown that $\operatorname{res}(H) \leq g(H) - n(H)$, if $t(H) \leq 3$. This generalizes our Corollary 6.30, because for a 3-generated numerical semigroup $t(H) \leq 2$, as follows from Theorem 5.11. In the same paper $\operatorname{res}(H)$ is compared with $\operatorname{bg}(H)$, the so-called birational Gorenstein colength, which was introduced by Kobayashi [45]. It is shown that if H is not symmetric, then $\operatorname{res}(H) \leq 2\operatorname{bg}(H) - 1$.

Appendix A
Basic Concepts in Homological Algebra

This chapter presents key concepts in homological algebra, including the homology functor, which analyses chain complexes, Hom and tensor, derived functors like Tor and Ext, double complexes and representable functors.

A.1 Homology Functors

Let R be a ring. A sequence

$$(M_\bullet, \partial_\bullet)\colon \quad \cdots \longrightarrow M_{i+1} \xrightarrow{\partial_{i+1}} M_i \xrightarrow{\partial_i} M_{i-1} \longrightarrow \cdots$$

of R-modules M_i and R-module homomorphisms $\partial_i : M_i \longrightarrow M_{i-1}$ is called a *complex* of R-modules, if $\partial_i \circ \partial_{i+1} = 0$ for all i.

The maps ∂_i are called the *differentials* of $(M_\bullet, \partial_\bullet)$. If it is clear which are the differentials, we simply write $M_\bullet$ instead of $(M_\bullet, \partial_\bullet)$. For all i we define the R-submodules $Z_i(M_\bullet) = \operatorname{Ker} \partial_i$ and $B_i(M_\bullet) = \operatorname{Im} \partial_{i+1}$ of M_i. Since $\partial_i \circ \partial_{i+1} = 0$, it follows that $B_i(M_\bullet) \subseteq Z_i(M_\bullet)$.

The elements of $B_i(M_\bullet)$ are called *i-boundaries* and the elements of $Z_i(M_\bullet)$ are called *i-cycles* of $M_\bullet$. The factor module

$$H_i(M_\bullet) - Z_i(M_\bullet)/B_i(M_\bullet)$$

is called the ith *homology module* of $M_\bullet$.

If $z \in Z_i(M_\bullet)$ is an i-cycle, then we write $[z] = z + B_i(M_\bullet)$ for the homology class of z in $H_i(M_\bullet)$. Hence we have $H_i(M_\bullet) = \{[z]\colon\ z \in Z_i(M_\bullet)\}$.

The complex $M_\bullet$ is called *exact*, if $B_i(M_\bullet) = Z_i(M_\bullet)$ for all i. In other words, $M_\bullet$ is an exact sequence if and only if $H_i(M_\bullet) = 0$ for all i. An exact sequence of the form $0 \to M \to N \to P \to 0$ is called a *short exact sequence*. Moreover,

J. Herzog et al., *Numerical Semigroups*, Compact Textbooks in Mathematics,
https://doi.org/10.1007/978-3-032-05424-1

a complex $\cdots \longrightarrow M_i \longrightarrow \cdots \longrightarrow M_1 \longrightarrow M_0 \longrightarrow 0$ is called *acyclic*, whenever $H_i(M_\bullet) = 0$ for all $i > 0$.

Exercise A.1
Let $M_\bullet$ be a complex, and let T be an exact covariant functor. Prove that $H_i(T(M_\bullet)) \cong T(H_i(M_\bullet))$.

Let $(M_\bullet, \partial_\bullet)$ and $(N_\bullet, \partial'_\bullet)$ be two complexes of R-modules. A *chain map* between $M_\bullet$ and $N_\bullet$ is a family $\varphi_\bullet = (\varphi_i)_{i\in\mathbb{Z}}$ where for all i, $\varphi_i : M_i \to N_i$ is an R-module homomorphism such that the following diagrams

$$\begin{array}{ccc} M_i & \xrightarrow{\partial_i} & M_{i-1} \\ \downarrow{\scriptstyle \varphi_i} & & \downarrow{\scriptstyle \varphi_{i-1}} \\ N_i & \xrightarrow{\partial'_i} & N_{i-1} \end{array}$$

commute. In other words, we have $\partial'_i \circ \varphi_i = \varphi_{i-1} \circ \partial_i$ for all i.

Let $z \in Z_i(M_\bullet)$. Then $\varphi_i(z) \in Z_i(N_\bullet)$, because $\partial'_i(\varphi_i(z)) = \varphi_{i-1}(\partial_i(z)) = 0$. Similarly, if $z \in B_i(M_\bullet)$, then $\varphi_i(z) \in B_i(N_\bullet)$. Therefore, $\varphi_\bullet$ induces for all i an R-module homomorphism

$$H_i(\varphi_\bullet)\colon H_i(M_\bullet) \longrightarrow H_i(N_\bullet),\ [z] \mapsto [\varphi_i(z)].$$

If $\psi_\bullet\colon N_\bullet \longrightarrow L_\bullet$ is another chain map, then $H_i(\psi_\bullet \circ \varphi_\bullet) = H_i(\psi_\bullet) \circ H_i(\varphi_\bullet)$. As you can see, we not only have homology modules of complexes but also induced homology maps of chain maps. Since $H_i(\mathrm{id}_{M_\bullet}) = \mathrm{id}_{H_i(M_\bullet)}$, it follows that H_i is a functor from the category of complexes of R-modules to the category of R-modules.

Let $\varphi_\bullet, \psi_\bullet\colon M_\bullet \to N_\bullet$ be two chain maps. Then $\varphi_\bullet$ and $\psi_\bullet$ are called *homotopic* if there exists a family $\eta_\bullet$ of R-module homomorphisms with $\eta_i : M_i \to N_{i+1}$ for all i, such that

$$\varphi_i - \psi_i = \partial'_{i+1} \circ \eta_i + \eta_{i-1} \circ \partial_i$$

for all i.

We write $\varphi_\bullet \sim \psi_\bullet$ if $\varphi_\bullet$ and $\psi_\bullet$ are homotopic. The family of maps $\eta_\bullet$ is called a *chain homotopy*.

$$\begin{array}{ccccccccc} \cdots & \longrightarrow & M_{i+1} & \longrightarrow & M_i & \xrightarrow{\partial_i} & M_{i-1} & \longrightarrow & \cdots \\ & & \downarrow\downarrow & \swarrow{\scriptstyle \eta_i} & \downarrow\downarrow & \swarrow{\scriptstyle \eta_{i-1}} & \downarrow\downarrow & & \\ \cdots & \longrightarrow & N_{i+1} & \xrightarrow[\partial'_{i+1}]{} & N_i & \longrightarrow & N_{i-1} & \longrightarrow & \cdots . \end{array}$$

Homotopic chain maps have the following nice property:

Theorem A.2
Let $\varphi_\bullet, \psi_\bullet \colon M_\bullet \to N_\bullet$ be homotopic chain maps. Then

$$H_i(\varphi_\bullet) = H_i(\psi_\bullet) \quad \text{for all} \quad i.$$

Proof. Let $z \in Z_i(M_\bullet)$. Then

$$\begin{aligned}(H_i(\varphi_\bullet) - H_i(\psi_\bullet))[z] &= H_i(\varphi_\bullet - \psi_\bullet)[z] = [(\varphi_i - \psi_i)(z)] \\ &= [\partial'_{i+1}(\eta_i(z)) + \eta_{i-1}(\partial_i(z))] = 0,\end{aligned}$$

since $\partial'_{i+1}(\eta_i(z)) \in B_i(N_\bullet)$ and since $\partial_i(z) = 0$. This shows that $H_i(\varphi_\bullet) = H_i(\psi_\bullet)$, as desired. □

Corollary A.3
Let $M_\bullet$ be a complex of R-modules. If $\mathrm{id}_{M_\bullet} \sim 0$, then $M_\bullet$ is exact.

Proof. The assumption implies that $\mathrm{id}_{H_i(M_\bullet)} = H_i(\mathrm{id}_{M_\bullet}) = H_i(0) = 0$ for all i. This is only possible if $H_i(M_\bullet) = 0$ for all i. □

A sequence of complexes and chain maps

$$0 \longrightarrow A_\bullet \xrightarrow{\varphi_\bullet} B_\bullet \xrightarrow{\psi_\bullet} C_\bullet \longrightarrow 0$$

is called a short exact sequence of complexes, if

$$0 \longrightarrow A_i \xrightarrow{\varphi_i} B_i \xrightarrow{\psi_i} C_i \longrightarrow 0$$

is a short exact sequence for all i.

Now, we come to the *fundamental theorem of homological algebra.*

Theorem A.4

Let $0 \longrightarrow A_\bullet \xrightarrow{\varphi_\bullet} B_\bullet \xrightarrow{\psi_\bullet} C_\bullet \longrightarrow 0$ be a short exact sequence of complexes of R-modules. Then we obtain the long exact sequence

$$\cdots \longrightarrow H_{i+1}(C_\bullet) \xrightarrow{\delta_{i+1}} H_i(A_\bullet) \xrightarrow{H_i(\varphi_\bullet)} H_i(B_\bullet) \xrightarrow{H_i(\psi_\bullet)} H_i(C_\bullet)$$
$$\xrightarrow{\delta_i} H_{i-1}(A_\bullet) \longrightarrow \cdots$$

of homology modules. The maps δ_i are called connecting homomorphisms.

Proof. We denote by $\partial_\bullet$ the differential of $A_\bullet$, by $\partial'_\bullet$ the differential of $B_\bullet$ and by $\partial''_\bullet$ the differential of $C_\bullet$.

We first define the connecting homomorphisms δ_i. Let $[z] \in H_i(C_\bullet)$. Then, $z \in Z_i(C_\bullet)$. Since $0 \longrightarrow A_i \xrightarrow{\varphi_i} B_i \xrightarrow{\psi_i} C_i \longrightarrow 0$ is exact, there exists $b \in B_i$ with $\psi_i(b) = z$. Then, $\partial'_i(b) \in B_{i-1}$ and $\psi_{i-1}(\partial'_i(b)) = \partial''_i(\psi_i(b)) = \partial''_i(z) = 0$. Hence, $\partial'_i(b) \in \operatorname{Ker}\psi_{i-1} = \operatorname{Im}\varphi_{i-1}$. Let $w \in A_{i-1}$ with $\varphi_{i-1}(w) = \partial'_i(b)$, see the following diagram

$$\begin{array}{ccccccccc}
0 & \longrightarrow & A_i & \xrightarrow{\varphi_i} & B_i & \xrightarrow{\psi_i} & C_i & \longrightarrow & 0 \\
 & & \downarrow{\scriptstyle \partial_i} & & \downarrow{\scriptstyle \partial'_i} & & \downarrow{\scriptstyle \partial''_i} & & \\
0 & \longrightarrow & A_{i-1} & \xrightarrow{\varphi_{i-1}} & B_{i-1} & \xrightarrow{\psi_{i-1}} & C_{i-1} & \longrightarrow & 0 \\
 & & \downarrow{\scriptstyle \partial_{i-1}} & & \downarrow{\scriptstyle \partial'_{i-1}} & & \downarrow{\scriptstyle \partial''_{i-1}} & & \\
0 & \longrightarrow & A_{i-2} & \xrightarrow{\varphi_{i-2}} & B_{i-2} & \xrightarrow{\psi_{i-2}} & C_{i-2} & \longrightarrow & 0.
\end{array}$$

We have $\varphi_{i-2}(\partial_{i-1}(w)) = \partial'_{i-1}(\varphi_{i-1}(w)) = \partial'_{i-1}(\partial'_i(b)) = 0$. Since φ_{i-2} is injective, it follows that $\partial_{i-1}(w) = 0$. Therefore, $w \in Z_{i-1}(A_\bullet)$, and we set $\delta_i([z]) = [w]$. That the so defined map δ_i is well defined is part of Exercise A.5.

We prove the exactness of the long sequence of homology modules only at the position $H_i(B_\bullet) \xrightarrow{H_i(\psi_\bullet)} H_i(C_\bullet) \xrightarrow{\delta_i} H_{i-1}(A_\bullet)$. The rest of the steps in the proof we leave to you as an exercise.

Let $[b] \in H_i(B_\bullet)$. Then $H_i(\psi_\bullet)([b]) = [z] \in H_i(C_\bullet)$, where $z = \psi_i(b)$. By definition, $\delta_i([z]) = [w]$, where $\varphi_{i-1}(w) = \partial'_i(b)$. Since $b \in Z_i(B_\bullet)$, it follows that $\partial'_i(b) = 0$. Now we use that φ_{i-1} is injective and deduce that $w = 0$. Therefore, $(\delta_i \circ H_i(\psi_\bullet))([b]) = \delta_i(H_i(\psi_\bullet)([b])) = \delta_i([z]) = 0$. This shows that $\operatorname{Im} H_i(\psi_\bullet) \subseteq \operatorname{Ker}\delta_i$.

Conversely, let $[z] \in H_i(C_\bullet)$ and assume that $\delta_i([z]) = [w] = 0$, where $\varphi_{i-1}(w) = \partial'_i(b)$ for some $b \in B_i$ with $\psi_i(b) = z$. Then, $w = \partial_i(v)$ for some $v \in A_i$. Therefore $\partial'_i(b - \varphi_i(v)) = \partial'_i(b) - \partial'_i(\varphi_i(v)) = \varphi_{i-1}(w) - \varphi_{i-1}(w) = 0$, and hence $b' := b - \varphi_i(v) \in Z_i(B_\bullet)$. Furthermore, $H_i(\psi_\bullet)([b']) = [\psi_i(b - \varphi_i(v))] = [\psi_i(b) - \psi_i(\varphi_i(v))] = [z - 0] = [z]$. □

Exercise A.5
(a) Show that δ_i is well defined, that is, δ_i does not depend on the various choices in its construction.

(b) Prove the exactness of the long homology sequence at the positions $H_i(A_{\bullet})$ and $H_i(B_{\bullet})$.

One calls a sequence of maps and R-modules

$$(M^{\bullet}, \partial^{\bullet}): \quad \cdots \longrightarrow M^{i-1} \xrightarrow{\partial^{i-1}} M^i \xrightarrow{\partial^i} M^{i+1} \longrightarrow \cdots$$

a *cocomplex*, if $\partial^i \circ \partial^{i-1} = 0$ for all i. The module $H^i(M^{\bullet}) = \operatorname{Ker} \partial^i / \operatorname{Im} \partial^{i-1}$ is called the *ith cohomology module* of $M^{\bullet}$. Cocomplexes appear naturally in the definition of Ext-modules, as we shall see in the next sections. It is clear that for cocomplexes the same rules hold as for complexes.

We call a complex $(M_{\bullet}, \partial_{\bullet})$ a complex of graded modules, if all M_i are graded R-modules and the differentials are graded R-module homomorphisms. It follows from Proposition 1.4 that the homology modules $H_i(M_{\bullet})$ are graded R-modules. In a similar way, one defines a cocomplex of graded modules.

A.2 Double Complexes

Double (co)complexes play an important role when we consider in Sect. A.4 the functors Tor and Ext, which are both functors in two variables.

A *double complex* $D_{\bullet,\bullet}$ is a diagram

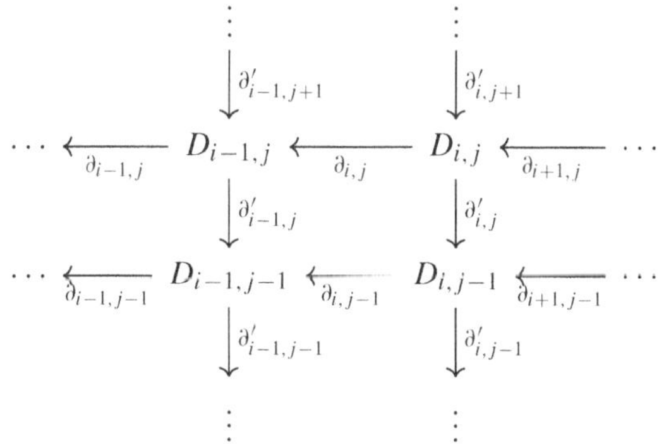

$$\begin{array}{ccccccc}
 & & \vdots & & \vdots & & \\
 & & \downarrow{\scriptstyle \partial'_{i-1,j+1}} & & \downarrow{\scriptstyle \partial'_{i,j+1}} & & \\
\cdots & \xleftarrow{\partial_{i-1,j}} & D_{i-1,j} & \xleftarrow{\partial_{i,j}} & D_{i,j} & \xleftarrow{\partial_{i+1,j}} & \cdots \\
 & & \downarrow{\scriptstyle \partial'_{i-1,j}} & & \downarrow{\scriptstyle \partial'_{i,j}} & & \\
\cdots & \xleftarrow{\partial_{i-1,j-1}} & D_{i-1,j-1} & \xleftarrow{\partial_{i,j-1}} & D_{i,j-1} & \xleftarrow{\partial_{i+1,j-1}} & \cdots \\
 & & \downarrow{\scriptstyle \partial'_{i-1,j-1}} & & \downarrow{\scriptstyle \partial'_{i,j-1}} & & \\
 & & \vdots & & \vdots & &
\end{array}$$

satisfying the following conditions:

1. For all i and j we have

$$\partial'_{i-1,j} \circ \partial_{i,j} = \partial_{i,j-1} \circ \partial'_{i,j}.$$

2. For all j, $(D_{\bullet,j}, \partial_{\bullet,j})$ is a complex, and for all i, $(D_{i,\bullet}, \partial'_{i,\bullet})$ is a complex.

A double cocomplex is defined in a similar way, but with all arrows reversed.

The complex $D_{\bullet,j}$ is called the *jth row complex* and the complex $D_{i,\bullet}$ is called the *ith column complex* of $D_{\bullet,\bullet}$.

Associated with the double complex $D_{\bullet,\bullet}$ is its *total complex* $(D_\bullet, d_\bullet)$ with

$$D_l = \bigoplus_{\substack{i,j \\ i+j=l}} D_{i,j},$$

where $d_l : D_l \to D_{l-1}$ is defined by

$$d_l(a) = \partial_{i,j}(a) + (-1)^i \partial'_{i,j}(a) \quad \text{for} \quad a \in D_{i,j} \quad \text{with} \quad i + j = l.$$

Exercise A.6

Show that $D_\bullet$ is indeed a complex. In other words, prove that $d_{l-1} \circ d_l = 0$ for all l.

We are primarily interested in first quadrant double (co)complexes. A double complex $D_{\bullet,\bullet}$ is called a *first quadrant* double complex, if $D_{i,j} = 0$ for $i < 0$ or $j < 0$. Hence such a double complex is of the shape

$$\begin{array}{ccccccccc}
 & & \vdots & & \vdots & & \vdots & & \\
 & & \downarrow & & \downarrow & & \downarrow & & \\
0 & \longleftarrow & D_{0,2} & \longleftarrow & D_{1,2} & \longleftarrow & D_{2,2} & \longleftarrow & \cdots \\
 & & \downarrow & & \downarrow & & \downarrow & & \\
0 & \longleftarrow & D_{0,1} & \longleftarrow & D_{1,1} & \longleftarrow & D_{2,1} & \longleftarrow & \cdots \\
 & & \downarrow & & \downarrow & & \downarrow & & \\
0 & \longleftarrow & D_{0,0} & \longleftarrow & D_{1,0} & \longleftarrow & D_{2,0} & \longleftarrow & \cdots \\
 & & \downarrow & & \downarrow & & \downarrow & & \\
 & & 0 & & 0 & & 0 & &
\end{array}$$

The chain maps $D_{i-1,\bullet} \xleftarrow{\partial_{i,\bullet}} D_{i,\bullet}$ induce module homomorphisms

$$H_0(D_{i-1,\bullet}) \xleftarrow{H_0(\partial_{i,\bullet})} H_0(D_{i,\bullet}).$$

Since $\partial_{i-1,\bullet} \circ \partial_{i,\bullet} = 0$ for all i, it follows that

$$0 \longleftarrow H_0(D_{0,\bullet}) \xleftarrow{H_0(\partial_{1,\bullet})} H_0(D_{1,\bullet}) \xleftarrow{H_0(\partial_{2,\bullet})} H_0(D_{2,\bullet}) \xleftarrow{H_0(\partial_{3,\bullet})} \cdots$$

is a complex. In order to simplify notation, we set $C_i = H_0(D_{i,\bullet})$ and $\psi_i = H_0(\partial_{i,\bullet})$ for all i. Then this complex becomes the complex

$$C_\bullet : 0 \longleftarrow C_0 \xleftarrow{\psi_1} C_1 \xleftarrow{\psi_2} C_2 \xleftarrow{\psi_3} \cdots$$

Similarly, one defines the complex $C'_\bullet$ with $C'_i = H_0(D_{\bullet,i})$ for all i.

Theorem A.7
Suppose all column complexes are acyclic, that is, $H_j(D_{i,\bullet}) = 0$ for all $i \geq 0$ and all $j > 0$. Then

$$H_i(D_\bullet) \cong H_i(C_\bullet) \quad \text{for all} \quad i \geq 0.$$

By symmetry, a similar statement holds for the row complexes. In particular, if all row and column complexes are acyclic, then $H_i(C_\bullet) \cong H_i(C'_\bullet)$ for all i.

Proof. We first define a map $\alpha\colon H_i(D_\bullet) \to H_i(C_\bullet)$. Let $[z] \in H_i(D_\bullet)$, where $z = (z_0, z_1, \ldots, z_i)$ is an i-cycle of $D_\bullet$ with $z_k \in D_{k,i-k}$ for $k = 0, \ldots i$. Note that z is an i-cycle, if and only if

$$\partial'_{k,i-k}(z_k) = (-1)^{k+1}\partial_{k+1,i-(k+1)}(z_{k+1}) \tag{A.1}$$

for $k = 0, \ldots, i-1$.

For each k, let $\epsilon_k\colon D_{k,0} \to H_0(D_{k,\bullet}) = C_k$ be the augmentation map of the complex $D_{k,\bullet}$. Then

$$\operatorname{Im} \partial'_{k,1} = \operatorname{Ker} \epsilon_k \quad \text{and} \quad \epsilon_{k-1} \circ \partial_{k,0} = \psi_k \circ \epsilon_k. \tag{A.2}$$

We set

$$\alpha([z]) = [\epsilon_i(z_i)].$$

For the map α to be well-defined, we have to show:
(1) $\epsilon_i(z_i)$ is an i-cycle of $C_\bullet$.
(2) If z' is an i-cycle of $D_\bullet$ with $[z] = [z']$, then $\alpha([z]) = \alpha([z'])$.

Using (A.1) and (A.2), we have $\psi_i(\epsilon_i(z_i)) = \epsilon_{i-1}(\partial_{i,0}(z_i)) = \epsilon_{i-1}((-1)^i\partial'_{i-1,1}(z_{i-1})) = 0$. This proves (1).

For the proof of (2), we note $z' = z + b$, where b is an i-boundary of $D_\bullet$. Let $b = (b_0, b_1, \ldots, b_i)$ with $b_k \in D_{k,i-k}$ for $k = 0, \ldots, i$. Then $z'_i = z_i + b_i$. Since b is an i-boundary, there exists $a = (a_0, \ldots, a_{i+1}) \in D_{i+1}$ with $d_i(a) = b$. This implies that $b_i = (-1)^i\partial'_{i,1}(a_i) + \partial_{i+1,0}(a_{i+1})$. Therefore, since $\epsilon_i(\partial'_{i,1}(a_i)) = 0$, it follows that $\epsilon_i(z'_i) = \epsilon_i(z_i) + \epsilon_i(\partial_{i+1,0}(a_{i+1})) = \epsilon_i(z_i) + \psi_{i+1}(\epsilon_{i+1}(a_{i+1}))$. This shows that $[\epsilon_i(z_i)] = [\epsilon_i(z'_i)]$, as desired.

Next we show that $\alpha\colon H_i(D_\bullet) \to H_i(C_\bullet)$ is surjective. Indeed, let $[\epsilon_i(z_i)] \in H_i(C_\bullet)$ with some $z_i \in D_{i,0}$. For $k = 1, \ldots, i$ we will construct by induction on k, elements $z_k \in D_{k,i-k}$ such that

$$\partial'_{i-k,k}(z_{i-k}) = (-1)^{i-(k-1)}\partial_{i-(k-1),k-1}(z_{i-(k-1)}).$$

Having constructed these z_k, we let $z = (z_0, \ldots, z_i)$. Then by (A.1), z is an i-cycle of $D_\bullet$ and $\alpha([z]) = [\epsilon_i(z_i)]$. This will then prove that α is surjective.

Now we start the induction proof: since $\epsilon_i(z_i)$ in an i-cycle of $C_\bullet$, it follows that $\epsilon_{i-1}(\partial_{i,0}(z_i)) = \psi_i(\epsilon_i(z_i)) = 0$. Therefore, since $\operatorname{Ker} \epsilon_{i-1} = \operatorname{Im} \partial'_{i-1,1}$, there exists $z_{i-1} \in D_{i-1,1}$ such that $\partial'_{i-1,1}(z_{i-1}) = (-1)^i \partial_{i,0}(z_i)$. This completes the proof of the induction begin.

Assume now that z_{i-l} has already been constructed for $l = 1, \ldots, k-1$. Our induction hypothesis implies that

$$\partial'_{i-(k-1),k-1}(z_{i-(k-1)}) = (-1)^{i+1-(k-1)}\partial_{i+1-(k-1),k-2}(z_{i+1-(k-1)}).$$

Therefore,

$$\begin{aligned}\partial'_{i-k,k-1}(\partial_{i-(k-1),k-1}(z_{i-(k-1)})) &= \partial_{i-(k-1),k-2}(\partial'_{i-(k-1),k-1}(z_{i-k+1}))\\ &= (-1)^{i+1-(k-1)}\partial_{i-(k-1),k-2}(\partial_{i+1-(k-1),k-2}(z_{i+1-(k-1)})) = 0.\end{aligned}$$

This shows that $\partial_{i-k+1,k-1}(z_{i-k+1}) \in \operatorname{Ker} \partial'_{i-k,k-1}$. Hence, since $D_{i-k,\bullet}$ is acyclic, there exists $z_{i-k} \in D_{i-k,k}$ such that $\partial'_{i-k,k}(z_{i-k}) = (-1)^{i-(k-1)}\partial_{i-(k-1),k-1}$ $(z_{i-(k-1)})$, as desired.

Finally we show that $\alpha\colon H_i(D_\bullet) \to H_i(C_\bullet)$ is injective. Indeed, let $z \in D_i$ be an i-cycle of $D_\bullet$ with $\alpha([z]) = 0$. We must show that $[z] = 0$. By induction on k we construct for $k = 0, \ldots, i+1$ elements $a_{i+1-k} \in D_{i+1-k,k}$ such that $d_{i+1}(a) = z$ for $a = (a_0, a_1, \ldots, a_{i+1})$. This then implies that $[z] = 0$.

To say that $d_{i+1}(a) = z$ is equivalent to saying that

$$\partial'_{i+1-(k+1),k+1}(a_{i+1-(k+1)}) = (-1)^{i+1-k}(\partial_{i+1-k,k}(a_{i+1-k}) - z_{i-k})$$

for $k = 0, \ldots, i$.

Now we start our construction of the a_{i+1-j}. Since $[\epsilon_i(z_i)] = 0$, there exists $a_{i+1} \in D_{i+1,0}$ with $\psi_i(\epsilon_{i+1}(a_{i+1})) = \epsilon_i(z_i)$. This implies that $\epsilon_i(\partial_{i+1,0}(b_{i+1})) = \epsilon_i(z_i)$. Therefore, $\partial_{i+1,0}(a_{i+1}) - z_i \in \operatorname{Ker} \partial'_{i,1}$, and hence there exists $a_i \in D_{i,1}$ such that $\partial'_{i,1}(a_i) = (-1)^{i+1}(\partial_{i+1,0}(a_{i+1}) - z_i)$.

Assume now that a_{i+1-j} has already been constructed for $j = 0, \ldots, k-1$.
We have

$$\partial'_{i+1-k,k-1}(\partial_{i+1-(k-1),k-1}(a_{i+1-(k-1)})) = \\ \partial_{i+1-(k-1),k-2}(\partial'_{i+1-(k-1),k-1}(a_{i+1-(k-1)})), \quad \text{(A.3)}$$

and by our induction hypothesis we have

$$\partial'_{i+1-(k-1),k-1}(a_{i+1-(k-1)}) = (-1)^{i+1-(k-2)}(\partial_{i+1-(k-2),k-2}(a_{i+1-(k-2)}) - z_{i+1-(k-1)}).$$

Therefore,

$$\partial_{i+1-(k-1),k-2}(\partial'_{i+1-(k-1),k-1}(a_{i+1-(k-1)})) = (-1)^{i+1-(k-1)}\partial_{i+1-(k-1),k-2}(z_{i+1-(k-1)}),$$

and hence (A.3) implies that

$$\partial'_{i+1-k,k-1}(\partial_{i+1-(k-1),k-1}(a_{i+1-(k-1)})) = (-1)^{i+1-(k-1)}\partial_{i+1-(k-1),k-2}(z_{i+1-(k-1)}). \quad (A.4)$$

On the other hand, since z is a cycle, we also have

$$\partial'_{i+1-k,k-1}(z_{i+1-k}) = (-1)^{i+1-(k-1)}\partial_{i+1-(k-1),k-2}(z_{i+1-(k-1)}).$$

This, together with (A.4), implies that

$$\partial'_{i+1-k,k-1}(\partial_{i+1-(k-1),k-1}(a_{i+1-(k-1)}) - z_{i+1-k}) = 0.$$

Hence, since $D_{i+1-k,\bullet}$ is acyclic there exists $a_{i+1-k} \in D_{i+1-k,k}$ with

$$\partial'_{i+1-k,k}(a_{i+1-k}) = (-1)^{i+1-(k-1)}(\partial_{i+1-(k-1),k-1}(a_{i+1-(k-1)}) - z_{i+1-k}).$$

This completes the induction step and finishes the proof of the theorem. □

We have a similar theorem for first quadrant double cocomplexes $D^{\bullet,\bullet}$. Their shape is as shown in the following diagram.

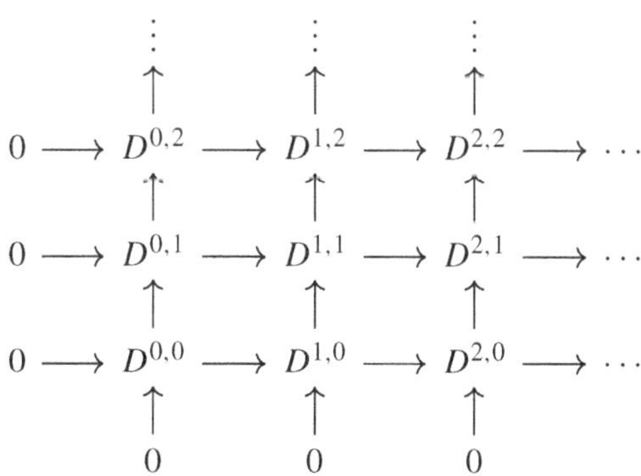

Let $C^i = H^0(D^{i,\bullet})$ and $C'^i = H^0(D^{\bullet,i})$. Then we obtain the cocomplexes

$$0 \to C^0 \to C^1 \to C^2 \to \cdots \quad \text{and} \quad 0 \to C'^0 \to C'^1 \to C'^2 \to \cdots,$$

and as in Theorem A.7 we see that $H^i(D^\bullet) \cong H^i(C^\bullet)$ for all i, if all the column cocomplexes $D^{i,\bullet}$ are acyclic, that is, if $H^j(D^{i,\bullet}) = 0$ for all $i \geq 0$ and all $j > 0$. If the row cocomplexes are acyclic as well, then

$$H^i(C^\bullet) \cong H^i(D^\bullet) \cong H^i(C'^\bullet) \quad \text{for all} \quad i. \tag{A.5}$$

We leave the details to the reader.

Of course, if we have double (co)complexes in the category GMod_R, then the homology modules and the homomorphisms between them are also graded.

A.3 Hom and Tensor

In what follows we review basic properties of the Hom and tensor functors.

We start with the Hom functor. Let R be an arbitrary commutative ring, and let M and N be R-modules. We denote by $\mathrm{Hom}_R(M, N)$, the set of all R-module homomorphisms from M to N. Let $r \in R$ and $\sigma, \tau \in \mathrm{Hom}_R(M, N)$. We define $r\sigma : M \to N$ by $m \mapsto r\sigma(m)$ for all $m \in M$ and $(\sigma + \tau): M \to N$ by $m \mapsto \sigma(m) + \tau(m)$ for all $m \in M$. It is easily checked that $r\sigma$ and $\sigma + \tau$ are R-module homomorphisms from M to N, and with this definition of addition and scalar multiplication, $\mathrm{Hom}_R(M, N)$ has the structure of an R-module.

Now, let $\varphi: M \to M'$ and $\psi: N \to N'$ be further R-module homomorphisms. Then these maps induce the R-module homomorphisms

$$\mathrm{Hom}_R(\varphi, N): \mathrm{Hom}_R(M', N) \to \mathrm{Hom}_R(M, N), \quad \beta \mapsto \beta \circ \varphi$$

and

$$\mathrm{Hom}_R(M, \psi): \mathrm{Hom}_R(M, N) \to \mathrm{Hom}_R(M, N'), \quad \alpha \mapsto \psi \circ \alpha.$$

Let $\varphi' : M' \to M''$ and $\psi': N' \to N''$ be further R-module homomorphisms. Then

$$\mathrm{Hom}_R(\varphi' \circ \varphi, N) = \mathrm{Hom}_R(\varphi, N) \circ \mathrm{Hom}_R(\varphi', N),$$

and

$$\mathrm{Hom}_R(M, \psi' \circ \psi) = \mathrm{Hom}_R(M, \psi') \circ \mathrm{Hom}_R(M, \psi).$$

We also have $\mathrm{Hom}_R(\mathrm{id}_M, N) = \mathrm{id}_{\mathrm{Hom}_R(M,N)} = \mathrm{Hom}_R(M, \mathrm{id}_N)$. We denote by Mod_R the category of all R-modules. These properties show that $\mathrm{Hom}_R(-, N)$ is a contravariant functor and $\mathrm{Hom}_R(M, -)$ is a covariant functor from Mod_R to Mod_R.

How much is the functor $\mathrm{Hom}_R(M, -)$ determined by the module M? The answer is given in

Proposition A.8
Let M and M' be R-modules, and suppose the functors $\operatorname{Hom}_R(M, -)$ and $\operatorname{Hom}_R(M', -)$ are functorially isomorphic. Then $M \cong M'$.

Proof. Let $\varphi\colon \operatorname{Hom}_R(M, -) \to \operatorname{Hom}_R(M', -)$ be the functorial isomorphism. Then $\alpha := \varphi_M(\mathrm{id}_M) \in \operatorname{Hom}_R(M', M)$, and we obtain the following commutative diagram

$$\begin{array}{ccc} \operatorname{Hom}_R(M, M') & \xrightarrow{\varphi_{M'}} & \operatorname{Hom}_R(M', M') \\ {\scriptstyle \operatorname{Hom}_R(M,\alpha)}\downarrow & & \downarrow{\scriptstyle \operatorname{Hom}_R(M',\alpha)} \\ \operatorname{Hom}_R(M, M) & \xrightarrow{\varphi_M} & \operatorname{Hom}_R(M', M). \end{array}$$

Let $\beta \in \operatorname{Hom}_R(M, M')$ with $\varphi_{M'}(\beta) = \mathrm{id}_{M'}$. Since the diagram is commutative, we then have

$$\varphi_M(\alpha \circ \beta) = \varphi_M(\operatorname{Hom}_R(M, \alpha)(\beta) = \varphi_M(\mathrm{id}_M).$$

It follows that $\alpha \circ \beta = \mathrm{id}_M$.

Similarly, we have a commutative diagram

$$\begin{array}{ccc} \operatorname{Hom}_R(M', M) & \xrightarrow{\varphi_M^{-1}} & \operatorname{Hom}_R(M, M) \\ {\scriptstyle \operatorname{Hom}_R(M',\beta)}\downarrow & & \downarrow{\scriptstyle \operatorname{Hom}_R(M,\beta)} \\ \operatorname{Hom}_R(M', M') & \xrightarrow{\varphi_{M'}^{-1}} & \operatorname{Hom}_R(M, M'). \end{array}$$

Hence

$$\varphi_{M'}^{-1}(\beta\circ\alpha) = \varphi_{M'}^{-1}(\operatorname{Hom}_R(M',\beta)(\alpha)) = \operatorname{Hom}_R(M,\beta)(\varphi_M^{-1}(\alpha)) = \beta = \varphi_{M'}^{-1}(\mathrm{id}_{M'}),$$

from which we deduce that $\beta \circ \alpha = \mathrm{id}_{M'}$. Therefore, $\beta\colon M \to M'$ is an isomorphism. $\square$

For an element $x \in R$ and an R-module M one has the R-module homomorphism

$$M \xrightarrow{\mu_x} M, \quad m \mapsto xm,$$

which is multiplication by x. Sometimes by misuse of notation we denote this map by $M \xrightarrow{x} M$. If N is another R-module, then it is straightforward to see that

$$\operatorname{Hom}_R(\mu_x, N) = \operatorname{Hom}_R(M, N) \xrightarrow{x} \operatorname{Hom}_R(M, N). \tag{A.6}$$

The same rule applies to $\operatorname{Hom}_R(M, \mu_x)$.

Our observation, which says that the multiplication map μ_x commutes with the Hom-functors, is a special case of a more general concept and result: A covariant functor $F\colon \mathrm{Mod}_R \to \mathrm{Mod}_R$ is called *R-linear*, if the map

$$F\colon \mathrm{Hom}_R(M,N) \longrightarrow \mathrm{Hom}_R(F(M),F(N)), \quad \varphi \mapsto F(\varphi),$$

is an R-module homomorphism. Similarly, a contravariant functor $F\colon \mathrm{Mod}_R \to \mathrm{Mod}_R$ is called R-linear, if the map $\mathrm{Hom}_R(M,N) \longrightarrow \mathrm{Hom}_R(F(N),F(M))$ is an R- module homomorphism.

Example A.9
For any R-module W, the functors $\mathrm{Hom}_R(W,-)$ and $\mathrm{Hom}_R(-,W)$ are R-linear.

The purpose of the next exercises is to show the left exactness of $\mathrm{Hom}_R(-,W)$ and $\mathrm{Hom}_R(W,-)$ for any R-module W.

Exercise A.10
Let $U \xrightarrow{\alpha} M \xrightarrow{\beta} N \longrightarrow 0$ be a complex of R-modules. Then $U \xrightarrow{\alpha} M \xrightarrow{\beta} N \longrightarrow 0$ is exact, if and only if

$$0 \longrightarrow \mathrm{Hom}_R(N,W) \xrightarrow{\mathrm{Hom}_R(\beta,W)} \mathrm{Hom}_R(M,W) \xrightarrow{\mathrm{Hom}_R(\alpha,W)} \mathrm{Hom}_R(U,W)$$

is exact for all R-modules W. Prove a similar statement for the functor $\mathrm{Hom}_R(W,-)$.

Exercise A.11
Let M be an R-module and I be an ideal of R. Prove that

$$\mathrm{Hom}_R(R/I,M) \cong 0 :_M I.$$

The functor $\mathrm{Hom}_R(-,W)$ is in general only left exact but not exact, as the following example demonstrates.

Example A.12
Let $R = K[x]$ be the polynomial ring over the field K, and let $\mathfrak{m} = (x)$ be the graded maximal ideal of R. The short exact sequence $0 \to \mathfrak{m} \to R \to R/\mathfrak{m} \to 0$ induces the exact sequence

$$0 \longrightarrow \mathrm{Hom}_R(R/\mathfrak{m},R) \longrightarrow \mathrm{Hom}_R(R,R) \longrightarrow \mathrm{Hom}_R(\mathfrak{m},R).$$

Suppose that $\mathrm{Hom}_R(-,R)$ is an exact functor. Then the map $\mathrm{Hom}_R(R,R) \longrightarrow \mathrm{Hom}_R(\mathfrak{m},R)$ is surjective, which means that for any $\varphi \in \mathrm{Hom}_R(\mathfrak{m},R)$, there exists $\psi \in \mathrm{Hom}_R(R,R)$ such that $\varphi = \psi|_{\mathfrak{m}}$, where $\psi|_{\mathfrak{m}}$ is the restriction of ψ to the submodule $\mathfrak{m}$.

Any $\psi \in \mathrm{Hom}_R(R,R)$ is equal to μ_r for some $r \in R$. So $\psi|_{\mathfrak{m}}(f) = rf$ for all $f \in \mathfrak{m}$. Thus for any $\varphi \in \mathrm{Hom}_R(\mathfrak{m},R)$, there exists $r \in R$ such that $\varphi(f) = rf$ for all $f \in \mathfrak{m}$. In particular, this assumption implies that $\varphi(\mathfrak{m}) \subseteq \mathfrak{m}$ for all $\varphi \in \mathrm{Hom}_R(\mathfrak{m},R)$. However, since the ideal $\mathfrak{m} = (x)$ is a free R-module, Proposition 1.9 shows that there exists $\varphi \in \mathrm{Hom}_R(\mathfrak{m},R)$ with $\varphi(x) = 1$, contradicting our assumption.

Exercise A.13

(a) Let N be an R-module and let F be a free R-module with basis $e_1, \ldots, e_r$. Show that $\mathrm{Hom}_R(F, N) \cong N^r$.

(b) Let R be a Noetherian ring, and let M and N be finitely generated R-modules. Show that $\mathrm{Hom}_R(M, N)$ is a finitely generated R-module.

Exercise A.14

(a) Let M and $N_1, \ldots, N_r$ be R-modules. Show that

$$\mathrm{Hom}_R(M, \bigoplus_{i=1}^{r} N_i) \cong \bigoplus_{i=1}^{r} \mathrm{Hom}_R(M, N_i).$$

(b) Let $\{M_i\}_{i\in\Lambda}$ be a family of R-modules and N be an R-module. Show that

$$\mathrm{Hom}_R(\bigoplus_{i\in\Lambda} M_i, N) \cong \prod_{i\in\Lambda} \mathrm{Hom}_R(M_i, N).$$

Hint: (a) If $\varphi_i \in \mathrm{Hom}_R(M, N_i)$ for $i = 1, \ldots, r$, then $\varphi \in \mathrm{Hom}_R(M, \bigoplus_{i=1}^{r} N_i)$, where $\varphi(x) = (\varphi_1(x), \ldots, \varphi_r(x))$ for all $x \in M$.

(b) Corresponding to any sequence $\{\varphi_i\} \in \bigoplus_{i\in\Lambda} \mathrm{Hom}_R(M_i, N)$, consider $\varphi : \bigoplus_{i\in\Lambda} M_i \to N$ with $\varphi(\{m_i\}) = \sum_{i\in\Lambda} \varphi_i(m_i)$

Theorem A.15

Let $0 \longrightarrow U \xrightarrow{\alpha} M \xrightarrow{\beta} N \longrightarrow 0$ be an exact sequence of R-modules. The following conditions are equivalent:

(i) There exists an R-module homomorphism $\gamma\colon N \to M$ such that $\beta \circ \gamma = \mathrm{id}_N$.

(ii) There exists an R-module homomorphism $\sigma : M \to U$ such that $\sigma \circ \alpha = \mathrm{id}_U$.

(iii) There exists an isomorphism $\varphi\colon U \oplus N \to M$ such that the following diagram

$$\begin{array}{ccccccccc}
0 & \longrightarrow & U & \xrightarrow{\alpha} & M & \xrightarrow{\beta} & N & \longrightarrow & 0 \\
 & & \uparrow{\scriptstyle \mathrm{id}_U} & & \uparrow{\scriptstyle \varphi} & & \uparrow{\scriptstyle \mathrm{id}_N} & & \\
0 & \longrightarrow & U & \xrightarrow{\iota} & U \oplus N & \xrightarrow{\pi} & N & \longrightarrow & 0,
\end{array}$$

is commutative. Here $\iota\colon U \to U \oplus N$ is the natural inclusion map and $\pi : U \oplus N \to N$ is the natural projection map.

If the equivalent conditions in Theorem A.15 hold, then the exact sequence $0 \longrightarrow U \stackrel{\alpha}{\longrightarrow} M \stackrel{\beta}{\longrightarrow} N \longrightarrow 0$ is called *split exact*.

Proof. (of Theorem A.15) (i) $\Rightarrow$ (ii): Let $m \in M$. Then $\beta(m - \gamma(\beta(m)) = \beta(m) - \beta(\gamma(\beta(m))) = \beta(m) - \beta(m) = 0$. Therefore, $m - \gamma(\beta(m)) \in \operatorname{Ker}\beta = \operatorname{Im}\alpha$. This together with the fact that α is injective, imply that there exists a unique element in U, which we denote by $\sigma(m)$, such that $\alpha(\sigma(m)) = m - \gamma(\beta(m))$. Now let $u \in U$. Since $\alpha(u) \in M$ and $\beta(\alpha(u)) = 0$, we have $\alpha(\sigma(\alpha(u))) = \alpha(u)$. Therefore, $\sigma(\alpha(u)) = u$ because α is injective. Note that σ is an R-module homomorphism. Indeed, for any $m, m' \in M$ we have $\alpha(\sigma(m + m')) = m + m' - \gamma(\beta(m + m')) = m - \gamma(\beta(m)) + m' - \gamma(\beta(m')) = \alpha(\sigma(m)) + \alpha(\sigma(m')) = \alpha(\sigma(m) + \sigma(m'))$. Since α is injective, we get $\sigma(m + m') = \sigma(m) + \sigma(m')$. Similarly, $\sigma(rm) = r\sigma(m)$ for any $r \in R$.

(ii) $\Rightarrow$ (i): Let $n \in N$. Then there exists $m \in M$ with $\beta(m) = n$. We set $\gamma(n) = m - \alpha(\sigma(m))$. It follows that $\beta(\gamma(n)) = \beta(m) = n$. The definition of $\gamma(n)$ does not depend on the choice of m. In fact, if $\beta(m') = n$ for some other $m' \in M$, then $m - m' \in \operatorname{Ker}\beta = \operatorname{Im}\alpha$. Therefore, there exists $u \in U$ with $\alpha(u) = m - m'$, and it follows that

$$
\begin{aligned}
&(m - \alpha(\sigma(m))) - (m' - \alpha(\sigma(m'))) = (m - m') - \alpha(\sigma(m - m')) = \\
&(m - m') - \alpha(\sigma(\alpha(u))) = (m - m') - \alpha(u) = (m - m') - (m - m') = 0.
\end{aligned}
$$

Thus γ is indeed well defined, and you may easily check that γ is an R-module homomorphism.

(i), (ii) $\Rightarrow$ (iii): Consider the R-module homomorphism $\varphi\colon U \oplus N \to M$ with $\varphi(u, n) = \alpha(u) + \gamma(n)$. Then, φ is injective. Because, if $\varphi(u, n) = 0$, then $\alpha(u) = -\gamma(n)$, and hence $0 = \beta(\alpha(u)) = -\beta(\gamma(n)) = -n$. Therefore, $n = 0$, and this implies that $\alpha(u) = 0$, which in turn implies that $u = 0$, since α is injective.

We also see that φ is surjective. Because, if $m \in M$, then $\beta o \gamma o \beta(m) = \beta(m)$. Hence $m - \gamma o \beta(m) \in \operatorname{Ker}\beta = \operatorname{Im}\alpha$. Hence $m - \gamma o \beta(m) = \alpha(u)$ for some $u \in U$. So $m = \alpha(u) + \gamma(\beta(m)) = \varphi(u, \beta(m))$. It is obvious that φ makes the diagram commutative.

(iii) $\Rightarrow$ (i), (ii): We let γ be the composition of the natural inclusion map $\iota\colon N \to U \oplus N$ with $\varphi\colon U \oplus N \to M$. Then $\beta \circ \gamma = \mathrm{id}_N$. Indeed,

$$
\beta \circ (\varphi \circ \iota) = (\beta \circ \varphi) \circ \iota = \pi \circ \iota = \mathrm{id}_N .
$$

Similarly, we let σ be the composition of $\varphi^{-1}\colon M \to U \oplus N$ with the natural projection map $U \oplus N \to U$. Then $\sigma \circ \alpha = \mathrm{id}_U$. □

Example A.1

The exact sequence $0 \longrightarrow U \stackrel{\alpha}{\longrightarrow} M \stackrel{\beta}{\longrightarrow} N \longrightarrow 0$ of finitely generated R-modules is split exact, if N is a free module.

Indeed, let $e_1, \dots, e_r$ be a basis of N, and choose $m_1, \dots, m_r \in M$ with $\beta(m_i) = e_i$ for all i. By Proposition 1.9, there exists an R-module homomorphism $\gamma : N \to M$ with $\gamma(e_i) = m_i$ for all i. It follows from the definition of γ that $\gamma \circ \beta = \mathrm{id}_M$. Hence our exact sequence splits.

Exercise A.16

Show that the exact sequence $0 \longrightarrow U \xrightarrow{\alpha} M \xrightarrow{\beta} N \longrightarrow 0$ of R-modules is split exact if and only if for any R-module W, the sequence

$$0 \to \mathrm{Hom}_R(N, W) \overset{\mathrm{Hom}_R(\beta, W)}{\longrightarrow} \mathrm{Hom}_R(M, W) \overset{\mathrm{Hom}_R(\alpha, W)}{\longrightarrow} \mathrm{Hom}_R(U, W) \to 0$$

is split exact.

Next, we study the tensor functor. Let R be any commutative ring and M and N be R-modules. We define the *tensor product* $M \otimes_R N$ as the R-module generated by the symbols $m \otimes n$ with $m \in M$ and $n \in N$, and the relations

$$\begin{aligned} r(m \otimes n) &= rm \otimes n = m \otimes rn \\ (m + m') \otimes n &= m \otimes n + m' \otimes n \quad \text{and} \\ m \otimes (n + n') &= m \otimes n + m \otimes n' \end{aligned} \tag{A.7}$$

for all $r \in R$, $m, m' \in M$ and $n, n' \in N$. When there is no confusion possible, we simply denote $M \otimes_R N$ by $M \otimes N$.

Let M, N and W be R-modules. A map $\alpha \colon M \times N \to W$ is called *R-bilinear* if for all $r \in R$, $m, m' \in M$ and $n, n' \in N$, we have

(1) $\alpha(rm, n) = \alpha(m, rn) = r\alpha(m, n)$
(2) $\alpha(m + m', n) = \alpha(m, n) + \alpha(m', n)$
(3) $\alpha(m, n + n') = \alpha(m, n) + \alpha(m, n')$.

The map $\alpha_{M,N} \colon M \times N \to M \otimes_R N$ defined by $\alpha_{M,N}(m, n) = m \otimes n$ is R-bilinear. Moreover, $M \otimes_R N$ satisfies the following universal property: for each R-bilinear map $\varphi \colon M \times N \to W$, there exists a unique R-module homomorphism $\varphi' \colon M \otimes_R N \to W$ such that the following diagram

$$\begin{array}{ccc} M \times N & \overset{\alpha_{M,N}}{\longrightarrow} & M \otimes_R N \\ & \searrow^{\varphi} & \downarrow \varphi' \\ & & W \end{array}$$

is commutative.

For given R-modules M, N and W, we denote by $\mathrm{Bihom}_R(M \times N, W)$ the set of R-bilinear maps $M \times N \to W$. The set $\mathrm{Bihom}_R(M \times N, W)$ has a natural R-module structure, and it follows from the universal property of the tensor product that the map

$$\mathrm{Hom}_R(M \otimes N, W) \to \mathrm{Bihom}_R(M \times N, W), \quad \psi \mapsto \psi \circ \alpha_{M,N} \tag{A.8}$$

is an isomorphism of R-modules.

Exercise A.17

Let M and N be R-modules. Let $\mathscr{G}$ be a system of generators of M and $\mathscr{H}$ be a system of generators of N. Show that the elements $m \otimes n$ with $m \in \mathscr{G}$ and $n \in \mathscr{H}$ generate $M \otimes N$, and conclude that $M \otimes N$ is a finitely generated R-module, if M and N are finitely generated R-modules.

Proposition A.18

Up to an isomorphism, the tensor product $M \otimes_R N$ of M and N is uniquely determined by its universal property.

Proof. Let $\alpha'_{M,N}: M \times N \to W$ be an R-bilinear map which has the universal property that is for all R-bilinear maps $\varphi: M \times N \to V$ there exists a unique R-module homomorphism $\varphi': W \to V$ with $\varphi \circ \alpha'_{M,N} = \varphi$.

By using the universal property of $M \otimes_R N$ and that of W, there exist R-module homomorphisms $\psi : M \otimes_R N \to W$ and $\varphi: W \to M \otimes_R N$ such that $\alpha'_{M,N} = \psi \circ \alpha_{M,N}$ and $\alpha_{M,N} = \varphi \circ \alpha'_{M,N}$. It follows that

$$\mathrm{id}_{M\otimes_R N} \circ \alpha_{M,N} = (\varphi \circ \psi) \circ \alpha_{M,N}.$$

The uniqueness condition in the universal property of $M \otimes_R N$ implies that $\mathrm{id}_{M\otimes_R N} = \varphi \circ \psi$. With the same argument one shows that $\mathrm{id}_W = \psi \circ \varphi$. Thus, $\psi : M \otimes_R N \to W$ is an isomorphism. □

We consider the following simple but important

Example A.19

Let N be an R-module. Then the map $R \otimes_R N \to N$ with $r \otimes n \mapsto rn$ is an isomorphism.

In fact, let $\alpha: R \times N \to N$ be given by $\alpha(r, n) = rn$ for all $r \in R$ and $n \in N$. Then α is an R-bilinear map. So by the universal property, there exists an R-module homomorphism $\varphi: R \otimes_R N \to N$ with $\varphi(r \otimes n) = rn$.

Now, define $\psi: N \to R \otimes_R N$ as $\psi(n) = 1 \otimes n$ for any $n \in N$. Then ψ is an R-module homomorphism. Indeed, for any $r \in R$ and $n, n' \in N$, we have $\psi(rn) = 1 \otimes rn = r \otimes n = r(1 \otimes n) = r\psi(n)$ and $\psi(n + n') = 1 \otimes (n + n') = 1 \otimes n + 1 \otimes n' = \psi(n) + \psi(n')$.

For any $r \in R$ and $n \in N$, $\psi o \varphi(r \otimes n) = \psi(rn) = 1 \otimes rn = r \otimes n$. So $\psi \circ \varphi = \mathrm{id}_{R\otimes_R N}$. Similarly, $\varphi \circ \psi = \mathrm{id}_N$. Thus φ is an isomorphism.

The following exercise generalizes Example A.19.

Exercise A.20

Let F be a free R-module with basis $e_1, \ldots, e_n$, and let $e_1^*, \ldots, e_n^*$ be its dual basis. This dual basis is a basis of $F^* = \operatorname{Hom}_R(F, R)$, see Exercise 2.4. Furthermore, let M be an R-module.

Show that the map

$$\operatorname{Hom}_R(F, M) \to F^* \otimes_R M, \quad \varphi \mapsto \sum_{i=1}^{n} e_i^* \otimes \varphi(e_i)$$

is an isomorphism, which does not depend on the choice of the basis $e_1, \ldots, e_n$.

Hint: let $f_1, \ldots, f_n$ be another basis of F. First show that if

$$\begin{pmatrix} e_1 \\ \vdots \\ e_n \end{pmatrix} = A \begin{pmatrix} f_1 \\ \vdots \\ f_n \end{pmatrix}, \quad \text{then} \quad \begin{pmatrix} f_1^* \\ \vdots \\ f_n^* \end{pmatrix} = A^T \begin{pmatrix} e_1^* \\ \vdots \\ e_n^* \end{pmatrix},$$

where A^T denotes the transpose of A.

Let R and R' be rings and $\varphi\colon R \to R'$ be an R-module homomorphism, and let M be an R-module and N be an R'-module. The R'-module N may be viewed as an R-module via φ with scalar multiplication $r \cdot n = \varphi(r)n$ for all $r \in R$ and $n \in N$. In particular, R' is an R-module via φ, and hence we can define the tensor products $R' \otimes_R M$ and $M \otimes_R R'$. Both tensor products can be given the structure of an R'-module. Indeed, if $x \in R'$, then the multiplication map $\mu_x\colon R' \to R'$ is an R-module homomorphism, and hence induces the R-module homomorphism

$$\mu_x \otimes \mathrm{id}_M\colon R' \otimes_R M \to R' \otimes_R M,\ r' \otimes m \mapsto xr' \otimes m \quad \text{for } r' \in R' \text{ and } m \in M,$$

which makes $R' \otimes_R M$ an R'-module. Similarly, $M \otimes_R R'$ is given an R'-module structure, and the map $R' \otimes_R M \to M \otimes_R R'$ with $r' \otimes m \mapsto m \otimes r'$ is an isomorphism of R'-modules.

Let $\varphi : M \to M'$ and $\psi : N \to N'$ be R-module homomorphisms. By the universal property of the tensor product we obtain the R-module homomorphism $\varphi \otimes_R N\colon M \otimes_R N \to M' \otimes_R N$ with $m \otimes n \mapsto \varphi(m) \otimes n$ and the R-module homomorphism $M \otimes_R \psi\colon M \otimes_R N \to M \otimes_R N'$ with $m \otimes n \mapsto m \otimes \psi(n)$. We have the following rules: $\mathrm{id}_M \otimes_R N = \mathrm{id}_{M \otimes_R N} = M \otimes_R id_N$, $(\varphi' \circ \varphi) \otimes_R N = (\varphi' \otimes_R N) \circ (\varphi \otimes_R N)$ for $\varphi'\colon M' \to M''$ and $M \otimes_R (\psi' \circ \psi) = (M \otimes_R \psi') \circ (M \otimes_R \psi)$ for $\psi'\colon N' \to N''$.

Exercise A.21
Show that $\varphi : M \to M'$ and $\psi : N \to N'$ induce a homomorphism $\varphi \otimes \psi : M \otimes_R N \to M' \otimes_R N', m \otimes n \mapsto \varphi(m) \otimes \psi(n)$, and

$$\varphi \otimes \psi = (\varphi \otimes_R N) \circ (M \otimes_R \psi).$$

Summarizing we obtain

Lemma A.22
The tensor product $\otimes_R$ is a covariant functor in two variables.

Exercise A.23
(a) Show that there is a unique R-module homomorphism $\alpha_{M,N} : M \otimes N \to N \otimes M$ with $\alpha(m \otimes n) = n \otimes m$ for all $m \in M$ and $n \in N$ and that φ is an isomorphism.

(b) Let $\varphi : M \to M'$ be an R-module homomorphism. Show that the following diagram

$$\begin{array}{ccc} M \otimes N & \xrightarrow{\alpha_{M,N}} & N \otimes M \\ {\scriptstyle \varphi\otimes N}\downarrow & & \downarrow{\scriptstyle N\otimes\varphi} \\ M' \otimes N & \xrightarrow{\alpha_{M',N}} & N \otimes M' \end{array}$$

is commutative.

Exercise A.24
The map

$$\sigma' : \mathrm{Bihom}_R(M \times N, W) \to \mathrm{Hom}_R(M, \mathrm{Hom}_R(N, W))$$

with $(\sigma'(\varphi)(m))(n) = \varphi(m, n)$ for all $m \in M$ and $n \in N$ is an isomorphism of R-modules.

Lemma A.25
Let M, N and W be R-modules. Then the map

$$\sigma : \mathrm{Hom}_R(M \otimes_R N, W) \to \mathrm{Hom}_R(M, \mathrm{Hom}_R(N, W))$$

with $(\sigma(\varphi)(m))(n) = \varphi(m \otimes n)$ for all $m \in M$ and $n \in N$, is an isomorphism of R-modules.

Proof. The composition of the map given in (A.8) with σ' defined in Exercise A.24 gives us the map σ and shows that σ is an isomorphism. □

Exercise A.26
Let $\{M_i\}_{i\in\Lambda}$ and N be R-modules. Show that

$$(\bigoplus_{i\in\Lambda} M_i) \otimes_R N \cong \bigoplus_{i\in\Lambda}(M_i \otimes_R N).$$

In Exercise A.23(a) we noted that the tensor product is commutative. In other words, for any two R-modules M and N we have $M \otimes_R N \cong N \otimes_R M$. We will see that the tensor product is also associative. Indeed, we have

Proposition A.27
Let L, M and N be R-modules. Then

$$L \otimes_R (M \otimes_R N) \cong (L \otimes_R M) \otimes_R N.$$

Proof. For the proof we use Lemma A.25, and obtain for all R-modules W.

$$\begin{aligned}&\mathrm{Hom}_R(L \otimes_R (M \otimes_R N), W) \cong \mathrm{Hom}_R(L, \mathrm{Hom}_R(M \otimes_R N, W)) \cong \\ &\mathrm{Hom}_R(L, \mathrm{Hom}_R(M, \mathrm{Hom}_R(N, W))) \cong \mathrm{Hom}_R(L \otimes_R M, \mathrm{Hom}_R(N, W)) \cong \\ &\mathrm{Hom}_R((L \otimes_R M) \otimes_R N, W).\end{aligned}$$

Since all the above isomorphisms are functorial, the desired conclusion follows from Proposition A.8. □

A more general version of Proposition A.27 is left to you as an exercise.

Exercise A.28
Let L be an R-module, let N be an S-module, and let M be an (R, S)-bimodule. Show the isomorphism $L \otimes_R (M \otimes_S N) \cong (L \otimes_R M) \otimes_S N$.

Hint: similar to an R-bilinear map, define a trilinear map from $L \times M \times N$ to any (R, S)-bimodule W. Show that both abelian groups $L \otimes_R (M \otimes_S N)$ and $(L \otimes_R M) \otimes_S N$ satisfy the corresponding universal property.

Proposition A.29

Let N be an R-module. The functor $-\otimes N$ is right exact. Indeed, if $U \xrightarrow{\beta} V \xrightarrow{\alpha} W \to 0$ is an exact sequence of R-modules, then the sequence

$$U \otimes_R N \xrightarrow{\beta\otimes N} V \otimes_R N \xrightarrow{\alpha\otimes N} W \otimes_R N \longrightarrow 0 \tag{A.9}$$

is also exact.

Proof. Since $\alpha \circ \beta = 0$, it follows that $(\alpha \otimes N) \circ (\beta \otimes N) = \alpha \circ \beta \otimes N = 0 \otimes N = 0$. Thus, the sequence (A.9) is a complex. In order to prove exactness, we consider for each R-module X the complex

$$\begin{aligned} 0 \to \operatorname{Hom}_R(W \otimes_R N, X) &\to \operatorname{Hom}_R(V \otimes_R N, X) \\ &\to \operatorname{Hom}_R(U \otimes_R N, X). \end{aligned} \tag{A.10}$$

By Lemma A.25, this complex is isomorphic to the complex

$$\begin{aligned} 0 \to \operatorname{Hom}_R(W, \operatorname{Hom}_R(N, X)) &\to \operatorname{Hom}_R(V, \operatorname{Hom}_R(N, X)) \\ &\to \operatorname{Hom}_R(U, \operatorname{Hom}_R(N, X)), \end{aligned}$$

which by the left exactness of Hom is exact, see Exercise A.10. This implies that the complex (A.10) is exact for all X. Hence, the desired conclusion follows from Exercise A.10. □

We say that an R-module M is *finitely presented*, if there exists an exact sequence $G \to F \to M \to 0$ for which F and G are finitely generated free R-modules. For example, if R is Noetherian, then M is finitely presented if and only if M is finitely generated. This follows from Theorem 1.17.

Proposition A.30

Let M and N be R-modules, and let $S \subseteq R$ be a multiplicatively closed set. Then

(a) $S^{-1}(M \otimes_R N) \cong S^{-1}M \otimes_{S^{-1}R} S^{-1}N$.
(b) If M is a finitely presented R-module, then

$$S^{-1} \operatorname{Hom}_R(M, N) \cong \operatorname{Hom}_{S^{-1}R}(S^{-1}M, S^{-1}N).$$

Proof. (a) Let $G \to F \to M \to 0$ be a free presentation of M. Then we obtain the exact sequence

$$G \otimes_R N \longrightarrow F \otimes_R N \longrightarrow M \otimes_R N \to 0. \tag{A.11}$$

We may assume that $F = R^s$ and $G = R^t$. Then it follows from Exercise A.26 and Example A.19 that $F \otimes_R N \cong N^s$ and $G \otimes_R N \cong N^t$. It follows that

$$S^{-1}(F \otimes_R N) \cong S^{-1}N^s \cong S^{-1}F \otimes_{S^{-1}R} S^{-1}N,$$

and similarly, $S^{-1}(G \otimes_R N) \cong S^{-1}G \otimes_{S^{-1}R} S^{-1}N$.

Since these isomorphisms are functorial and localization is an exact functor, the exact sequence (A.11) yields the exact sequence

$$S^{-1}G \otimes_{S^{-1}R} S^{-1}N \to S^{-1}F \otimes_{S^{-1}R} S^{-1}N \to S^{-1}(M \otimes_R N) \to 0. \tag{A.12}$$

We also have the exact sequence $S^{-1}G \to S^{-1}F \to S^{-1}M \to 0$ of $S^{-1}R$-modules. This gives us the exact sequence

$$S^{-1}G \otimes_{S^{-1}R} S^{-1}N \to S^{-1}F \otimes_{S^{-1}R} S^{-1}N \to S^{-1}M \otimes_{S^{-1}R} S^{-1}N \to 0 \tag{A.13}$$

A comparison of the exact sequence (A.12) with the exact sequence (A.13) yields that $S^{-1}(M \otimes_R N) \cong S^{-1}M \otimes_{S^{-1}R} S^{-1}N$.

(b) Let $G \to F \to M \to 0$ be a finite presentation of M. Then we obtain the exact sequence

$$0 \longrightarrow \operatorname{Hom}_R(M, N) \longrightarrow \operatorname{Hom}_R(F, N) \longrightarrow \operatorname{Hom}_R(G, N). \tag{A.14}$$

We may assume that $F = R^s$ and $G = R^t$. Then it follows from Exercise A.13(a) that $\operatorname{Hom}_R(F, N) \cong N^s$ and $\operatorname{Hom}_R(G, N) \cong N^t$. It follows that

$$S^{-1}\operatorname{Hom}_R(F, N) \cong S^{-1}N^s \cong \operatorname{Hom}_{S^{-1}R}(S^{-1}F, S^{-1}N),$$

and similarly, $S^{-1}\operatorname{Hom}_R(G, N) \cong \operatorname{Hom}_{S^{-1}R}(S^{-1}G, S^{-1}N)$. Since we have this isomorphisms, an analogous argument that was applied in the proof of (a) yields that $S^{-1}\operatorname{Hom}_R(M, N) \cong \operatorname{Hom}_{S^{-1}R}(S^{-1}M, S^{-1}N)$. □

Exercise A.31

Let $I, J \subsetneq R$ be ideals of R, and let M be an R-module. Show that $R/I \otimes_R M \cong M/IM$. Conclude that $R/I \otimes_R R/J \cong R/(I + J)$.

Hint: Use the right exactness of the tensor product and show that the image $I \otimes_R M \to R \otimes_R M \cong M$ is the submodule IM of M.

▶ **Remark A.32** (a) Similarly as for the Hom-functor we have that $0 \to U \otimes_R W \to M \otimes_R W \to N \otimes_R W \to 0$ is exact for all R-modules W, if $0 \to U \to M \to N \to 0$ is a split exact sequence of R-modules. In particular, this is the case if N is a free R-module.

(b) If F is a free R-module and $0 \to U \xrightarrow{\alpha} M \to N \to 0$ is a short exact sequence of R-modules, then $0 \to U \otimes_R F \to M \otimes_R F \to N \otimes_R F \to 0$ is exact. Indeed, if $F = \bigoplus_{i \in \Lambda} R$, then the map $\alpha \otimes F : U \otimes_R F \to M \otimes_R F$ is injective if and only if the map $\overline{\alpha} : \bigoplus_{i \in \Lambda} U \to \bigoplus_{i \in \Lambda} M$ with $\overline{\alpha}(\{u_i\}) = \{\alpha(u_i)\}$ is injective (see Exercise A.26). Clearly if α is injective, then $\overline{\alpha}$ is injective.

A.4 The Functors Tor and Ext

Let M be an R-module. By Proposition 1.9 there exists a free R-module F_0 and a surjective homomorphism $\epsilon\colon F_0 \longrightarrow M$. Similarly, for $\operatorname{Ker} \epsilon$ there is a free presentation that is a surjective graded R-module homomorphism $F_1 \to \operatorname{Ker} \epsilon$. Composing the morphism $F_1 \to \operatorname{Ker} \epsilon$ with the inclusion map $\operatorname{Ker} \epsilon \hookrightarrow F_0$, we obtain the exact sequence $F_1 \longrightarrow F_0 \longrightarrow M \longrightarrow 0$. Proceeding in this way, we obtain an exact sequence

$$\cdots \longrightarrow F_i \longrightarrow \cdots \longrightarrow F_1 \longrightarrow F_0 \longrightarrow M \longrightarrow 0, \tag{A.15}$$

where all F_i are free R-modules. Replacing M by 0 in this exact sequence gives the complex

$$F_\bullet\colon \cdots \longrightarrow F_i \longrightarrow \cdots \longrightarrow F_1 \longrightarrow F_0 \longrightarrow 0 \tag{A.16}$$

of free R-modules with $H_i(F_\bullet) = 0$ for $i > 0$ and $H_0(F_\bullet) \cong M$. A complex of free modules as in (2.10) is called a *free resolution* of M.

Let M and N be R-modules, and let $F_\bullet$ be a free R-resolution of M. For all $i \geq 0$ we define

$$\operatorname{Tor}_i^R(M, N) = H_i(F_\bullet \otimes_R N) \quad \text{and} \quad \operatorname{Ext}_R^i(M, N) = H^i(\operatorname{Hom}_R(F_\bullet, N)).$$

Proposition A.33
The definitions of $\operatorname{Tor}_i^R(M, N)$ and $\operatorname{Ext}_R^i(M, N)$ do not depend on the particular chosen free R-resolution $F_\bullet$ of M. That is, if $G_\bullet$ is another free R-resolution of M, then

$$H_i(F_\bullet \otimes_R N) \cong H_i(G_\bullet \otimes_R N) \quad \text{and} \quad H^i(\operatorname{Hom}_R(F_\bullet, N)) \cong H^i(\operatorname{Hom}_R(G_\bullet, N)).$$

More precisely, if $\varphi_\bullet\colon\ F_\bullet \to G_\bullet$ is a chain map with $H_0(\varphi_\bullet) = \mathrm{id}_M$, then for each i, both maps, $H_i(\varphi_\bullet \otimes N)$ and $H^i(\operatorname{Hom}_R(\varphi_\bullet, N)$, are isomorphisms.

Proof. Like in the graded case, see Theorem 2.16(a), there are chain maps $\varphi_\bullet\colon F_\bullet \to G_\bullet$ and $\psi_\bullet\colon G_\bullet \to F_\bullet$ with $H_0(\varphi_\bullet) = H_0(\psi_\bullet) = \mathrm{id}_M$, and similarly as in Theorem 2.16(b), $\psi_\bullet \circ \varphi_\bullet \sim \mathrm{id}_{F_\bullet}$ and $\varphi_\bullet \circ \psi_\bullet \sim \mathrm{id}_{G_\bullet}$. This implies that

$$(\psi_\bullet \otimes_R N) \circ (\varphi_\bullet \otimes_R N) \sim \mathrm{id}_{F_\bullet} \otimes_R N \quad \text{and} \quad (\varphi_\bullet \otimes_R N) \circ (\psi_\bullet \otimes_R N) \sim \mathrm{id}_{G_\bullet} \otimes_R N.$$

It follows from Theorem A.2 that

$$H_i(\psi_\bullet \otimes_R N) \circ H_i(\varphi_\bullet \otimes_R N) = \mathrm{id}_{H_i(F_\bullet \otimes_R N)} \quad \text{and}$$
$$H_i(\varphi_\bullet \otimes_R N) \circ H_i(\psi_\bullet \otimes_R N) = \mathrm{id}_{H_i(G_\bullet \otimes_R N)}\,.$$

This shows that $H_i(\varphi_\bullet \otimes N)\colon H_i(F_\bullet \otimes_R N) \to H_i(G_\bullet \otimes_R N)$ is an isomorphism. The second isomorphism is proved similarly. □

Exercise A.34

(a) Let F be a free R-module. Show that

$$\mathrm{Tor}_i^R(F, N) = \mathrm{Ext}_R^i(F, N) = 0 \quad \text{for all} \quad i > 0$$

and all R-modules N.

(b) Let $M = M_1 \oplus M_2$. Show that $\mathrm{Tor}_i^R(M, N) \cong \mathrm{Tor}_i^R(M_1, N) \oplus \mathrm{Tor}_i^R(M_2, N)$ for all i. A similar statement holds for $\mathrm{Ext}_R^i(-, N)$,

Let M and N be R-modules, let $(F_\bullet, \partial_\bullet)$ be a free resolution of M, and let $(G_\bullet, \partial'_\bullet)$ be a free resolution of N. We form the complex $(F_\bullet \otimes G_\bullet, d_\bullet)$ of free R-modules with

$$(F_\bullet \otimes G_\bullet)_i = \bigoplus_{j=0}^{i} F_j \otimes G_{i-j} \quad \text{and} \quad d_i(f \otimes g) = \partial_j(f) \otimes g + (-1)^j f \otimes \partial'_{i-j}(g)$$

for $f \otimes g \in F_j \otimes G_{i-j}$ and all i and j.

Exercise A.35

Show that $F_\bullet \otimes G_\bullet$ is a complex. In other words, prove that $d_{i+1} \circ d_i = 0$ for all i.

The complex $F_\bullet \otimes G_\bullet$ is the total complex of the double complex with column complexes $(F_j \otimes G_\bullet, \mathrm{id}_{F_j} \otimes \partial'_\bullet)$ and the row complexes $(F_\bullet \otimes G_i, \partial_\bullet \otimes \mathrm{id}_{G_i})$. This complex is called the *tensor product of the complexes* $F_\bullet$ and $G_\bullet$. Applying Theorem A.7 to $F_\bullet \otimes G_\bullet$ allows us to compute $\mathrm{Tor}_i^R(M, N)$ in terms of $G_\bullet$, as well. We see this in

Theorem A.36
Let M and N be R-modules, and let $F_{\bullet}$ be a free R-resolution of M and $G_{\bullet}$ be a free R-resolution of N. Then for all $i \geq 0$, there exists an isomorphism

$$\mathrm{Tor}_i^R(M, N) \cong H_i(F_{\bullet} \otimes N) \cong H_i(F_{\bullet} \otimes G_{\bullet}) \cong H_i(M \otimes G_{\bullet})$$

of graded R-modules.

Proof. Since F_j is a free module, the column complexes $F_j \otimes G_{\bullet}$ are all acyclic. Since $H_0(F_j \otimes G_{\bullet}) \cong F_j \otimes N$, the complex

$$0 \leftarrow F_0 \otimes N \leftarrow F_1 \otimes N \leftarrow F_2 \otimes N \leftarrow \cdots$$

is the complex $C_{\bullet}$ of Theorem A.7. Hence, Theorem A.7 implies that

$$\mathrm{Tor}_i^R(M, N) \cong H_i(F_{\bullet} \otimes N) \cong H_i(C_{\bullet}) \cong H_i(F_{\bullet} \otimes G_{\bullet}),$$

where the first isomorphism results from our definition of Tor. Similarly, we see that $H_i(F_{\bullet} \otimes G_{\bullet}) \cong H_i(M \otimes G_{\bullet})$. □

Let $\varphi\colon M \to M'$ be an R-module homomorphism. Let $F_{\bullet}$ be a free resolution of M, $F'_{\bullet}$ be a free resolution of M', and let $\varphi_{\bullet}\colon F_{\bullet} \to F'_{\bullet}$ by a lifting of φ. Given this, we set

$$\mathrm{Tor}_i(\varphi, N) = H_i(\varphi_{\bullet} \otimes N) : \mathrm{Tor}_i^R(M, N) \to \mathrm{Tor}_i^R(M', N).$$

Note that this definition does not depend on the particular lifting chosen for φ. Indeed, if $\psi_{\bullet}$ is another lifting of φ, then Theorem 2.16 tells us that $\varphi_{\bullet}$ and $\psi_{\bullet}$ are homotopic, and this induces a homotopy between $\varphi_{\bullet} \otimes N$ and $\psi_{\bullet} \otimes N$. Then, by Theorem A.2, we have $H_i(\varphi_{\bullet} \otimes N) = H_i(\psi_{\bullet} \otimes N)$ for all i, as desired.

But what happens if we choose different resolutions of M and M'? This will also be no problem. Say, $G_{\bullet}$ is another resolution of M and $G'_{\bullet}$ is another resolution of M'. Then there is a chain map $\alpha_{\bullet}\colon G_{\bullet} \to F_{\bullet}$ with $H_0(\alpha_{\bullet}) = \mathrm{id}_M$ and a chain map $\beta_{\bullet}\colon F'_{\bullet} \to G'_{\bullet}$ with $H_0(\beta_{\bullet}) = \mathrm{id}_{M'}$, and we let $\varphi'_{\bullet} = \beta_{\bullet} \circ \varphi_{\bullet} \circ \alpha_{\bullet}$.

Then $\varphi'_{\bullet}\colon G_{\bullet} \to G'_{\bullet}$ is a chain map with $H_0(\varphi'_{\bullet}) = \varphi$. With this choice of $\varphi'_{\bullet}$ we obtain for all i the following commutative diagram

$$\begin{array}{ccc}
H_i(F_{\bullet} \otimes N) & \xrightarrow{H_i(\varphi_{\bullet} \otimes N)} & H_i(F'_{\bullet} \otimes N) \\
{\scriptstyle H_i(\alpha_{\bullet} \otimes N)}\uparrow & & \downarrow{\scriptstyle H_i(\beta_{\bullet} \otimes N)} \\
H_i(G_{\bullet} \otimes N) & \xrightarrow{H_i(\varphi'_{\bullet} \otimes N)} & H_i(G'_{\bullet} \otimes N).
\end{array}$$

By Proposition A.33 the vertical maps in the diagram are isomorphisms, while the horizontal maps give us $\mathrm{Tor}_i^R(\varphi, N)$ defined via the corresponding resolutions.

The commutativity of the diagram shows that the definition of $\mathrm{Tor}_i^R(\varphi, N)$ is compatible with the various presentations of $\mathrm{Tor}_i^R(M, N)$ and $\mathrm{Tor}_i^R(M', N)$ in terms of resolutions and hence is well defined.

It also follows that $\mathrm{Tor}_i^R(\mathrm{id}_M, N) = \mathrm{id}_{\mathrm{Tor}_i^R(M,N)}$ and that for any two R-module homomorphisms $\varphi\colon M \to M'$ and $\psi : M' \to M''$ we have

$$\mathrm{Tor}_i^R(\psi \circ \varphi, N) = \mathrm{Tor}_i^R(\psi, N) \circ \mathrm{Tor}_i^R(\varphi, N).$$

This shows that $\mathrm{Tor}_i^R(-, N)$ is a covariant functor. In the same way one shows that $\mathrm{Ext}^i_R(-, N)$ is a contravariant functor.

Sometimes it is needed to compute the map $\mathrm{Tor}_i^R(\varphi, N)$ by choosing a free resolution of N. Suppose we choose a free resolution $G_\bullet$ of N. Then for each i, the map $\varphi\colon M \to M'$ induces the R-module homomorphism $H_i(\varphi \otimes G_\bullet)$, and we obtain a commutative diagram

$$\begin{array}{ccc} H_i(M \otimes G_\bullet) & \xrightarrow{H_i(\varphi\otimes G_\bullet)} & H_i(M' \otimes G_\bullet) \\ \downarrow & & \downarrow \\ H_i(F_\bullet \otimes N) & \xrightarrow{H_i(\varphi_\bullet\otimes N)} & H_i(F'_\bullet \otimes N), \end{array}$$

where the vertical maps are isomorphisms, see Theorem A.36. This shows that $\mathrm{Tor}_i^R(\varphi, N) = H_i(\varphi \otimes G_\bullet)$, if we compute $\mathrm{Tor}_i^R(M, N)$ and $\mathrm{Tor}_i^R(M', N)$ via the free resolution $G_\bullet$.

In a similar way as we defined $\mathrm{Tor}_i^R(\varphi, N)$ we define $\mathrm{Tor}_i^R(M, \psi)$ for an R-module homomorphism $\psi\colon N \to N'$. This makes $\mathrm{Tor}_i^R(M, -)$ a functor in the second variable, as well.

Exercise A.37

Let M be an R-module. Prove that there exists a functorial isomorphism $\mathrm{Tor}_i^R(M, -) \cong \mathrm{Tor}_i^R(-, M)$. Indeed, show that if $\varphi : M \to M'$ is an R-module homomorphism, and N is an R-module, then the following diagram commutes

$$\begin{array}{ccc} \mathrm{Tor}_i^R(M, N) & \longrightarrow & \mathrm{Tor}_i^R(N, M) \\ {\scriptstyle \mathrm{Tor}_i^R(\varphi,N)}\downarrow & & \downarrow{\scriptstyle \mathrm{Tor}_i^R(N,\varphi)} \\ \mathrm{Tor}_i^R(M', N) & \longrightarrow & \mathrm{Tor}_i^R(N, M'). \end{array}$$

Hint: use Theorem A.36.

We have seen that Tor_i^R is a functor in both variables. The same is true for Ext^i_R. Indeed, let $\varphi : N \to N'$ be an R-module homomorphism, and let $F_\bullet$ be a free R-resolution of M. Then

$$\mathrm{Hom}_R(F_\bullet, N)\colon 0 \to \mathrm{Hom}_R(F_0, N) \to \mathrm{Hom}_R(F_1, N) \to \mathrm{Hom}_R(F_2, N) \cdots$$

is a cocomplex. Let

$$\mathrm{Hom}_R(F_\bullet, \varphi) = (\mathrm{Hom}_R(F_i, \varphi)\colon \mathrm{Hom}_R(F_i, N) \to \mathrm{Hom}_R(F_i, N'))_{i\geq 0}.$$

Then $\mathrm{Hom}_R(F_\bullet, \varphi)\colon \mathrm{Hom}_R(F_\bullet, N) \to \mathrm{Hom}_R(F_\bullet, N')$ is a cochain map. We set $\mathrm{Ext}^i_R(M, \varphi) = H^i(\mathrm{Hom}_R(F_\bullet, \varphi))$. With this definition, $\mathrm{Ext}^i_R(M, -)$ is a covariant functor.

Proposition A.38
Let R be a ring, and let M and N be R-modules. Then we have

(a) $\mathrm{Tor}^R_0(M, N) \cong M \otimes_R N$ and $\mathrm{Ext}^0_R(M, N) \cong \mathrm{Hom}_R(M, N)$.
(b) If R is Noetherian and M and N are finitely generated R-modules, then $\mathrm{Tor}^R_i(M, N)$ and $\mathrm{Ext}^i_R(M, N)$ are also finitely generated R-modules for all i.
(c) If R is a graded K-algebra and $M, N \in \mathcal{M}_R$, then $\mathrm{Tor}^R_i(M, N)$ and $\mathrm{Ext}^i_R(M, N)$ also belong to $\mathcal{M}_R$ for all i.

Proof. (a) Let $F_\bullet$ be a free R-resolution of M with the differential map $\partial_\bullet$. Then the sequence $F_1 \xrightarrow{\partial_1} F_0 \to M \to 0$ is exact. Since tensor is a right exact functor, it follows that $F_1 \otimes_R N \xrightarrow{\partial_1 \otimes_R N} F_0 \otimes_R N \to M \otimes_R N \to 0$ is also exact. Therefore,

$$\begin{aligned} H_0(F_\bullet \otimes_R N) &= \mathrm{Ker}(\partial_0 \otimes_R N)/\,\mathrm{Im}(\partial_1 \otimes_R N) \\ &= (F_0 \otimes_R N)/\,\mathrm{Im}(\partial_1 \otimes_R N) \cong M \otimes_R N. \end{aligned}$$

Similarly one shows that $H^0(\mathrm{Hom}_R(F_\bullet, N)) \cong \mathrm{Hom}_R(M, N)$ by using the left exactness of $\mathrm{Hom}_R(-, N)$.

We only give the proof of (c), because the proof of (b) follows similarly.

(c) Let $F_\bullet$ be a graded free R-resolution of M. Note that $F_i \otimes_R N \in \mathcal{M}_R$ and $\partial_i \otimes_R N$ is a morphism in $\mathcal{M}_R$. Therefore, $H_i(F_\bullet \otimes_R N) = \mathrm{Ker}(\partial_i \otimes_R N)/\,\mathrm{Im}(\partial_{i+1} \otimes_R N)$ is an object in $\mathcal{M}_R$, see Proposition 1.4. The same arguments apply for $H^i(\mathrm{Hom}_R(F_\bullet, N))$. □

We say that an R-module M is *finitely presented*, if there exists an exact sequence $G \to F \to M \to 0$ for which F and G are finitely generated free R-modules. For example, if R is Noetherian, then M is finitely presented if and only if M is finitely generated. This follows from Theorem 1.17.

One property that is useful for later applications, is the fact that Tor and Ext localize. More precisely, we have

Proposition A.39
Let R be a ring, let $S \subseteq R$ be a multiplicatively closed subset of R, and let M and N be R-modules. Then

$$S^{-1}\operatorname{Tor}_i^R(M, N) \cong \operatorname{Tor}_i^{S^{-1}R}(S^{-1}M, S^{-1}N)$$

for all i.

Moreover, suppose that M is a finitely presented R-module. Then

$$S^{-1}\operatorname{Ext}_R^i(M, N) \cong \operatorname{Ext}_{S^{-1}R}^i(S^{-1}M, S^{-1}N)$$

for all i.

Proof. Let $F_\bullet$ be a free R-resolution of M. Since localization is an exact functor, it follows that $S^{-1}F_\bullet$ is a free $S^{-1}R$-resolution of $S^{-1}M$, and that localization commutes with taking (co)homology. Thus, together with Exercise A.1 and the fact that tensor commutes with localization, we obtain that

$$\begin{aligned} S^{-1}\operatorname{Tor}_i^R(M, N) &\cong S^{-1}H_i(F_\bullet \otimes_R N) \cong H_i(S^{-1}(F_\bullet \otimes_R N)) \\ &\cong H_i(S^{-1}F_\bullet \otimes_{S^{-1}R} S^{-1}N) \cong \operatorname{Tor}_i^{S^{-1}R}(S^{-1}M, S^{-1}N). \end{aligned}$$

To prove the second isomorphism, let $G \to F \to M \to 0$ be a finite presentation of M. Then we obtain the exact sequence

$$0 \longrightarrow \operatorname{Hom}_R(M, N) \longrightarrow \operatorname{Hom}_R(F, N) \longrightarrow \operatorname{Hom}_R(G, N). \qquad \text{(A.17)}$$

We may assume that $F = R^s$ and $G = R^t$. Then $\operatorname{Hom}_R(F, N) \cong N^s$ and $\operatorname{Hom}_R(G, N) \cong N^t$, since finite direct products are direct sums. (At this point we use the finite presentation of M.) It follows that

$$S^{-1}\operatorname{Hom}_R(F, N) \cong S^{-1}N^s \cong \operatorname{Hom}_{S^{-1}R}(S^{-1}F, S^{-1}N),$$

and similarly, $S^{-1}\operatorname{Hom}_R(G, N) \cong \operatorname{Hom}_{S^{-1}R}(S^{-1}G, S^{-1}N)$.

Since these isomorphisms are functorial and localization is an exact functor, the exact sequence (A.17) yields the exact sequence

$$\begin{aligned} 0 \to S^{-1}\operatorname{Hom}_R(M, N) &\to \operatorname{Hom}_{S^{-1}R}(S^{-1}F, S^{-1}N) \qquad \text{(A.18)} \\ &\to \operatorname{Hom}_{S^{-1}R}(S^{-1}G, S^{-1}N). \end{aligned}$$

We also have the exact sequence $S^{-1}G \to S^{-1}F \to S^{-1}M \to 0$ of $S^{-1}R$-modules. This gives us the exact sequence

$$\begin{aligned} 0 \to \operatorname{Hom}_{S^{-1}R}(S^{-1}M, S^{-1}N) &\to \operatorname{Hom}_{S^{-1}R}(S^{-1}F, S^{-1}N) \qquad \text{(A.19)} \\ &\to \operatorname{Hom}_{S^{-1}R}(S^{-1}G, S^{-1}N). \end{aligned}$$

A comparison of the exact sequence (A.18) with the exact sequence (A.19) yields that $\operatorname{Hom}_{S^{-1}R}(S^{-1}M, S^{-1}N) \cong S^{-1}\operatorname{Hom}_R(M, N)$. Hence

$$\begin{aligned} S^{-1}\operatorname{Ext}^i_R(M, N) &\cong S^{-1}H_i(\operatorname{Hom}_R(F_\bullet, N)) \cong H_i(S^{-1}(\operatorname{Hom}_R(F_\bullet, N))) \\ &\cong H_i(\operatorname{Hom}_{S^{-1}R}(S^{-1}F_\bullet, S^{-1}N)) \cong \operatorname{Ext}^i_{S^{-1}R}(S^{-1}M, S^{-1}N). \end{aligned}$$

□

We derive the long exact sequences for Ext and Tor for modules over a commutative ring. This is described in

Theorem A.40

Let W be an R-module, and let $0 \to U \xrightarrow{\varphi} M \xrightarrow{\psi} N \to 0$ be an exact sequence of R-modules. Then one has the long exact sequence

$$\begin{aligned} \cdots &\to \operatorname{Tor}^R_{i+1}(N, W) \to \operatorname{Tor}^R_i(U, W) \to \operatorname{Tor}^R_i(M, W) \to \operatorname{Tor}^R_i(N, W) \to \cdots \\ &\to \operatorname{Tor}^R_1(N, W) \to U \otimes_R W \to M \otimes_R W \to N \otimes_R W \to 0, \end{aligned}$$

and the long exact sequence

$$\begin{aligned} 0 &\to \operatorname{Hom}_R(N, W) \to \operatorname{Hom}_R(M, W) \to \operatorname{Hom}_R(U, W) \to \operatorname{Ext}^1_R(N, W) \to \cdots \\ &\to \operatorname{Ext}^i_R(N, W) \to \operatorname{Ext}^i_R(M, W) \to \operatorname{Ext}^i_R(U, W) \to \operatorname{Ext}^{i+1}_R(N, W) \to \cdots. \end{aligned}$$

Proof. Let $F_\bullet$ be a free R-resolution of U and $H_\bullet$ be a free R-resolution of N. First we are going to construct a free R-resolution $G_\bullet$ of M and chain maps $\varphi_\bullet\colon F_\bullet \to G_\bullet$ and $\psi_\bullet\colon G_\bullet \to H_\bullet$ such that $0 \to F_\bullet \xrightarrow{\varphi_\bullet} G_\bullet \xrightarrow{\psi_\bullet} H_\bullet \to 0$ is an exact sequence of complexes and that $H_0(\varphi_\bullet) = \varphi$ and $H_0(\psi_\bullet) = \psi$. Having this, $0 \to F_i \otimes_R W \to G_i \otimes_R W \to H_i \otimes_R W \to 0$ is exact for all i. Thus, we obtain a short exact sequence of complexes $0 \to F_\bullet \otimes_R W \to G_\bullet \otimes_R W \to H_\bullet \otimes_R W \to 0$. The corresponding long exact sequence as shown in Theorem A.4 gives us the long exact Tor-sequence.

Similarly, we obtain the long exact Ext-sequence by observing that

$$0 \to \operatorname{Hom}_R(H_\bullet, W) \to \operatorname{Hom}_R(G_\bullet, W) \to \operatorname{Hom}_R(F_\bullet, W) \to 0$$

is an exact sequence of complexes.

It remains to construct the exact sequence $0 \to F_\bullet \to G_\bullet \to H_\bullet \to 0$ of complexes. For each i we let $G_i = F_i \oplus H_i$ and $\varphi_i\colon F_i \to F_i \oplus H_i$ be the map with $\varphi_i(f) = (f, 0)$ for all $f \in F_i$ and $\psi_i\colon F_i \oplus H_i \to H_i$ be the map with $\psi_i(f, h) = h$ for all $(f, h) \in F_i \oplus H_i$. Next we have to define the differential of $G_\bullet$. Let $\partial_\bullet$ be the differential of $F_\bullet$ and $\partial''_\bullet$ be the differential of $H_\bullet$. We construct the differential

$\partial'_\bullet = (\partial'_i)_{i\geq 0}$ by induction on i. Let $F_0 \xrightarrow{\epsilon} U \to 0$ and $H_0 \xrightarrow{\epsilon''} N \to 0$ be the augmentation maps. We construct $\epsilon' \colon G_0 = F_0 \oplus H_0 \to M$ such that the diagram

$$\begin{array}{ccccccccc}
0 & \longrightarrow & F_0 & \longrightarrow & F_0 \oplus H_0 & \longrightarrow & H_0 & \longrightarrow & 0 \\
 & & \downarrow{\scriptstyle \epsilon} & & \downarrow{\scriptstyle \epsilon'} & & \downarrow{\scriptstyle \epsilon''} & & \\
0 & \longrightarrow & U & \xrightarrow{\ \varphi\ } & M & \xrightarrow{\ \psi\ } & N & \longrightarrow & 0
\end{array}$$

is commutative and ϵ' is an epimorphism.

Let $\mathscr{B}$ be a basis of F_0 and $\mathscr{B}'$ be a basis of H_0. For each element $h \in \mathscr{B}'$ choose an element $m_h \in M$ which is mapped onto $\epsilon''(h) \in N$ by the homomorpism $M \to N$. Then for $f \in \mathscr{B}$ and $h \in \mathscr{B}'$, we set $\epsilon'(f, h) = \varphi \circ \epsilon(f) + m_h$. Since the set $\{(f, h) \colon f \in \mathscr{B}, h \in \mathscr{B}'\}$ is a basis of G_0, the map $\epsilon' \colon G_0 \to M$ is well defined. We leave it to you to check that ϵ' is surjective and the resulting diagram is commutative.

Now, suppose ∂'_j has already been constructed for $j = 0, \ldots, i$, where $\partial'_0 = \epsilon'$. Then we have a commutative diagram with exact rows and columns

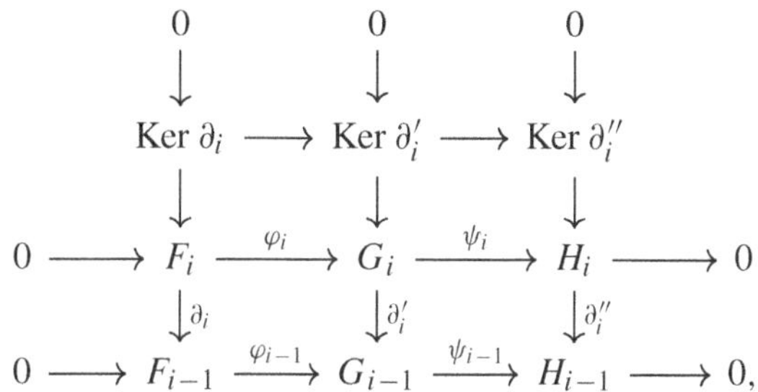

where the map $\operatorname{Ker} \partial_i \to \operatorname{Ker} \partial'_i$ is the restriction of φ_i to $\operatorname{Ker} \partial_i$, and the map $\operatorname{Ker} \partial'_i \to \operatorname{Ker} \partial''_i$ is the restriction of ψ_i to $\operatorname{Ker} \partial'_i$. By induction hypothesis assume that $0 \to \operatorname{Ker} \partial_{i-1} \to \operatorname{Ker} \partial'_{i-1} \to \operatorname{Ker} \partial''_{i-1} \to 0$ is exact. Then replacing the bottom row of the above diagram by this short exact sequence and applying the Snake Lemma, we see that $0 \to \operatorname{Ker} \partial_i \to \operatorname{Ker} \partial' \to \operatorname{Ker} \partial''_i \to 0$ is exact. Then we consider the diagram

$$\begin{array}{ccccccccc}
0 & \longrightarrow & F_{i+1} & \longrightarrow & F_{i+1} \oplus H_{i+1} & \longrightarrow & H_{i+1} & \longrightarrow & 0 \\
 & & \downarrow & & & & \downarrow & & \\
0 & \longrightarrow & \operatorname{Ker} \partial_i & \longrightarrow & \operatorname{Ker} \partial'_i & \longrightarrow & \operatorname{Ker} \partial''_i & \longrightarrow & 0 \\
 & & \downarrow & & & & \downarrow & & \\
 & & 0 & & & & 0 & &
\end{array}$$

with exact rows and columns, and construct $F_{i+1} \oplus H_{i+1} \to \operatorname{Ker} \partial'_i$ as we did it in the first step to make the resulting diagram commutative. Then ∂'_{i+1} is the composition of $F_{i+1} \oplus H_{i+1} \to \operatorname{Ker} \partial'_i$ with the inclusion map $\operatorname{Ker} \partial'_i \hookrightarrow G_i$. □

In analogy to Theorem A.40 we have long exact sequences in the second variable for the Tor- and Ext-functors.

Theorem A.41
Under the assumptions of Theorem A.40 one has the long exact sequence

$$\begin{aligned}\cdots \to \operatorname{Tor}_{i+1}^R(W,N) \to \operatorname{Tor}_i^R(W,U) \to \operatorname{Tor}_i^R(W,M) \to \operatorname{Tor}_i^R(W,N) \to \cdots\\ \to \operatorname{Tor}_1^R(W,N) \to W\otimes U \to W\otimes M \to W\otimes N \to 0,\end{aligned}$$

and the long exact sequence

$$\begin{aligned}0 \to \operatorname{Hom}_R(W,U) \to \operatorname{Hom}_R(W,M) \to \operatorname{Hom}_R(W,N) \to \operatorname{Ext}_R^1(W,U) \to \cdots\\ \to \operatorname{Ext}_R^i(W,U) \to \operatorname{Ext}_R^i(W,M) \to \operatorname{Ext}_R^i(W,N) \to \operatorname{Ext}_R^{i+1}(W,U) \to \cdots.\end{aligned}$$

Proof. Let $F_\bullet$ be a free R-resolution of W. Because of the right exactness of the tensor product, we obtain the short exact sequence $F_\bullet\otimes U \to F_\bullet\otimes M \to F_\bullet\otimes N \to 0$ of complexes. Actually, the complex homomorphism $F_\bullet\otimes U \to F_\bullet\otimes M$ is injective, since for each i we have an exact sequence

$$\cdots \longrightarrow \operatorname{Tor}_1^R(F_i,N) \longrightarrow F_i\otimes U \longrightarrow F_i\otimes M \longrightarrow \cdots$$

and $\operatorname{Tor}_1^R(F_i,N) = 0$, by Exercise A.34. Thus, we indeed have the short exact sequence

$$0 \longrightarrow F_\bullet\otimes U \longrightarrow F_\bullet\otimes M \longrightarrow F_\bullet\otimes N \to 0,$$

which gives rise to the long exact Tor-sequence.

The proof for the long exact Ext-sequence is similar. We only need to know that

$$0 \longrightarrow \operatorname{Hom}_R(F,U) \longrightarrow \operatorname{Hom}_R(F,M) \longrightarrow \operatorname{Hom}_R(F,N) \longrightarrow 0$$

is exact, when F is free. We know that the functor $\operatorname{Hom}_R(W,-)$ is left exact for any R-module W. Thus, we need only to show that $\operatorname{Hom}_R(F,M) \to \operatorname{Hom}_R(F,N)$ is surjective. Indeed, let $\varphi \in \operatorname{Hom}_R(F,N)$ and let $\mathscr{B}$ be a basis of F. For each $f \in \mathscr{B}$ choose $m_f \in M$ which is mapped by $M \to N$ onto $\varphi(f) \in N$. We let $\psi \in \operatorname{Hom}_R(F,M)$ be the map with $\psi(f) = m_f$ for all $f \in \mathscr{B}$. Then $\operatorname{Hom}_R(F,M) \to \operatorname{Hom}_R(F,N)$ maps ψ to φ. □

Lemma A.42
Let M and N be R-modules and suppose that there exists an element $x \in \operatorname{Ann}(M)$ which is regular on N. Then $\operatorname{Hom}_R(M, N) = 0$.

Proof. Let $\varphi \in \operatorname{Hom}_R(M, N)$. Then $x\varphi(M) = \varphi(xM) = 0$. Since x is regular on N, we conclude that $\varphi(M) = 0$. □

Proposition A.43
Let M and N be R-modules and assume that $\mathbf{f} = f_1, \ldots, f_r$ is an N-sequence in $\operatorname{Ann}(M)$. Then

$$\operatorname{Ext}^i_R(M, N) \cong \operatorname{Ext}^{i-r}_R(M, N/(\mathbf{f})N).$$

Proof. We prove the assertion by induction on r. If $r = 0$, then there is nothing to prove. Let $r > 0$. Set $\bar{N} = N/f_1N$. By Theorem A.41, the short exact sequence $0 \to N \xrightarrow{f_1} N \to \bar{N} \to 0$ induces the exact sequence

$$\operatorname{Ext}^{i-1}_R(M, N) \to \operatorname{Ext}^{i-1}_R(M, \bar{N}) \to \operatorname{Ext}^i_R(M, N) \xrightarrow{\varphi} \operatorname{Ext}^i_R(M, N).$$

By induction hypothesis, $\operatorname{Ext}^{i-1}_R(M, N) \cong \operatorname{Ext}^{i-r}_R(M, N/(f_1, \ldots, f_{r-1})N)$. Since f_n is regular on $N/(f_1, \ldots, f_{n-1})N$, by Lemma A.42, $\operatorname{Ext}^{i-1}_R(M, N) = 0$.

It follows from Proposition A.48 that $\operatorname{Ext}^i_R(M, -)$ is an R-linear functor. This implies that φ is the multiplication map by f_1. Since $f_1M = 0$, we have $f_1 \operatorname{Ext}^i_R(M, N) = 0$, which means that $\varphi = 0$. This together with our induction hypothesis implies that

$$\begin{aligned}\operatorname{Ext}^i_R(M, N) \cong \operatorname{Ext}^{i-1}_R(M, \bar{N}) &\cong \operatorname{Ext}^{i-r}_R(M, \bar{N}/(f_2, \ldots, f_n)\bar{N}) \\ &\cong \operatorname{Ext}^{i-r}_R(M, N/(\mathbf{f})N).\end{aligned}$$

□

A.5 Left and Right Derived Functors

Let Mod_R be the category of R-modules, and let $F\colon \mathrm{Mod}_R \to \mathrm{Mod}_R$ be an additive functor. When F is a right exact covariant functor, the left derived functors of F are defined and when F is left exact, the right derived functors of F are defined.

We start with a right exact covariant additive functor F. For $i = 0, 1, \dots$ the *left derived functors* $L_i F\colon \mathrm{Mod}_R \to \mathrm{Mod}_R$ of F satisfy the following properties:

(1) $L_0 F = F$.
(2) $L_i F(P) = 0$ for all $i > 0$, if P is projective.
(3) For every short exact sequence $0 \to U \xrightarrow{\varphi} M \xrightarrow{\psi} N \to 0$, there is a long exact sequence

$$\cdots \to L_{i+1}F(N) \xrightarrow{\delta_{i+1}} L_i F(U) \xrightarrow{L_i F(\varphi)} L_i F(M) \xrightarrow{L_i F(\psi)} L_i F(N) \to \cdots$$
$$\to L_1 F(N) \xrightarrow{\delta_1} F(U) \xrightarrow{F(\varphi)} F(M) \xrightarrow{F(\psi)} F(N) \to 0.$$

(4) The connecting homomorphisms δ_i are natural in the following sense: if

$$\begin{array}{ccccccccc} 0 & \longrightarrow & U & \longrightarrow & M & \longrightarrow & N & \longrightarrow & 0 \\ & & \downarrow{\alpha} & & \downarrow{\beta} & & \downarrow{\gamma} & & \\ 0 & \longrightarrow & U' & \longrightarrow & M' & \longrightarrow & N' & \longrightarrow & 0, \end{array}$$

is a commutative diagram with exact rows, then

$$\begin{array}{ccc} L_{i+1}F(N) & \xrightarrow{\delta_{i+1}} & L_i F(U) \\ {\scriptstyle L_{i+1}F(\gamma)}\downarrow & & \downarrow{\scriptstyle L_i F(\alpha)} \\ L_{i+1}F(N') & \xrightarrow{\delta'_{i+1}} & L_i F(U') \end{array}$$

is commutative for all $i > 0$.

The above properties (1) to (4) determine left derived functors of F uniquely up to functorial isomorphism.

Let us check that $\mathrm{Tor}_i(-, W) = L_i F(-)$, where $F(-) = - \otimes W$. Condition (1) is shown in Proposition A.38(a) and condition (3) is shown in Theorem A.40. Now let P be a projective module. Then P is a direct summand of a free module F. Thus we may write $F = P \oplus Q$. By Exercise A.34 for $i > 0$ we have $0 = \mathrm{Tor}_i^R(F, W) = \mathrm{Tor}_i^R(P, W) \oplus \mathrm{Tor}_i^R(Q, W)$. This implies that $\mathrm{Tor}_i^R(P, W) = 0$ for $i > 0$ and proves (2).

When $F\colon \mathrm{Mod}_R \to \mathrm{Mod}_R$ is a right exact covariant additive functor, the left derived functors $L_i F$ are obtained by choosing for each $M \in \mathrm{Mod}_R$ a projective resolution $P_\bullet$ and setting $L_i F(M) = H_i(F(P_\bullet))$.

Next we consider a left exact, contravariant additive functors $F : \mathrm{Mod}_R \to \mathrm{Mod}_R$. For such functors the right derived functors $R^i F$ exist and are characterized by the properties (1) to (4) with the arrows reversed between the derived functors in the diagrams.

An important example is the functor $F(-) = \operatorname{Hom}_R(-, W)$ with $W \in \operatorname{Mod}_R$. For this functor we have $R^i F(-) = \operatorname{Ext}^i_R(-, W)$ for all i, see Exercise A.45. We use this fact to prove the following result of Rees.

Theorem A.44
Let M and N be R-modules, and let $x \in R$ be a regular element on R and N with $xM = 0$. Then

$$\operatorname{Ext}^{i+1}_R(M, N) \cong \operatorname{Ext}^i_{R/xR}(M, N/xN)$$

for all $i \geq 0$.

Moreover, if R is graded, $M, N \in \mathcal{M}_R$ and x is homogeneous of degree a, then $\underline{\operatorname{Ext}}^{i+1}_R(M, N) \cong \underline{\operatorname{Ext}}^i_{R/xR}(M, N/xN)(a)$ for all $i \geq 0$.

Proof. Set $\bar{R} = R/xR$ and $\bar{N} = N/xN$. Since $xM = 0$, we may view M as an $\bar{R}$-module. Conversely, any $\bar{R}$-module W may be viewed as an R-module with $xW = 0$. We show that the functors $\operatorname{Ext}^{i+1}_R(-, N)\colon \operatorname{Mod}_{\bar{R}} \to \operatorname{Mod}_{\bar{R}}$ are the right derived functors of $\operatorname{Hom}_{\bar{R}}(-, \bar{N})$.

By Theorem A.40, the exact sequence $0 \to N \xrightarrow{x} N \to \bar{N} \to 0$ induces the exact sequence

$$\operatorname{Hom}_R(M, N) \to \operatorname{Hom}_R(M, \bar{N}) \to \operatorname{Ext}^1_R(M, N) \xrightarrow{x} \operatorname{Ext}^1_R(M, N)$$

We observe that $\operatorname{Hom}_R(M, N) = 0$. Indeed, let $\varphi \in \operatorname{Hom}_R(M, N)$. Then $\operatorname{Im}\varphi \subseteq N$ and $x \operatorname{Im}\varphi = 0$. Since x is regular of N, it follows that $\operatorname{Im}\varphi = 0$. Now, since x annihilates $\operatorname{Ext}^1_R(M, N)$, we deduce from the above exact sequence that $\operatorname{Hom}_{\bar{R}}(M, \bar{N}) \cong \operatorname{Ext}^1_R(M, N)$. This proves property (1) for the right derived functors of $\operatorname{Hom}_{\bar{R}}(-, \bar{N})$.

Let P be a projective $\bar{R}$-module. Then there is a free $\bar{R}$-module F for which P is a direct summand. The exact sequence $0 \to R \xrightarrow{x} R \to \bar{R} \to 0$ yields to a free resolution of F. Hence $\operatorname{pd}(F) \leq 1$ as an R-module. It follows that $\operatorname{Ext}^{i+1}_R(F, N) = 0$ for $i > 0$, and since P is a direct summand of F, we also have $\operatorname{Ext}^{i+1}_R(P, N) = 0$ for $i > 0$. This proves property (2).

Property (3) follows from Theorem A.40.

The proof of the graded case is essentially the same. The shift by a results from the fact that under the assumptions of the theorem, the exact sequence $0 \to N(-a) \xrightarrow{x} N \to \bar{N} \to 0$ implies that $\underline{\operatorname{Hom}}_{\bar{R}}(M, \bar{N})(a) \cong \underline{\operatorname{Ext}}^1_R(M, N)$, see Exercise 2.3. The rest of the proof is as before. $\square$

Exercise A.45
Show that the right derived functors of $\operatorname{Hom}_R(-, W)$ are the functors $\operatorname{Ext}^i_R(-, W)$, for $i \geq 0$.

Finally we consider a left exact, covariant additive functor $F : \operatorname{Mod}_R \to \operatorname{Mod}_R$. An example of interest for us is the functor $F(-) = \operatorname{Hom}_R(W, -)$. Another important example is the section functor $\Gamma_{\mathfrak{m}}$ which will be considered in Sect. 3.2 where we study the canonical module.

Now we are ready to define the right derived functors $R^i F$ of a covariant left exact functor F, and to prove their existence. They should satisfy the properties (1)-(4) similar to those for left derived covariant functors but with the obvious modifications. In (2) the projective module has to be replaced by an injective module. In other words, we want that $R^i F(E) = 0$ for $i > 0$, if E is injective. In (3) the long exact sequence should begin as follows:

$$0 \to F(U) \to F(M) \to F(N) \to R^1 F(U) \to R^1 F(M) \to R^1 F(N) \to \cdots .$$

In order to prove the existence of the right derived functor of F we choose for each $M \in \operatorname{Mod}_R$ an injective resolution $E^\bullet$, and set $R^i F(M) = H^i(F(E^\bullet))$. Let $\varphi : N \to M$ be an R-module homomorphism and $L^\bullet$ be an injective resolution of N. By using the extension property for morphisms into injective modules, one can extend $\varphi : N \to M$ to a cocomplex homomorphism $\varphi^\bullet : L^\bullet \to E^\bullet$ and for each i one defines the R-module homomorphism $R^i F(\varphi) : R^i F(N) \to R^i F(M)$ as $R^i F(\varphi) = H^i(F(\varphi^\bullet))$. The so defined functors satisfy the axioms of the right derived functors of F. We leave it to the reader to check this.

Proposition A.46
Let $W \in \operatorname{Mod}_R$. Then $R^i \operatorname{Hom}(W, -) = \operatorname{Ext}^i_R(W, -)$ for all i.

Proof. Let $M \in \operatorname{Mod}_R$, and $E^\bullet$ be an injective resolution of M. We show that $\operatorname{Ext}^i_R(W, M) \cong H^i(\operatorname{Hom}_R(W, E^\bullet))$ for all i.

Let $P_\bullet$ be a projective resolution of W. Then we obtain a first quadrant double cocomplex $D^{\bullet,\bullet}$ of the form

$$
\begin{array}{ccccccccc}
 & & 0 & & 0 & & 0 & & \\
 & & \uparrow & & \uparrow & & \uparrow & & \\
0 & \longrightarrow & \mathrm{Hom}_R(P_2, E^0) & \longrightarrow & \mathrm{Hom}_R(P_2, E^1) & \longrightarrow & \mathrm{Hom}_R(P_2, E^2) & \longrightarrow & \cdots \\
 & & \uparrow & & \uparrow & & \uparrow & & \\
0 & \longrightarrow & \mathrm{Hom}_R(P_1, E^0) & \longrightarrow & \mathrm{Hom}_R(P_1, E^1) & \longrightarrow & \mathrm{Hom}_R(P_1, E^2) & \longrightarrow & \cdots \\
 & & \uparrow & & \uparrow & & \uparrow & & \\
0 & \longrightarrow & \mathrm{Hom}_R(P_0, E^0) & \longrightarrow & \mathrm{Hom}_R(P_0, E^1) & \longrightarrow & \mathrm{Hom}_R(P_0, E^2) & \longrightarrow & \cdots \\
 & & \uparrow & & \uparrow & & \uparrow & & \\
 & & 0 & & 0 & & 0 & &
\end{array}
$$

For each i, the ith row cocomplex $\mathrm{Hom}_R(P_i, E^\bullet)$ is acyclic and with $H^0(\mathrm{Hom}_R(P_i, E^\bullet)) = \mathrm{Hom}_R(P_i, M)$, inducing the cocomplex $\mathrm{Hom}_R(P_\bullet, M)$ with $H^i(\mathrm{Hom}_R(P_\bullet, M)) = \mathrm{Ext}^i_R(W, M)$, and for each j, the jth column cocomplex $\mathrm{Hom}_R(P_\bullet, E^j)$ is acyclic with $H^0(\mathrm{Hom}_R(P_\bullet, E^j)) = \mathrm{Hom}_R(W, E^j)$, inducing the cocomplex $\mathrm{Hom}_R(W, E^\bullet)$. Therefore, (A.5) implies that

$$\mathrm{Ext}^i_R(W, M) \cong H^i(D^{\bullet,\bullet}) \cong H^i(\mathrm{Hom}_R(W, E^\bullet)),$$

as desired. □

Summarizing we have

Corollary A.47
Let $M, N \in \mathrm{Mod}_R$, let $P_\bullet$ be a projective resolution of M, and let $E^\bullet$ be an injective resolution of N. Then

$$\mathrm{Ext}^i_R(N, M) \cong H^i(\mathrm{Hom}_R(P_\bullet, N)) \cong H^i(\mathrm{Hom}_R(M, E^\bullet)).$$

We conclude this section with the following result.

Proposition A.48
Let $F\colon \mathrm{Mod}_R \to \mathrm{Mod}_R$ be a right exact R-linear functor. Then its left derived functors $L_i F$ are also R-linear. A similar statement holds for left exact R-linear functors.

Proof. We give the proof for the case that F is right exact and covariant. The other cases are proved similarly. Let $M, N \in \text{Mod}_R$, $\varphi, \psi \in \text{Hom}_R(M, N)$, and let $x, y \in R$. Let $P_\bullet$ be a projective resolution of M, and $Q_\bullet$ be a projective resolution of N. There exist chain maps $\varphi_\bullet, \psi_\bullet\colon P_\bullet \to Q_\bullet$ with $H_0(\varphi_\bullet) = \varphi$ and $H_0(\psi_\bullet) = \psi$. It follows that $x\varphi_\bullet + y\psi_\bullet\colon P_\bullet \to Q_\bullet$ is a chain map with $H_0(x\varphi_\bullet + y\psi_\bullet) = x\varphi + y\psi$. Let $z \in F(P_i)$ be a cycle of the complex $F(P_\bullet)$ and $[z] \in H_i(F(P_\bullet)) = L_iF(M)$ be its homology class. Then

$$\begin{aligned} L_iF(x\varphi + y\psi)([z]) &= [(x\varphi_i + y\psi_i)(z)] = [x\varphi_i(z) + y\psi_i(z)] \\ &= x[\varphi_i(z)] + y[\psi_i(z)] = xL_iF(\varphi)([z]) + yL_iF(\psi)([z]) \\ &= (xL_iF(\varphi) + yL_iF(\psi))([z]). \end{aligned}$$

This shows that $L_iF(x\varphi + y\psi) = xL_iF(\varphi) + yL_iF(\psi)$, as desired. □

Now we come to a concept which plays a role in the proof of Grothendieck's local duality theorem in Chap. 3. A contravariant functor $F\colon \mathcal{M}_R \to \mathcal{M}_R$ is called a *representable* functor, if there exists an R-module $W \in \mathcal{M}_R$ such that $F(-) \cong \underline{\text{Hom}}_R(-, W)$. In this case $W \cong F(R)$.

We have the following important characterization of representable functors.

Proposition A.49

Let $F\colon \mathcal{M}_R \to \mathcal{M}_R$ be a contravariant functor. The following conditions are equivalent:

(i) F is representable.
(ii) F is R-linear and left exact, it commutes with finite direct summands and $F(R(a)) = F(R)(-a)$ for all $a \in \mathbb{Z}$.

Proof. (i) ⇒ (ii): Since F is representable, there exists $W \in \mathcal{M}_R$ such that $F(-) \cong \underline{\text{Hom}}_R(-, W)$. Since $\underline{\text{Hom}}_R(-, W)$ satisfies the conditions in (ii), then F does it as well.

(ii) ⇒ (i): We set $W = F(R)$ and show that there is a functorial isomorpism $F(M) \cong \underline{\text{Hom}}_R(M, W)$ for $M \in \mathcal{M}_R$. We choose a graded free presentation

$$H \xrightarrow{\varphi} G \longrightarrow M \longrightarrow 0$$

of M with $G = \bigoplus_{i=1}^n R(-a_i)$ and $H = \bigoplus_{j=1}^m R(-b_j)$. Notice that by Proposition 1.9 such presentation always exist.

Due to the conditions on F given in (ii), and due to the properties of the functor $\underline{\mathrm{Hom}}_R(-, W)$, we have the functorial isomorphisms

$$F(G) \cong \bigoplus_{i=1}^{n} F(R(-a_i)) \cong \bigoplus_{i=1}^{n} F(R)(a_i) \cong \bigoplus_{i=1}^{n} W(a_i)$$
$$\cong \underline{\mathrm{Hom}}_R(\bigoplus_{i=1}^{n} R(-a_i), W) = \underline{\mathrm{Hom}}_R(G, W).$$

Similarly, $F(H) \cong \underline{\mathrm{Hom}}_R(H, W)$. Thus, we obtain a diagram

$$\begin{array}{ccccccc} 0 & \longrightarrow & F(M) & \longrightarrow & F(G) & \xrightarrow{F(\varphi)} & F(H) \\ & & & & \downarrow & & \downarrow \\ 0 & \longrightarrow & \underline{\mathrm{Hom}}_R(M, W) & \longrightarrow & \underline{\mathrm{Hom}}_R(G, W) & \xrightarrow{\underline{\mathrm{Hom}}_R(\varphi, W)} & \underline{\mathrm{Hom}}_R(H, W) \end{array}$$

with vertical functorial isomorphisms. Since F is R-linear, this diagram is commutative (see Exercise A.50) and hence induces a functorial isomorphism $F(M) \to \underline{\mathrm{Hom}}_R(M, W)$. □

Exercise A.50
Let $F: \mathscr{M}_R \to \mathscr{M}_R$ be a contravariant R-linear functor, which commutes with finite direct summands and for which $F(R(-a)) = F(R)(-a)$, and let

$$\bigoplus_{j=1}^{m} R(-b_j) = H \xrightarrow{\varphi} G = \bigoplus_{i=1}^{n} R(-a_i)$$

be a morphism in the category $\mathscr{M}_R$, that is, a graded R-module homomorphism of degree 0. Let $h_1, \ldots, h_m$ be the canonical basis of H, let $g_1, \ldots, g_n$ be the canonical basis of G and let $C = (c_{ij})$ be the $n \times m$-matrix representing φ with respect to these bases. Then we obtain the morphism

$$F(G) = \bigoplus_{i=1}^{n} F(R)(a_i) \xrightarrow{F(\varphi)} \bigoplus_{j=1}^{m} F(R)(b_j) = F(H).$$

Show that for each homogeneous element $w = (w_1, \ldots, w_n)$ we have

$$F(\varphi)(w) = (\sum_{j=1} c_{1j} w_j, \ldots, \sum_{j=1} c_{mj} w_j),$$

and deduce from this fact that $F(\varphi) = \underline{\mathrm{Hom}}_R(\varphi, F(R))$.

Notation

$\mathbf{a}^{\mathsf{T}}$	transpose of the vector $\mathbf{a}$
$\langle a, bH\rangle$	gluing of H induced by (a, b)
$\langle a_1, \dots, a_m\rangle$	numerical semigroup generated by $a_1, \dots, a_m$
$\mathscr{A}_R$	category of Artinian graded R-modules
$\mathrm{Ann}(M)$	Annihilator of M
$\mathrm{Ap}_H(a)$	Apéry set of H with respect to a
$\mathrm{Ass}(M)$	set of associated prime ideals of M
$a(R)$	a-invariant of R
$\beta_{i,j}(M)$	(i, j)-th graded Betti number of M
$B_i(M_\bullet)$	i-boundaries of complex $M_\bullet$
$\mathrm{Bihom}_R(M \times N, W)$	set of R-bilinear maps from $M \times N$ to W
C_H	normalized canonical ideal of H
$C^\bullet$	modified Čech complex
$\mathrm{char}(K)$	characteristic of the field K
$D_{\bullet,\bullet}$	double complex
$\mathrm{depth}\, M$	depth of M
$\dim_K(V)$	K-vector space dimension of V
$\dim M$	Krull dimension of M
$\underline{\mathrm{Ext}}$	the restriction of Ext-functor to the category GMod_R
$e_{j_1} \wedge \cdots \wedge e_{j_i}$	wedge product
$e(M)$	multiplicity of module M
$e(H)$	multiplicity of semigroup H
$E(M)$	injective hull of M
$\underline{E}(M)$	injective hull of M in the category GMod_R
$\mathrm{emb}\, R$	embedding dimension of R
$\mathrm{emb}\, H$	embedding dimension of semigroup H
$\mathscr{F}_R$	category the graded R-modules M with $\dim_K M_i < \infty$ for all i

J. Herzog et al., *Numerical Semigroups*, Compact Textbooks in Mathematics,
https://doi.org/10.1007/978-3-032-05424-1

$F(H)$	Frobenius number of H
$f_{\mathbf{v}}$	binomial attached to vector $\mathbf{v}$
$\gcd(u, v)$	greatest common divisor of u and v
$\underline{\Gamma}_{\mathfrak{m}}$	section functor in the category GMod_R
$\mathscr{G}(H)$	set of gaps of H
$g(H)$	genus of H
$\mathrm{grade}(I, M)$	grade of I with respect to M
GMod_R	category of graded R-modules
$\mathrm{gr}_{\mathfrak{m}}(M)$	associated graded module of M
$\underline{H}^i_{\mathfrak{m}}$	ith local cohomology functor in the category GMod_R
$H(M, i)$	dimension of the i-th graded component of M
$\mathrm{height}\, I$	height of I
$\mathrm{height}_M\, I$	M-height of I
$H_i(M_\bullet)$	ith-homology module of complex $M_\bullet$
$\mathrm{Hilb}_M(t)$	Hilbert series of M
$\mathrm{HilbS}_M(t)$	Hilbert-Samuel series of M
$\mathrm{Hom}_R(M, N)$	set of all R-module homomorphisms from M to N
$\underline{\mathrm{Hom}}_R(M, N)$	direct sum of the graded R-module homomorphisms from M to N
$I - J$	difference of relative semigroup ideals
$I^{(k)}$	kth symbolic power of I
$I_s(A)$	ideal generated by all s-minors of the matrix A
$I : J$	colon ideal of I with respect to J
$\mathrm{id}(M)$	injective dimension of M
$\underline{\mathrm{id}}(M)$	graded injective dimension of M
$\mathrm{Im}\,\varphi$	image of φ
I_H	defining ideal of semigroup ring $K[H]$
$\mathscr{K}$	category of $\mathbb{Z}$-graded K-vector spaces
$\mathrm{Ker}\,\varphi$	kernel of φ
$K_\bullet(\mathbf{f}; M)$	Koszul complex of M with respect to $\mathbf{f} = f_1, \ldots, f_m$
$K[t_1^{\pm 1}, \ldots, t_d^{\pm 1}]$	Laurent polynomial ring over K in the variables $t_1, \ldots, t_d$
$K[x_1, \ldots, x_n]$	polynomial ring over K
$K[H]$	semigroup ring of H
$\bigwedge^s \varphi$	wedge product of φ
$\bigwedge^\bullet F$	exterior algebra of F
$\ell(M)$	length of M
$\mathrm{lcm}(u, v)$	least common multiple of u and v
M_P	localization with respect to prime ideal P
$\mu(M)$	minimal number of a homogeneous set of generators of M
$\mu_i(P, M)$	ith Bass number of M with respect to P.
$\mathrm{Min}(M)$	set of minimal prime ideals of M
$\mathscr{M}_R$	category of finitely generated graded R-modules
Mod_R	category of all R-modules
$M \otimes N$	tensor product of M and N in Mod_R
$M \underline{\otimes} N$	tensor product of M and N in GMod_R

$\mathcal{N}(H)$	set of non-gaps of H
$n(H)$	number of non-gaps of H
$[n]$	set of integers $\{1, \dots, n\}$
N^*	submodule generated by all homogeneous elements of N
ω_R	canonical module of R
Ω_H	canonical ideal of H
$\operatorname{pd} M$	projective dimension of M
$\underline{\operatorname{pd}}_R(M)$	projective dimension of M in $\mathcal{M}_R$
$\operatorname{PF}(H)$	set of pseudo-Frobenius numbers of H
$Q(R)$	field of fractions of R
$\operatorname{rank} M$	rank of M
$\sqrt{I}$	radical of I
$r(R)$	Cohen-Macaulay type of R
$\rho_H(a)$	number of factorizations of a
$\operatorname{res}(H)$	residuum of H
$S^{-1}M$	localization of M with respect to S
$\operatorname{Soc}(M)$	socle of M
$\operatorname{sgn}(\pi)$	signature of permutation π
$\operatorname{Spec}(R)$	set of prime ideals of R
$\operatorname{Supp}(M)$	support of M
$t(H)$	type of semigroup H
$\operatorname{trdeg}(L/K)$	transcendence degree of L/K
Tor	Tor-functor
$\underline{\operatorname{Tor}}$	restriction of Tor-functor to GMod_R
$\operatorname{tr}(\Omega_H)$	trace of the canonical ideal of H
$V^\vee$	graded dual of graded vector space V
$\mathbb{Z}_{\geq 0}$	set of nonnegative integers
$\mathbb{Z}^{d \times n}$	set of $d \times n$-matrices $A = (a_{ij})_{\substack{1 \le i \le d \\ 1 \le j \le n}}$ with each $a_{ij} \in \mathbb{Z}$
$Z(M)$	set of zerodivisors of M
$Z_i(M_\bullet)$	i-cycles of complex $M_\bullet$

References

1. R. Apéry, Sur les branches superlinéaires des courbes algébriques. C. R. Acad. Sci. Paris **222**, 1198–1200 (1946)
2. A. Assi, M. D'Anna, P.A. García-Sánchez, *Numerical Semigroups and Applications*, 2nd edn., RSME Springer Series, 3 (Springer, Cham, 2020)
3. M.F. Atiyah, L.G. Macdonald, *Introduction to Commutative Algebra* (Addison-Wesley Publishing Company, 1969)
4. J. Backelin, On the number of semigroups of natural numbers. Math. Scand. **66**, 197–215 (1990)
5. V. Barucci, D.E. Dobbs, M. Fontana, Maximality properties in numerical semigroups and applications to one-dimensional analytically irreducible local domains. Mem. Amer. Math. Soc. **125** (1997)
6. V. Barucci, R. Fröberg, One-dimensional almost Gorenstein rings. J. Algebra **188**, 418–442 (1997)
7. V. Barucci, F. Khouja, On the class semigroup of a numerical semigroup. Semigroup Forum **92**, 377–392 (2016)
8. R. Berger, Über eine Klasse unvergabelter lokaler Ringe. Math. Ann. **146**, 98–102 (1962)
9. A. Brauer, On a problem of partitions. Am. J. Math. **64**, 299–312 (1942)
10. A. Brauer, J.E. Shockley, On a problem of Frobenius. J. Reine Angew. Math. **211**, 215–220 (1962)
11. H. Bresinsky, On prime ideals with generic zero $x_i = t^{n_i}$. Proc. Am. Math. Soc. **47**, 329–332 (1975)
12. H. Bresinsky, Symmetric semigroups of integers generated by 4 elements. Manuscripta Math. **17**, 205–219 (1975)
13. M.P. Brodmann, R.Y. Sharp, *Local Cohomology: An Algebraic Introduction with Geometric Applications*, vol. 136 (Cambridge University Press, London, Cambridge, New York, 2012)
14. W. Bruns, P. Garcia-Sanchez, C. O'Neill, D. Wilburne, Wilf's conjecture in fixed multiplicity. Internat. J. Algebra Comput. **30**, 861–882 (2020)
15. W. Bruns, J. Herzog, *Cohen–Macaulay Rings*. Revised edition (Cambridge University Press, 1998)
16. W. Bruns, B. Ichim, C. Söger, U. von der Ohe, Normaliz. Algorithms for rational cones and affine monoids. https://www.normaliz.uni-osnabrueck.de

J. Herzog et al., *Numerical Semigroups*, Compact Textbooks in Mathematics, https://doi.org/10.1007/978-3-032-05424-1

17. H. Cartan, S. Eilenberg, *Homological Algebra* (Princeton University Press, 1956)
18. S.T. Chapman, P.A. García-Sánchez, D. Llena, V. Ponomarenko, J.C. Rosales, The catenary and tame degree in finitely generated commutative cancellative monoids. Manuscripta Math. **120**, 253–264 (2006)
19. S.T. Chapman, R. Hoyer, N. Kaplan, Delta Sets of Numerical Monoids are Eventually Periodic. Aequationes Math. **77**, 273–279 (2009)
20. H. Dao, T. Kobayashi, R. Takahashi, Trace of canonical modules, annihilator of Ext, and classes of rings close to being Gorenstein. J. Pure Appl. Algebra **225** (2021)
21. M. Delgado, P.A. García-Sánchez, J.J. Morais, *numericalsgps*, a package of the GAP System for Computational Discrete Algebra. Version 4.7.9 of 29-Nov-2015 (free software, GPL)
22. C. Delorme, Sous-monoïdes d'intersection complète de N. Ann. Scient. École Norm. Sup. **4**, 145–154 (1976)
23. G. Denham, Short generating functions for some semigroup algebras. Electron. J. Combin. **10** (2003)
24. S. Ding, A note on the index of Cohen-Macaulay local rings. Comm. Algebra **21**, 53–71 (1993)
25. P. Eakin, A. Sathaye, Prestable ideals. J. Algebra **41**, 439–454 (1976)
26. D. Eisenbud, *Commutative Algebra with a View Toward Algebraic Geometry*. Graduate Texts in Mathematics (Springer, New York, 1995)
27. S. Eliahou, Wilf's conjecture and Macaulay's theorem. J. Eur. Math. Soc. **20**, 2105–2129 (2018)
28. J. Elias, The conjecture of Sally on the Hilbert function for curve singularities. J. Algebra **160**, 42–49 (1993)
29. K. Eto, Almost Gorenstein monomial curves in affine four space. J. Algebra **488**, 362–387 (2017)
30. R. Fröberg, C. Gottlieb, R. Häggkvist, On numerical semigroups. Semigroup Forum **35**, 63–83 (1986)
31. A. Geroldinger, F. Halter-Koch, Non-unique factorizations of algebraic integers. Funct. Approx. Comment. Math. **39**, part 1, pp. 49–60 (2008)
32. P. Gimenez, I. Sengupta, H. Srinivasan, Minimal free resolutions for certain affine monomial curves, in *Commutative Algebra and its Connections to Geometry: Pan-American Advanced Studies Institute August 3–14, 2009*, ed. by A. Corso, C. Polini (Universidade Federal De Pernambuco, Olinda, Brazil (Contemporary Mathematics), 2011), pp. 87–95
33. S. Goto, R. Takahashi, N. Taniguchi, Almost Gorenstein rings–towards a theory of higher dimension. J. Pure Appl. Algebra **219**, 2666–2712 (2015)
34. J. Herzog, Generators and relations of abelian semigroups and semigroup rings. Louisiana State University. LSU Digital Commons, LSU Historical Dissertations and Theses (1969)
35. J. Herzog, Generators and relations of abelian semigroups and semigroup rings. Manuscripta Math. **3**, 175–193 (1970)
36. J. Herzog, T. Hibi, H. Ohsugi, *Binomial Ideals*. Graduate Texts in Mathematics, vol. 279 (Springer, Cham, 2018)
37. J. Herzog, T. Hibi, D.I. Stamate, The trace of the canonical module. Israel J. Math. **233**, 133–165 (2019)
38. J. Herzog, T. Hibi, D.I. Stamate, Canonical trace ideal and residue for numerical semigroup rings. Semigroup Forum **103**, 550–566 (2021)
39. J. Herzog, S. Kumashiro, Upper bound on the colength of the trace of the canonical module in dimension one. Arch. Math. (Basel) **119** (2022)
40. J. Herzog, E. Kunz, Der kanonische Modul eines Cohen-Macaulay Rings. Lecture Notes in Mathematics vol. 238 (Springer, 1971)
41. J. Herzog, E. Kunz, *Die Werthalbgruppe eines lokalen Rings der Dimension* 1 (Springer-Verlag, Sitzungsberichte der Heidelberger Akademie der Wissenschaften, 1971)
42. J. Herzog, K. Watanabe, Almost symmetric numerical semigroups. In Semigroup Forum **98**, 589–630 (2019)
43. C. Huneke, A. Vraciu, Rings that are almost Gorenstein. Pacific J. Math. **225**, 85–102 (2006)

44. S.M. Johnson, A linear diophantine problem. Canadian J. Math. **12**, 390–398 (1960)
45. T. Kobayashi, Local rings with self-dual maximal ideal. Illinois J. Math. **64**, 349–373 (2020)
46. V. Kodiyalam, Asymptotic behaviour of Castelnuovo-Mumford regularity. Proc. Am. Math. Soc. **128**, 407–411 (2000)
47. E. Kunz, The value-semigroup of a one-dimensional Gorenstein ring. Proc. Am. Soc. **25**, 748–751 (1970)
48. E. Kunz, Über die Klassifikation numerischer Halbgruppen. Regensburger Mathematische Schriften **11** (1987)
49. E. Kunz, *Introduction to Commutative Algebra and Algebraic Geometry* (Birkhäuser, New York, 2013)
50. F.S. Macaulay, The algebraic theory of modular systems. Revised reprint of the 1916 original. With an introduction by Paul Roberts. Cambridge Mathematical Library (Cambridge University Press, Cambridge, 1994)
51. H. Matsumura, *Commutative Ring Theory* (Cambridge University Press, 1986)
52. A. Moscariello, On the type of an almost Gorenstein monomial curve. J. Algebra **456**, 266–277 (2016)
53. H. Nari, Symmetries on almost symmetric numerical semigroups. Semigroup Forum **86**, 140–154 (2013)
54. H. Nari, T. Numata, K.I. Watanabe, Genus of numerical semigroups generated by three elements. J. Algebra **358**, 67–73 (2012)
55. A. Nijenhuis, H.S. Wilf, Representations of integers by linear forms in nonnegative integers. J. Number Theory **4**, 98–106 (1972)
56. L. O'Carroll, A uniform Artin-Rees theorem and Zariski's main lemma on holomorphic functions. Invent. Math. **90**, 674–682 (1987)
57. A. Oneto, F. Strazzanti, G. Tamone, One-dimensional Gorenstein local rings with decreasing Hilbert function. J. Algebra **489**, 91–114 (2017)
58. F. Orecchia, One-dimensional local rings with reduced associated graded ring and their Hilbert functions. Manuscripta Math. **32**, 391–405 (1980)
59. D.P. Patil, Minimal sets of generators for the relation ideals of certain monomial curves. Manuscripta Math. **80**, 239–248 (1993)
60. I. Peeva, *Graded Syzygies. Algebra and Applications*, vol. 14 (Springer-Verlag London, Ltd., London, 2011)
61. J.L. Ramírez Alfonsín, *The diophantine Frobenius problem*, vol. 30 (Oxford University Press on Demand, 2005)
62. L. Rédei, *Theorie der endlich erzeugbaren kommutativen Halbgruppen*, vol. 41 (Physica-Verlag, 1963)
63. J.C. Rosales, *Semigrupos numéricos*, Tesis Doctoral (Universidad de Granada, Spain, 2001)
64. J.C. Rosales, P.A. García-Sánchez, Numerical semigroups with embedding dimension three. Arch. Math. **83**, 488–496 (2004)
65. J.C. Rosales, P.A. García-Sánchez, *Numerical Semigroups*, vol. 20 (Springer, 2009)
66. J.C. Rosales, P.A. García-Sánchez, J.I. García García, Atomic commutative monoids and their elasticity. Semigroup Forum **68**, 64–86 (2004)
67. J.C. Rosales, P.A. García-Sánchez, J.I. García García, J.A. Madrid, Fundamental gaps in numerical semigroups. J. Pure Appl. Algebra **189**, 301–313 (2004)
68. J. Rotman, *An Introduction to Homological Algebra*, 2nd edn., Universitext (Springer, New York, 2009)
69. J.D. Sally, *Numbers of Generators of Ideals in Local Rings* (Marcel Dekker Inc., New York-Basel, 1978)
70. E.S. Selmer, On a linear Diophantine problem of Frobenius. J. Reine Angew. Math. **293**(294), 1–17 (1977)
71. R.Y. Sharp, *Steps in Commutative Algebra*, 2nd edn. (Cambridge University Press, 2000)

72. R.P. Stanley, *Combinatorics and Commutative Algebra*, 2nd edn. Progress in Mathematics, vol. 41 (Birkhäuser Boston, Inc., Boston, MA, 1996)
73. J.J. Sylvester, On subinvariants, i.e. Semi-Invariants to Binary Quantics of an Unlimited Order. Am. J. Math. **5**, 79–136 (1882)
74. A. Tripathi, The coin exchange problem for arithmetic progressions. Am. Math. Monthly **101**, 779–781 (1994)
75. A. Tripathi, On a variation of the coin exchange problem for arithmetic progression. Electron. J. Combin. Number Theory **3**, #A01 4 (2003)
76. A. Tripathi, S. Vijay, On a generalization of the coin exchange problem for three variables. J. Integer Seq. **9**, 1–8 (2006)
77. K. Watanabe, Some examples of one dimensional Gorenstein rings. Nagoya Math. J. **49**, 101–109 (1973)
78. K. Watanabe, A short proof of Bresinski's theorem on Gorenstein semigroup rings generated by four elements, in *Numerical Semigroups*, ed. by V. Barucci, S. Chapman, M. D'Anna, R. Fröberg, pp. 367–374, INdAM Series, vol. 40 (Springer, 2018)
79. C.A. Weibel, *An Introduction to Homological Algebra*. Cambridge Studies in Advanced Mathematics, vol. 38 (Cambridge University Press, London, Cambridge, New York, 1994)
80. H.S. Wilf, A circle-of-lights algorithm for the "money-changing problem". Am. Math. Monthly **85**, 562–565 (1978)
81. Wolfram Research, Inc. (www.wolfram.com), Mathematica Online, Champaign, IL (2020)
82. A. Zhai, Fibonacci-like growth of numerical semigroups of a given genus. Semigroup Forum **86**, 634–662 (2013)

Index

J. Herzog et al., *Numerical Semigroups*, Compact Textbooks in Mathematics,
https://doi.org/10.1007/978-3-032-05424-1

The manufacturer's authorised representative in the EU is Springer Nature Customer Service Centre GmbH, Europaplatz 3, 69115 Heidelberg, Germany. If you have any concerns regarding our products, please contact ProductSafety@springernature.com

Printed and bound by CPI Group (UK) Ltd, Croydon, CR0 4YY

07/07/2026

02160921-0001